Thermomechanical Processing
of Steels

Thermomechanical Processing of Steels

Editors

Jose M. Rodriguez-Ibabe
Pello Uranga

MDPI • Basel • Beijing • Wuhan • Barcelona • Belgrade • Manchester • Tokyo • Cluj • Tianjin

Editors
Jose M. Rodriguez-Ibabe
CEIT and Universidad de Navarra-Tecnun
Spain

Pello Uranga
CEIT and Universidad de Navarra-Tecnun
Spain

Editorial Office
MDPI
St. Alban-Anlage 66
4052 Basel, Switzerland

This is a reprint of articles from the Special Issue published online in the open access journal *Metals* (ISSN 2075-4701) (available at: https://www.mdpi.com/journal/metals/special_issues/thermomechanical_processing).

For citation purposes, cite each article independently as indicated on the article page online and as indicated below:

LastName, A.A.; LastName, B.B.; LastName, C.C. Article Title. *Journal Name* **Year**, *Article Number*, Page Range.

ISBN 978-3-03943-354-4 (Hbk)
ISBN 978-3-03943-355-1 (PDF)

© 2020 by the authors. Articles in this book are Open Access and distributed under the Creative Commons Attribution (CC BY) license, which allows users to download, copy and build upon published articles, as long as the author and publisher are properly credited, which ensures maximum dissemination and a wider impact of our publications.

The book as a whole is distributed by MDPI under the terms and conditions of the Creative Commons license CC BY-NC-ND.

Contents

About the Editors vii

Pello Uranga and José María Rodríguez-Ibabe
Thermomechanical Processing of Steels
Reprinted from: *Metals* **2020**, *10*, 641, doi:10.3390/met10050641 1

Liangyun Lan, Zhiyuan Chang and Penghui Fan
Exploring the Difference in Bainite Transformation with Varying the Prior Austenite Grain Size in Low Carbon Steel
Reprinted from: *Metals* **2018**, *8*, 988, doi:10.3390/met8120988 5

Yi Shao, Xiaohua Li, Junjie Ma, Chenxi Liu and Zesheng Yan
Microstructure Formation of Low-Carbon Ferritic Stainless Steel during High Temperature Plastic Deformation
Reprinted from: *Metals* **2019**, *9*, 463, doi:10.3390/met9040463 19

Unai Mayo, Nerea Isasti, Jose M. Rodriguez-Ibabe and Pello Uranga
Interaction between Microalloying Additions and Phase Transformation during Intercritical Deformation in Low Carbon Steels
Reprinted from: *Metals* **2019**, *9*, 1049, doi:10.3390/met9101049 35

Jintao Wang, Shouping Liu and Xiaoyu Han
Study on ς Phase in Fe–Al–Cr Alloys
Reprinted from: *Metals* **2019**, *9*, 1092, doi:10.3390/met9101092 53

Ying Han, Jiaqi Sun, Yu Sun, Jiapeng Sun and Xu Ran
Tensile Properties and Microstructural Evolution of an Al-Bearing Ferritic Stainless Steel at Elevated Temperatures
Reprinted from: *Metals* **2020**, *10*, 86, doi:10.3390/met10010086 63

Leire García-Sesma, Beatriz López and Beatriz Pereda
Effect of High Ti Contents on Austenite Microstructural Evolution During Hot Deformation in Low Carbon Nb Microalloyed Steels
Reprinted from: *Metals* **2020**, *10*, 165, doi:10.3390/met10020165 81

Silvia Mancini, Luigi Langellotto, Paolo Emilio Di Nunzio, Chiara Zitelli and Andrea Di Schino
Defect Reduction and Quality Optimization by Modeling Plastic Deformation and Metallurgical Evolution in Ferritic Stainless Steels
Reprinted from: *Metals* **2020**, *10*, 186, doi:10.3390/met10020186 105

Perla Julieta Cerda Vázquez, José Sergio Pacheco-Cedeño, Mitsuo Osvaldo Ramos-Azpeitia, Pedro Garnica-González, Vicente Garibay-Febles, Joel Moreno-Palmerin, José de Jesús Cruz-Rivera and José Luis Hernández-Rivera
Casting and Constitutive Hot Flow Behavior of Medium-Mn Automotive Steel with Nb as Microalloying
Reprinted from: *Metals* **2020**, *10*, 206, doi:10.3390/met10020206 123

Johannes Webel, Adrian Herges, Dominik Britz, Eric Detemple, Volker Flaxa, Hardy Mohrbacher and Frank Mücklich
Tracing Microalloy Precipitation in Nb-Ti HSLA Steel during Austenite Conditioning
Reprinted from: *Metals* **2020**, *10*, 243, doi:10.3390/met10020243 137

Caleb A. Felker, John. G. Speer, Emmanuel De Moor and Kip O. Findley
Hot Strip Mill Processing Simulations on a Ti-Mo Microalloyed Steel Using Hot Torsion Testing
Reprinted from: *Metals* **2020**, *10*, 334, doi:10.3390/met10030334 . . . . **159**

Irina Pushkareva, Babak Shalchi-Amirkhiz, Sébastien Yves Pierre Allain, Guillaume Geandier, Fateh Fazeli, Matthew Sztanko and Colin Scott
The Influence of Vanadium Additions on Isothermally Formed Bainite Microstructures in Medium Carbon Steels Containing Retained Austenite
Reprinted from: *Metals* **2020**, *10*, 392, doi:10.3390/met10030392 . . . . **179**

About the Editors

Jose M. Rodriguez-Ibabe, Professor, completed his Ph.D. in Engineering (1984) at the University of Navarra (Spain). He is a professor of Materials Science and Metallurgy (University of Navarra). He joined Ceit Research Institute in 1984, as a researcher in the Materials and Manufacturing Division. Its field of research is steel metallurgy and, more specifically, the thermomechanical processing of high value-added steels. He has published 127 papers in journals and about 250 contributions at national and international conferences. He has written 6 books and has graduated 19 doctoral students. He has received the following seven international awards: Meritorious Award (2000, Iron and Steel Society, USA); Vanadium Award (2001, Institute of Material, UK); Charles Hatchett Award (2003, Institute of Materials, Minerals and Mining, UK); Gilbert R. Speich Award (three times: 2012, 2013 and 2016, AIST, USA); Henry Meyers Award (2018, ABMM, Brazil). He is currently Executive President of Ceit Research Institute, located in San Sebastián (Spain).

Pello Uranga, Associate Professor, is the Associate Director of the Materials and Manufacturing Division at CEIT and an associate professor at Tecnun, the School of Engineering at the University of Navarra. He obtained his Ph.D. degree in Materials Engineering in 2002 from the University of Navarra. Currently, his research activity is mainly focused on the thermomechanical processing and the microstructural evolution modeling of steels. He has published over 150 technical papers in international journals and conferences, receiving five international awards. He is an active member of AIST (Association of Iron and Steel Technology) and TMS (The Mineral, Metals and Materials Society) professional associations, as well as the university program evaluator for ANECA (Spain) and ABET (USA).

Editorial

Thermomechanical Processing of Steels

Pello Uranga [1,2,*] and José María Rodríguez-Ibabe [1,2]

1. CEIT-Basque Research and Technology Alliance (BRTA), Materials and Manufacturing Division, M. Lardizabal 15, 20018 Donostia-San Sebastián, Basque Country, Spain
2. Mechanical and Materials Engineering Department, Universidad de Navarra-Tecnun, M. Lardizabal 13, 20018 Donostia-San Sebastián, Basque Country, Spain
* Correspondence: puranga@ceit.es; Tel.: +34-943-212800

Received: 8 April 2020; Accepted: 13 May 2020; Published: 15 May 2020

1. Introduction

The combination of hot working technologies with a thermal path, under controlled conditions (i.e., thermomechanical processing) provides opportunities to achieve required mechanical properties at lower costs. The replacement of conventional rolling plus post-rolling heat treatments by integrated controlled forming and cooling strategies implies important reductions in energy consumption, increases in productivity and more compact facilities in the steel industry. The metallurgical challenges that this integration implies, though, are relevant and impressive developments that have been achieved over the last 40 years. The development of new steel grades and processing technologies devoted to thermomechanically-processed products is increasing and their implementation is being expended to higher value added products and applications.

The achievement of mechanical properties and process stability during a thermomechanical controlled process (TMCP) depends on the chemical composition, process parameter control, and optimization, as well as post-forming cooling strategy and thermal treatments. Therefore, this Special Issue combines contributions to different fields, topics, steel grades, and forming technologies applying TMCP processes to steels.

In addition to the metallurgical peculiarities and relationships between chemical composition, process, and final properties, the impact of advanced characterization techniques and innovative modeling strategies provides new tools to achieve further deployment of TMCP technologies.

This Special Issue on Thermomechanical Processing of Steels gathers papers written by experts from the steel industry and academia summarizing their latest developments and achievements in the field.

2. Contributions

The present Special Issue on Thermomechanical Processing of Steels includes eleven research papers [1–11]. A wide range of steel grades are covered in these papers. Although most of the papers deal with low carbon microalloyed grades [1,3,6,9,10], several papers study ferritic stainless steels [2,5,7] and others focus on grades such as Fe–Al–Cr alloys [4], medium-Mn Nb microalloyed steels [8], and medium carbon V microalloyed grades [11].

While some papers cover specific performance problems and optimization alternatives for industrial processing conditions [3,7,8], most of the papers deal with more fundamental analyses of physical metallurgy mechanisms and microstructural evolution [1–7,9–11].

Chang et al. [1] simulate a welding thermal cycle technique to generate different sizes of prior austenite grains. Dilatometry tests, in situ laser scanning confocal microscopy, and transmission electron microscopy are used to investigate the role of prior austenite grain size on bainite transformation in low carbon steel. In addition to the Hall–Petch strengthening effect, the carbon segregation at the fine austenite grain boundaries is probably another factor that decreases the Bs temperature as a result of the increase in interfacial energy of nucleation.

Shao et al. [2] study the effects of deformation temperature, deformation reduction, and deformation rate on microstructural formation, ferritic and martensitic phase transformation, stress–strain behaviors, and micro-hardness in low-carbon ferritic stainless steels. The increase in deformation temperature promotes the formation of fine equiaxed dynamic strain-induced transformation ferrite and suppresses martensitic transformation. The increase in deformation can effectively promote the transformation of dynamic strain-induced transformation (DSIT) ferrite, and decrease the martensitic transformation rate, which is caused by the work hardening effect on the metastable austenite.

Mayo et al. [3] analyze laboratory thermomechanical simulations reproducing intercritical rolling conditions performed in plain low-carbon and NbV-microalloyed steels. Using the electron backscattered diffraction (EBSD) technique, the discretization between intercritically deformed ferrite and new ferrite grains formed after deformation is achieved. The austenite conditioning before intercritical deformation in the Nb-bearing steel affects the balance of final precipitates by modifying the size distributions and origin of the Nb (C, N). This fact could modify the substructure in the intercritically deformed grains.

Wang et al. [4] propose a method of using the second phase to control the grain growth in Fe–Al–Cr alloys, in order to obtain better mechanical properties. In Fe–Al–Cr alloys, austenitic transformation occurs by adding austenitizing elements, leading to the formation of the second phase and segregation at the grain boundaries, which hinders grain growth. The nucleation of σ phase in Fe–Al–Cr alloy was controlled by the ratio of nickel to chromium. When the Ni/Cr (eq) ratio of alloys was more than 0.19, σ phase could nucleate in Fe–Al–Cr alloy. The relationship between austenitizing and nucleation of FeCr(σ) phase is given by thermodynamic calculation.

Han et al. [5] study the influence of temperature and strain rate on the hot tensile properties of 0Cr18AlSi ferritic stainless steel, a potential structural material in the ultra-supercritical generation industry. This work provides a deep understanding of the hot deformation behavior and its mechanism of the Al-bearing ferritic stainless steel and thus provides a basal design consideration for its extensive application. Both yield strength and ultimate tensile strength increase with the increase in strain rate. At high temperatures and low strain rates, prolonged necking deformation is observed, which determines the ductility of the steel to some extent.

García-Sesma et al. [6] focused their paper on the study of hot working behavior of Ti–Nb microalloyed steels with high Ti contents (>0.05%). After analyzing the torsion tests, it was observed that the 0.1% and 0.15% Ti additions resulted in retarded softening kinetics at all temperatures. This retardation can be mainly attributed to the solute drag effect exerted by Ti in solid solution. The precipitation state of the steels after reheating and after deformation was characterized and the applicability of existing microstructural evolution models was also evaluated.

Mancini et al. [7] analyze the manufacturing of ferritic stainless steel flat bars. In their paper, the origin of some edge defects occurring during hot rolling of flat bars of this grade is analyzed and thermomechanical and microstructural calculations are carried out to enhance the quality of the finished products by reducing the jagged borders defect on the hot rolled bars. Results show that the defect is caused by processing conditions that trigger an anomalous heating which, in turn, induces an uncontrolled grain growth on the edges. The work-hardened and elongated grains do not recrystallize during hot deformation. Consequently, they tend to squeeze out the surrounding softer and recrystallized matrix towards the edges of the bar where the fractures that characterize the surface defect occur.

Cerda Vázquez et al. [8] propose a novel medium-Mn steel (6.5 wt.% Mn) microstructure with 0.1 wt.% Nb designed using Thermo-Calc and JMatPro thermodynamic simulation software. The as-cast microstructure consists mainly of a mixture of martensite, ferrite, and a low amount of austenite, while the microstructure in the homogenization condition corresponds to martensite and retained austenite, which is verified by X-ray diffraction tests. In order to design further production stages of the steel, the homogenized samples were subjected to hot compression testing to determine their plastic flow behavior.

Webel et al. [9] combine results from various characterization methods for tracing microalloy precipitation after simulating different austenite TMCP conditions in a Gleeble thermo-mechanical simulator. Atom probe tomography (APT), scanning transmission electron microscopy in a focused ion beam equipped scanning electron microscope (STEM-on-FIB), and electrical resistivity measurements provide complementary information on the precipitation status and correlate with each other. Precipitates that formed during cooling or isothermal holding could be distinguished from strain-induced precipitates by corroborating STEM measurements with APT results, because APT specifically allowed obtaining detailed information about the chemical composition of precipitates as well as the elemental distribution.

Felker et al. [10] assess the influence of thermomechanical processing on the evolution of austenite and the associated final ferritic microstructures. Hot strip mill processing simulations were performed on a low-carbon, titanium-molybdenum microalloyed steel using hot torsion testing to investigate the effects of extensive differences in austenite strain accumulation on austenite morphology and microstructural development after isothermal transformation. Greater austenite shear strain accumulation resulted in greater refinement of both the prior austenite and polygonal ferrite grain sizes. Further, polygonal ferrite grain diameter distributions were narrowed, and the presence of hard, secondary phase constituents was minimized, with greater amounts of austenite strain accumulation. The results indicate that extensive austenite strain accumulation before decomposition is required to achieve desirable, ferritic microstructures.

Finally, Pushkareva et al. [11] investigate V additions on isothermally formed bainite in medium carbon steels containing retained austenite using in-situ high-energy X-ray diffraction (HEXRD) and ex-situ electron energy loss spectroscopy (EELS) and energy dispersive X-ray analysis (EDX) techniques in the transmission electron microscope (TEM). No significant impact of V in solid solution on the bainite transformation rate, final phase fractions or on the width of bainite laths was seen for transformations in the range 375–430 °C. A beneficial refinement of blocky martensite-austenite (MA) and a corresponding size effect induced enhancement in austenite stability were found at the lowest transformation temperature. Overall, V additions result in a slight increase in strength levels.

3. Conclusions and Outlook

The development of thermomechanical processing has been impressive in the last years. Nowadays, the application of thermomechanical processes, reducing the cost and time needed for final heat treatments is extended to a wide variety of steel grades and products, as shown in the papers gathered in the current Special Issue. These developments, though, require a high level of intensive research and transferability to the industry and understanding of basic mechanisms, which with a proper processing control, will ensure the reliability of these processing routes under the most challenging operational conditions. The interest and high level of contributions published in this Special Issue ensure that the link between research and industry will not break anytime soon.

As Guest Editors, we would like to express our sincere thanks to all the authors for submitting their manuscripts and sharing their latest developments. We also would like to encourage them and the rest of the community to continue researching and publishing in steel-related topics, as their relevance to industry and society is vital for progress in the future.

Conflicts of Interest: The author declares no conflicts of interest.

References

1. Chang, Z.; Fan, P. Exploring the Difference in Bainite Transformation with Varying the Prior Austenite Grain Size in Low Carbon Steel. *Metals* **2018**, *8*, 988. [CrossRef]
2. Shao, Y.; Li, X.; Ma, J.; Liu, C.; Yan, Z. Microstructure Formation of Low-Carbon Ferritic Stainless Steel during High Temperature Plastic Deformation. *Metals* **2019**, *9*, 463. [CrossRef]
3. Mayo, U.; Isasti, N.; Rodriguez-Ibabe, J.; Uranga, P. Interaction between Microalloying Additions and Phase Transformation during Intercritical Deformation in Low Carbon Steels. *Metals* **2019**, *9*, 1049. [CrossRef]
4. Wang, J.; Liu, S.; Han, X. Study on σ Phase in Fe–Al–Cr Alloys. *Metals* **2019**, *9*, 1092. [CrossRef]
5. Han, Y.; Sun, J.; Sun, Y.; Sun, J.; Ran, X. Tensile Properties and Microstructural Evolution of an Al-Bearing Ferritic Stainless Steel at Elevated Temperatures. *Metals* **2020**, *10*, 86. [CrossRef]
6. García-Sesma, L.; López, B.; Pereda, B. Effect of High Ti Contents on Austenite Microstructural Evolution During Hot Deformation in Low Carbon Nb Microalloyed Steels. *Metals* **2020**, *10*, 165. [CrossRef]
7. Mancini, S.; Langellotto, L.; Di Nunzio, P.; Zitelli, C.; Di Schino, A. Defect Reduction and Quality Optimization by Modeling Plastic Deformation and Metallurgical Evolution in Ferritic Stainless Steels. *Metals* **2020**, *10*, 186. [CrossRef]
8. Cerda Vázquez, P.; Pacheco-Cedeño, J.; Ramos-Azpeitia, M.; Garnica-González, P.; Garibay-Febles, V.; Moreno-Palmerin, J.; Cruz-Rivera, J.; Hernández-Rivera, J. Casting and Constitutive Hot Flow Behavior of Medium-Mn Automotive Steel with Nb as Microalloying. *Metals* **2020**, *10*, 206. [CrossRef]
9. Webel, J.; Herges, A.; Britz, D.; Detemple, E.; Flaxa, V.; Mohrbacher, H.; Mücklich, F. Tracing Microalloy Precipitation in Nb-Ti HSLA Steel during Austenite Conditioning. *Metals* **2020**, *10*, 243. [CrossRef]
10. Felker, C.; Speer, J.; De Moor, E.; Findley, K. Hot Strip Mill Processing Simulations on a Ti-Mo Microalloyed Steel Using Hot Torsion Testing. *Metals* **2020**, *10*, 334. [CrossRef]
11. Pushkareva, I.; Shalchi-Amirkhiz, B.; Allain, S.; Geandier, G.; Fazeli, F.; Sztanko, M.; Scott, C. The Influence of Vanadium Additions on Isothermally Formed Bainite Microstructures in Medium Carbon Steels Containing Retained Austenite. *Metals* **2020**, *10*, 392. [CrossRef]

© 2020 by the authors. Licensee MDPI, Basel, Switzerland. This article is an open access article distributed under the terms and conditions of the Creative Commons Attribution (CC BY) license (http://creativecommons.org/licenses/by/4.0/).

Article

Exploring the Difference in Bainite Transformation with Varying the Prior Austenite Grain Size in Low Carbon Steel

Liangyun Lan [1,2,*], Zhiyuan Chang [3] and Penghui Fan [1]

1. School of Mechanical Engineering and Automation, Northeastern University, Shenyang 110819, China; 1600335@stu.neu.edu.cn
2. Key Laboratory of Vibration and Control of Aero-Propulsion Systems, Ministry of Education of China, Northeastern University, Shenyang 110819, China
3. State key Laboratory of Rolling Technology and Automation, Northeastern University, Shenyang 110819, China; changzyhpu@163.com
* Correspondence: lanly@me.neu.edu.cn; Tel.: +86-248-367-4140

Received: 2 November 2018; Accepted: 21 November 2018; Published: 24 November 2018

Abstract: The simulation welding thermal cycle technique was employed to generate different sizes of prior austenite grains. Dilatometry tests, in situ laser scanning confocal microscopy, and transmission electron microscopy were used to investigate the role of prior austenite grain size on bainite transformation in low carbon steel. The bainite start transformation (Bs) temperature was reduced by fine austenite grains (lowered by about 30 °C under the experimental conditions). Through careful microstructural observation, it can be found that, besides the Hall–Petch strengthening effect, the carbon segregation at the fine austenite grain boundaries is probably another factor that decreases the Bs temperature as a result of the increase in interfacial energy of nucleation. At the early stage of the transformation, the bainite laths nucleate near to the grain boundaries and grow in a "side-by-side" mode in fine austenite grains, whereas in coarse austenite grains, the sympathetic nucleation at the broad side of the pre-existing laths causes the distribution of bainitic ferrite packets to be interlocked.

Keywords: low carbon steel; prior austenite grain boundary; carbon segregation; Bs temperature

---

1. Introduction

Prior austenite grain size is a vitally important parameter to influence phase transformation behavior in steels, because to a great extent, it determines the number of nucleation sites and the space of the growth of products. For martensite transformation, it is well recognized that fine prior austenite grain size decreases the martensite start transformation (Ms) temperature, as the Hall–Petch strengthening effect for fine austenite grains improves resistance to the plastic deformation of martensite transformation [1]. Meanwhile, it further refines the martensitic block width and packet size, or likely forms a phenomenon of variant selection in the crystallography, which plays a critical role in tailoring the final mechanical properties [2–5].

The effect of prior austenite grain size on bainite transformation is also of significance, although the related results seem to be in dispute [6–11]. Some studies considered that fine austenite grains accelerate the kinetics of bainite transformation, as the increase in the number of grain boundaries provides more preferable sites for nucleation of ferritic laths. By contrast, others found that the bainite formation is insensitive to the prior grain size or even that the growth of bainite is retarded by fine prior grains. The bainite start transformation (Bs) temperature is found to be lowered with decreasing austenite grain size under the continuous cooling condition, a very similar trend to that found with the Ms temperature. It is taken for granted that they have a similar mechanism for the decrease in transformation temperature [10,11]. However, it should be noted that the martensite transformation is an athermal transformation, whereas the bainite transformation is, sometimes, regarded as a kind of thermally activated nucleation and growth reaction—it does not seem very convincing that the Hall–Petch strengthening effect in fine austenite grains is considered as a sole factor to decrease the Bs temperature. Recently, some studies reported that the variant distribution of bainite is limited in fine austenite grains compared with large austenite grains [12–14]. This phenomenon may be related to the difference in the temperature range of bainite transformation with varying prior grain size [13]. It is thus worth further exploring what else, besides the Hall–Petch strengthening effect, is likely to play a role in retarding the bainite transformation for fine austenite grains.

In this work, welding thermal simulation was employed to generate different sizes of prior austenite grains and to prove the effect of prior grain size on phase transformation, because the inhomogeneous austenite grain size is representative of the heat-affected zone (HAZ). In situ observation using high-temperature laser scanning confocal microscopy (LSCM) and partial bainite transformation tests were carried out to further reveal the microstructural characters at the early stage of the transformation with different prior austenite grain sizes.

2. Experimental Method

The chemical composition of low carbon steel in this study was 0.053 C, 0.22 Si, 1.35 Mn, 0.082 (Nb + V + Ti), 0.0012 B, 0.23 Cr, 0.37 Mo balanced by Fe.

Specimens with dimensions of φ 6 × 50 mm were cut from the hot rolled steel plate along the transversal direction. The HAZ simulations were carried out using a thermo-mechanical simulator and a two dimensional Rykalin mathematical model was adopted to control the testing temperature. A Pt-10% Rh thermocouple was welded at the middle of the cylinders to monitor the temperature change during heat treatment. A linear variable displacement transducer type dilatometer was used to measure the relative change in diameter of these samples.

Specimens were heated to a peak temperature at a rate of 130 °C/s and held for 2 s. The peak temperature was set at 1350 and 1100 °C to obtain widely varying austenite grain sizes, which simulates the thermal cycles at coarse grained HAZ (CGHAZ) and fine grained HAZ (FGHAZ), respectively. The samples were then subjected to various cooling rates that were defined as the cooling time from 800 to 500 °C (the $t_{8/5}$ time varies from 5 to 600 s) until the phase transformation has completed. Figure 1 shows the measured thermal cycle curves. Dilatation curves were obtained to determine the phase change temperatures. Continuous cooling transformation (CCT) diagrams for the CGHAZ and FGHAZ were constructed together to exhibit the effect of prior austenite grain size on phase transformation, although the CCT diagram of the CGHAZ has been reported elsewhere [13].

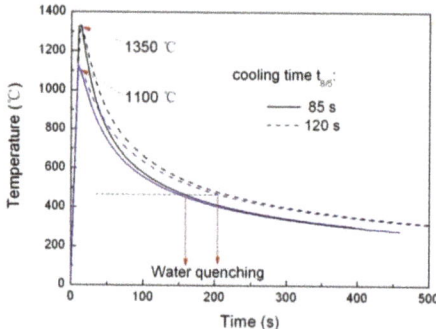

Figure 1. Typical thermal cycle curves with different peak temperatures and cooling times (water quenching process signified schematically with red arrows).

The partially transformed bainite microstructure was expected to be retained using an interrupted cooling method by water quenching. Based on the CCT diagrams, the partial bainite transformation samples were quenched in water immediately after the welding thermal cycle temperature was lowered to 470, 500, or 550 °C, as signified with red arrows in Figure 1.

For in situ observation of phase transformation, two samples were cut directly from the welding thermal simulation samples, machined into disks (6 mm diameter and 3 mm thickness), and then mechanically polished. These prepared samples were austenitized at 1300 °C and 1100 °C to simulate the CGHAZ and FGHAZ thermal cycles, respectively. The linear cooling rate of 2.5 °C/s was employed, which is approximately equal to the average cooling rate of the $t_{8/5}$ = 120 s cooling time. The real-time information of phase transformation was recorded using high-temperature LSCM.

Metallographic specimens were cut from the heat treatment region of interest and then polished by conventional techniques. Microstructural characteristics were examined using an optical microscope(Leica DMIRM, Leica company, Weitz, Germany), a field-emission scanning electron microscope equipped with electron probe microanalysis (SEM, ULTRA 55, ZEISS, Jena, Germany), and a transmission electron microscope (TEM, FEI Tecnai G^2F20, FEI company, Hillsboro, OR, USA). Vickers hardness measurements were conducted using an FM 700 hardness-testing machine (Hardness tester, FM-700, FUTURE-TECH company, Kawasaki-City, Japan) employing a 0.5 N load with 15 s dwell time. The average hardness values for each heat treatment condition were calculated based on at least 10 repeat tests.

3. Results and Discussion

On the basis of the dilatation curves (some typical examples were given in the literature [11,13]), the transformation start/finish temperatures could be confirmed with a normal extrapolation method [15]. The CCT diagrams for CGHAZ and FGHAZ are constructed together in Figure 2a. The transformation start temperature gradually lowers and the transformation finish temperature first increases, but then decreases with the decrease in cooling time. The bainite microstructure could be formed within a wide range of cooling rates, as marked in the diagram, which are attributed to the fact that the combined addition of multi-microalloys, such as Mo, Nb, Ti, and V, into steel extends the range of bainite transformation [16]. As expected, FGHAZ always has a lower Bs temperature than CGHAZ for any given cooling rate (highlighted with gray background), although the discrepancy in bainite transformation finish temperature seems to be very insignificant. The largest difference in the Bs temperature between CGHAZ and FGHAZ is about 30 °C at the middle cooling rate.

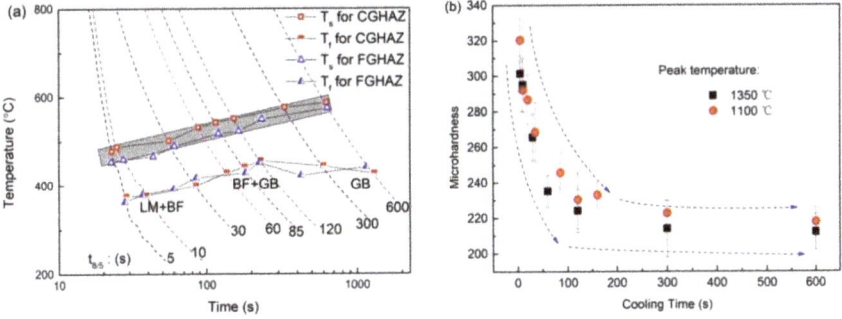

Figure 2. (a) Continuous cooling transformation (CCT) diagrams with different peak temperatures and (b) the change of Vickers hardness as a function of cooling time (CGHAZ: coarse grained heat-affected zone, FGHAZ: fine grained HAZ, LM: lath martensite, BF: bainitic ferrite, GB: granular ferrite).

Figure 2b shows the change of hardness with cooling time. As the cooling time increases, the hardness decreases rapidly and then gradually levels off. Meanwhile, at a given cooling rate, FGHAZ always seems to have a slightly higher hardness than CGHAZ. This further proves the difference in transformation behaviors (Figure 2a) because the lower temperature transformation microstructure normally results in a higher hardness.

Figure 3 represents three groups of typical microstructures for CGHAZ and FGHAZ after they were subjected to the $t_{8/5}$ times of 5, 85, and 600 s, respectively. The average grain size of FGHAZ measured with the line intercept method ranges from about 16.8 µm to 29.5 µm, while CGHAZ has an average grain size higher than 300 µm. At the shortest cooling time, the microstructure is predominantly lath martensite, accompanied by very little bainitic ferrite, regardless of peak temperature (Figure 3a,d). Comparing Figure 3a and d, the packet size of martensite in the morphology seems to be refined by the fine prior grain size (Figure 3e). Because of the hierarchical structure of martensite [2–5], the block size and lath width may be also decreased in FGHAZ.

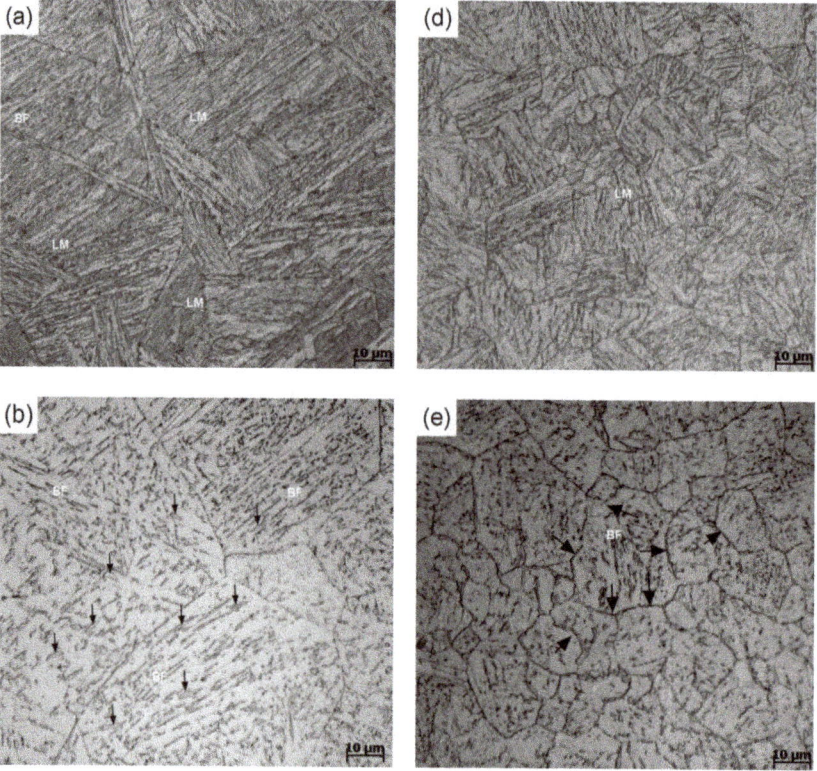

Figure 3. *Cont.*

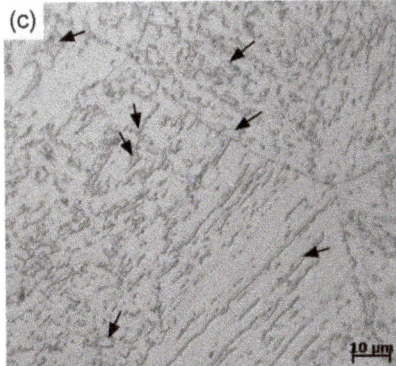 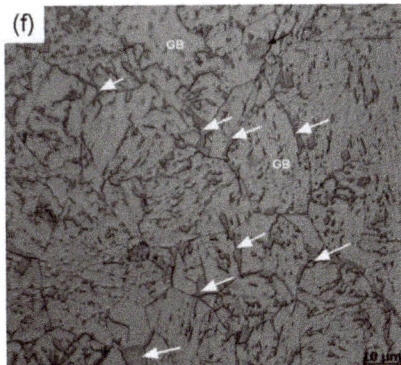

Figure 3. Optical microstructure with different peak temperatures: (**a**–**c**) 1350 °C and (**d**–**f**) 1100 °C, and different cooling times: $t_{8/5}$ (**a**,**d**) 5 s, (**b**,**e**) 85 s, and (**c**,**f**) 600 s. (LM: lath martensite, BF: bainitic ferrite, GB: granular ferrite, and some martensite–austenite (MA) constituents were signified with arrows).

At the cooling time (85 s), CGHAZ and FGHAZ microstructures are mainly characterized by bainitic ferrite (Figure 3b,e). However, the martensite–austenite (MA) constituents that are formed below the Ms temperature as a result of incomplete bainite transformation [17] have different sizes and distribution sites, although most of them have a similar short and slender shape. For example, the MA constituents in CGHAZ are present inside the grains and their distribution approximately describes some sites of parallel lath boundaries [17]; while in FGHAZ, most of them appear along the prior austenite grain boundaries (signified with arrows in Figure 3e), which seems to make the grain boundaries coarser. The distribution of MA constituents implies that the redistribution of carbon atoms is different during the transformation for CGHAZ and FGHAZ, because notable enrichment of carbon was observed within the MA constituents using atom probe tomography (APT) [18].

As the cooling time increases to 600 s, the predominant microstructure for CGHAZ and FGHAZ becomes granular bainite (Figure 3c,f). The MA constituents have a massive shape, mainly because the uphill diffusion of carbon atoms becomes much easier at the lower cooling rate condition [19]. However, for CGHAZ, most massive MA constituents distribute uniformly in the matrix, except that a few of the MA constituents appear along the prior austenite grain boundaries, whereas in FGHAZ, they seem to have a larger average size and are mainly attached to the grain boundaries, as signified by arrows in Figure 3c,f. Because CGHAZ and FGHAZ have similar microstructures, the difference of MA constituents should be mainly attributed to the number of grain boundaries (i.e., prior austenite grain size), as the grain boundary has a higher diffusivity for interstitial atoms (including C, B, and H) than the grain interior and also act as a sink for these atoms [20–22].

Using the high temperature LSCM observation, the real-time features of bainite at the early stage of the transformation were recorded, as shown in Figures 4 and 5. The phase transformation does not occur when the temperature decreases to 634 °C (Figure 4a). The first ferritic lath can be found at 632 °C and its nucleation site is exactly on the grain boundary, as signified with an arrow in Figure 4b. However, this temperature is higher than that detected using a dilatometer (Figure 2a). This difference is probably acceptable as the dilatometer sensitivity is about 10 percent volume fraction transformed and the free surface effect of the LSCM sample can enhance the transformation temperature [23]. With a further slightly decreasing temperature (629 °C), several ferritic laths form, simultaneously attached to the grain boundaries (Figure 4c). Meanwhile, it is very interesting to find that a newly formed ferritic lath sympathetically nucleates on the pre-existing ferritic lath and grows toward the other direction, as marked with a cycled white arrow. This mechanism is a necessity that gives rise to the interlocked distribution of bainitic ferrite packets and partitions the coarse austenite grains into several finer and separate regions [24]. Nevertheless, other grains still do not occur in phase transformation, which means the bainite formation is inhomogeneous at the grain scale. This agrees well with the results by Sainis et al. [25], who found that the rate of nucleation varies markedly between different austenite grains.

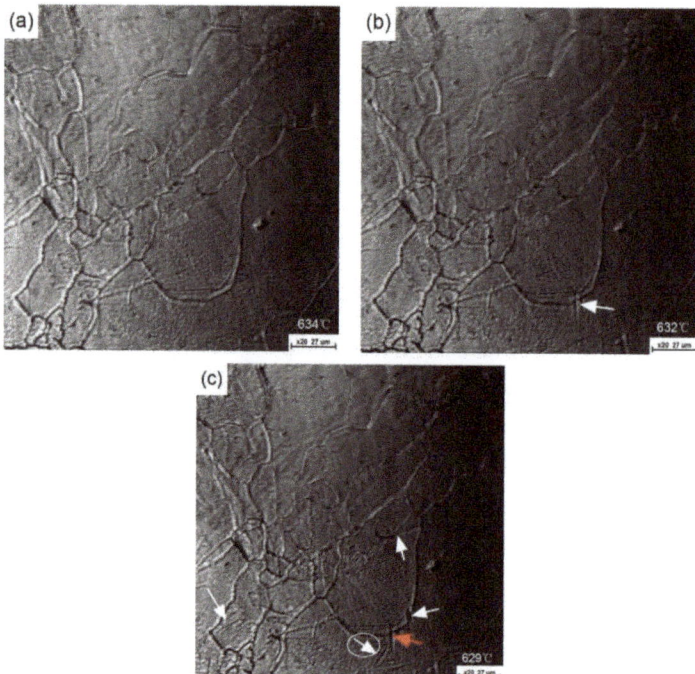

Figure 4. In situ observation of phase transformation with the CGHAZ thermal cycle: (**a**) 634 °C, (**b**) 632 °C, and (**c**) 629 °C. (The white arrows show that the ferritic laths form at the moment of real-time photographing for each image, the red arrows show the pre-existing ferritic laths.).

For the FGHAZ thermal cycle, the austenite grains are smaller and their size is more uniform (Figure 5a) compared with that of CGHAZ. When the temperature decreases to 610 °C, no obvious phase transformation takes place (Figure 5b). The ferritic laths can first be found at 605 °C, and their nucleation sites are also derived from the austenite grain boundaries (arrowed in Figure 5c), although the image quality was not as good as before. The growth rate of these ferritic laths seems to be very fast, leading to their length across almost the prior austenite grain in no time. Subsequently, the newly formed bainitic laths appear in a "side-by-side" mode relative to the pre-existing laths, rather than an interlocked mode as in CGHAZ with the decrease in temperature (Figure 5d), indicating that this mode may make only a packet structure in the morphology formed inside each fine prior austenite grain [14]. According to these real-time micrographs, it is confirmed that the fine austenite grain size decreases the Bs temperature by about 30 °C, which is in good agreement with the dilatation tests.

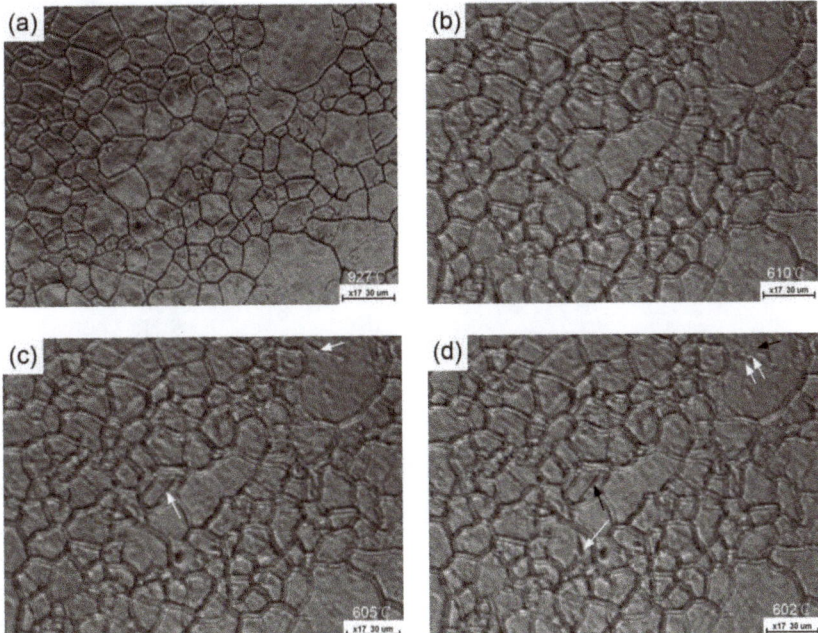

Figure 5. In situ observation of phase transformation with the FGHAZ thermal cycle: (**a**) 927 °C, (**b**) 610 °C, (**c**) 605 °C, and (**d**) 602 °C. (The white arrows show that the ferritic laths form at the moment of real-time photographing for each image, the black arrows show the pre-existing ferritic laths.)

Figure 6 shows the partially transformed bainite microstructure after the samples were subjected to partial welding thermal cycle cooling, followed by water quenching. At the very beginning of phase transformation, the lath morphology of ferrite forms and the nucleation of these ferritic laths are attached to the grain boundary (Figure 6a). With the progress of bainite transformation (Figure 6b), the packet structure of bainite can clearly be found as the carbides or MA constituents decorate the parallel lath boundaries. Several primary ferritic laths nucleate at different sites of the grain boundary inside one large prior austenite grain (marked with black arrows in Figure 6b), and many secondary laths sympathetically nucleate on the pre-existing ferritic laths (signified with white arrows), which agrees well with the LSCM results (Figure 4c). Through close observation of the nucleation sites of the primary laths, no MA constituents can be found, implying that in this case, the carbon atoms do not enrich at the nucleation sites of the prior austenite grain boundaries. On the contrary, most of carbon atoms should be expelled into the lath boundaries during transformation.

Only a small amount of bainite microstructures can be found for the FGHAZ sample (Figure 6c) when the quenching temperature decreases to 500 °C. The prior austenite grain boundaries seem to be particularly evident (arrowed in Figure 6c) compared with these in Figure 6a using the same etchant solution. The lath boundaries of bainite in the vicinity of prior austenite grain boundaries are ambiguous as almost no any carbide was formed to decorate these lath boundaries in these micro-regions (Figure 6d). An electron probe microanalysis on carbon distribution along the AB line (Figure 6d) across a prior austenite grain boundary semi-quantitatively shows that the prior austenite grain boundary with MA constituent contains much higher carbon concentration compared with the grain interior (Figure 6e).

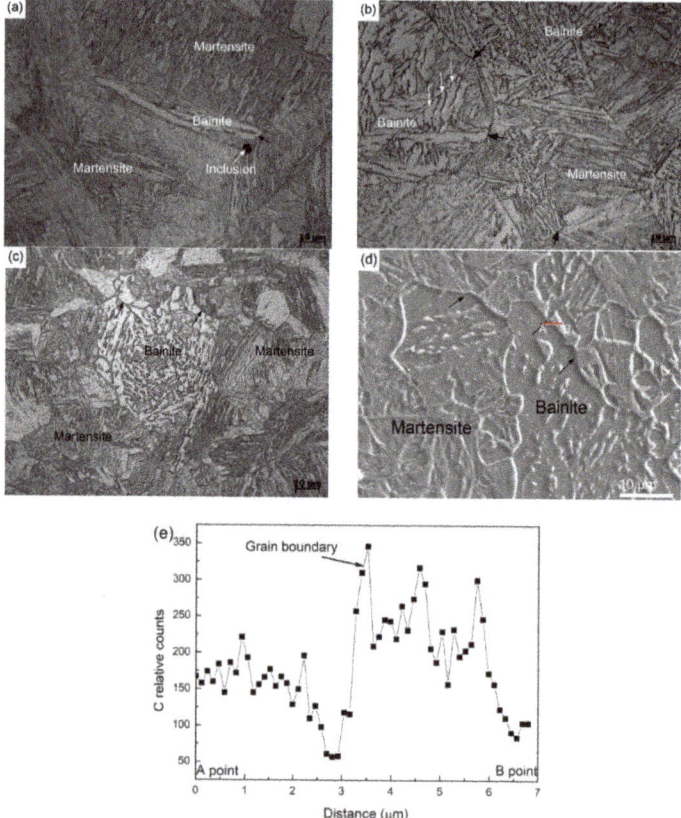

Figure 6. Partially transformed bainite microstructure obtained by water quenching: (**a,b**) peak temperature of 1350 °C with 85 °C/s cooling rate and quenching temperatures of 550 °C and 500 °C, respectively; (**c,d**) peak temperature of 1100 °C with 120 °C/s cooling rate and quenching temperatures of 500 °C and 470 °C, respectively; (**e**) the carbon distribution along the AB line across one grain boundary shown in (**d**).

Several TEM images were joined together in Figure 7 to show the morphology of bainite formed inside a fine prior austenite grain. Each of ferritic laths seem to be composed of several sub-units (one of which was delineated) and very few of MA constituents can be formed along the lath boundaries. These features are different from the morphology of bainite formed in coarse prior austenite grains (the lath boundaries are always decorated by the slender MA constituents shown in Figure 5d of the literature [13]). Based on the bainite transformation theory, these sub-units formed with a limited size may be attributed to the fact that the increase in strength of austenite due to a relatively lower transformation temperature or the limited space of ferritic growth impedes the plastic accommodation of remaining austenite for transformation strain [26].

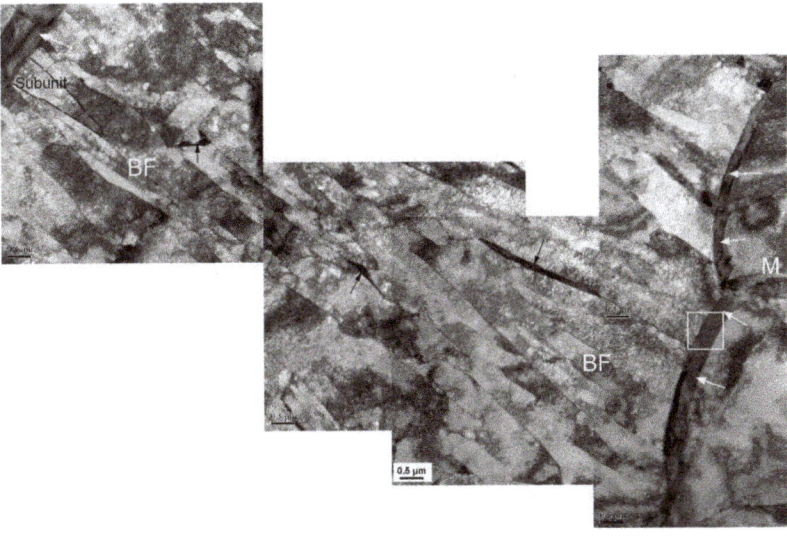

Figure 7. Transmission electron microscope (TEM) images showing the fine structure and grain boundary character of partially transformed microstructure using the same sample of Figure 6d (BF: bainitic ferrite, M: martensite, grain boundary MA constituents are signified with white arrows and fine MA constituents are marked with black arrows).

Most interestingly, the prior grain boundaries are decorated by many long strip MA constituents with an average width of about 0.4 µm, as signified with white arrows (Figure 7). The magnified image in Figure 8 shows they have a character of high density dislocation martensite, which implies that the austenite along the grain boundaries is stabilized by carbon enrichment at these micro-zones, and most of them transform into martensite during final quenching. From the morphology of ferritic lath point of view, these carbon atoms enriched at the grain boundaries are considered to occur in super-cooled austenite at very early stages of the transformation or even prior to the transformation. Assuming that these segregated carbon atoms are expelled from the product phase during the transformation, they should be enriched in front of the phase interface and a quasi-eutectoid structure is likely to be formed [27], as presented in CGHAZ.

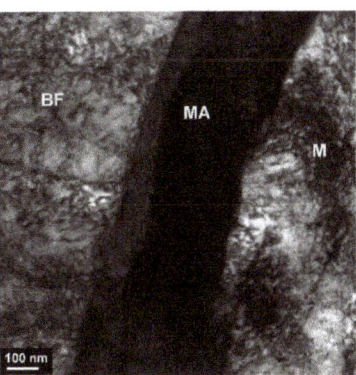

Figure 8. The magnified image showing the fine structure of the grain boundary MA constituent corresponding to the zone squared in Figure 7 (BF: bainite ferrite, MA: grain boundary MA constituent, M: martensite that transformed from remaining austenite of neighboring grain).

4. Overall Discussion

The size and distribution of MA constituents are different between CGAHZ and FGHAZ, regardless of similar microstructures under any cooling time condition (e.g., Figures 3 and 6). This means that the prior austenite grain size to some extent influences the redistribution of carbon atoms before and after the transformation, as schematically shown in Figure 9. The distribution of carbon atoms should be very uniform in CGHAZ grains at elevated temperatures (Figure 9a), as the solid solubility of carbon in austenite increases with the temperature. Additionally, the number of vacancies generated inside the grains increases with the temperature, and they bind with carbon atoms to form stable carbon–vacancy pairs [20,28]. As a result, the carbon atoms uniformly distribute in super-cooled austenite for CGHAZ. During the transformation, the excess carbon atoms for BCC crystals are expelled in front of the phase interface into the lath boundaries (Figure 9b).

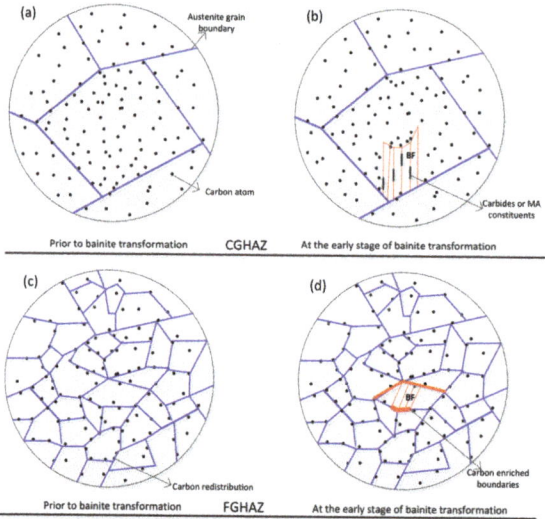

Figure 9. Carbon redistribution in super-cooled austenite before and after the onset of bainite transformation for CGHAZ (a,b) and FGHAZ (c,d).

The carbon atoms segregated along the prior austenite grain boundaries for FGHAZ (Figure 9c) may be derived from the following two reasons. First, the grain boundary acts as a sink for vacancies. The higher the density of grain boundaries, the easier the grain boundary segregation of carbon atoms occurs because of the equilibrium and non-equilibrium segregation mechanism [29]. Second, the trend of carbon diffusion into the grain boundaries may be enhanced as a result of the lowering solid solubility of carbon in super-cooled austenite. However, the distance of carbon diffusion in FGHAZ is much shorter than in CGHAZ, and the carbon diffusivity along the grain boundary is higher than inside the grains [22], which gives rise to the presence of carbon enrichment at the grain boundaries for the FGHAZ (Figure 9c), and the grain boundaries were finally decorated with the MA constituents (Figure 9d). Timokhina et al. [30] found that two types of the remaining austenite with different lattice parameters are present as a result of different carbon content at the beginning of the transformation using an in situ neutron diffraction technique. This means that the carbon redistribution does occur at the very early stage of the transformation, which seems to be in support of our result that the carbon atoms may be segregated at the prior grain boundaries prior to the transformation, although the present steel used has a much lower carbon content.

Zhang et al. [20] considered that the prior austenite grain boundaries, which are high-energy regions with disordered atomic arrangements, could accommodate a number of carbon atoms and thus decrease the overall energy. In this sense, the difference in the carbon concentration of the grain boundary will change the overall boundary energy, which further influences the drive force of grain boundary nucleation. Therefore, carbon segregation at the grain boundaries in FGHAZ mentioned above is probably responsible for the difference in Bs temperature. More details are discussed below.

According to the classical nucleation theory, the free energy ΔG for nucleation at prior austenite grain boundaries is given by [31,32]

$$\Delta G = \Delta G_{chem} + [\Delta G_{str} + \Delta G_{int}] \quad (1)$$

where ΔG_{chem} is the molar chemical Gibbs energy difference between the parent and product phase, ΔG_{str} is the summation of molar strain energy resulting from the volume misfit between ferrite and austenite (i.e., transformation strain), and ΔG_{int} is the molar interfacial energy for the new phase formed. The latter two items are the resistance to phase transformation. If the transformation strain can be accommodated by the plastic deformation of austenite, the required strain energy for the creation of a new nucleus is much lower. The grain boundary nucleation thus occurs under the low transformation driving force condition, as presented in CGHAZ. However, the plastic accommodation in fine austenite grains is largely constrained by the Hall–Petch strengthening effect. As a result, it is always considered to be the main factor to decrease the Bs and Ms temperatures [1,3,10,23], because a larger super-cooled temperature is required to proceed to the phase transformation.

Very little strain energy seems to be required at the early stage of the bainite transformation as the transformation strain is insignificant at this point [30]. Here, the change of interfacial energy probably plays a vital role in decreasing Bs temperature as a result of carbon segregation at the grain boundary. The heterogeneous nucleation attached to the grain boundary is mainly attributed to the fact that the prior grain boundary energy decreases the interfacial energy of new-phase nucleation. Song et al. [32] showed that the maximum reduction of interfacial energy can be achieved when the embryo forms at the high-energy austenite grain boundary. However, the austenite grain boundary energy will be reduced if the excess carbon atoms are segregated on them, leading to the fact that the energy assisting the new-phase nucleation is lowered. Madariaga et al. [33] also considered that grain boundary segregation has an influence on new-phase nucleation through the associated change of the interfacial energy and, presumably, of the structure of the boundary itself. On the basis of the partially transformed microstructure (Figure 8), the width of carbon enriched micro-zones at grain boundaries is around 0.4 μm in FGHAZ. This means that the actual nucleation sites of ferritic laths are some distance (maybe about 0.2 μm) away from the grain boundaries. The nucleation of ferritic laths might not even take advantage of the prior austenite grain boundary energy. Therefore, it can be concluded that compared with CGHAZ, a larger driving force of transformation is needed to offset the increase in interfacial energy of nucleation for FGHAZ, which is embodied in the lower in Bs temperature.

5. Conclusions

The effect of austenite grain size on the bainite transformation, especially on the Bs temperature, was investigated under the simulation welding thermal cycle condition. Through detailed microstructural characterization and in situ observation of the early stage of phase transformation, the main findings can be given below.

(1) The predominantly bainite microstructure can be obtained over a very wide range of cooling rates for this studied steel. The hardness decreases rapidly and then gradually levels off with the decrease in cooling rate and FGHAZ always results in a slightly higher hardness than the CGHAZ at any given cooling rate.
(2) For a given cooling condition, FGHAZ always has a lower Bs temperature than CGHAZ, which is proven by in situ LSCM observation and partial bainite transformation tests. The prior austenite grain boundaries were always decorated by many long strip MA constituents in FGHAZ, indicating that the carbon segregation occurred at the early stage of bainite transformation.

Besides the Hall–Petch strengthening effect in FGHAZ, the increasing interfacial energy for the grain boundary nucleation due to carbon enrichment is probably another factor that lowers the Bs temperature.

(3) A large bainitic ferrite packet can be transformed only in fine austenite grains because the ferritic laths grow in a "side-by-side" mode, although each of the laths may consist of several sub-units. In contrast, the interlocked distribution of bainitic ferrite packets can frequently be found in coarse austenite grains, which is attributed to the mechanism of sympathetic nucleation.

Author Contributions: Conceptualization, L.L.; Methodology, L.L.; Software, Z.C.; Validation, L.L., Z.C. and P.F.; Formal Analysis, L.L.; Investigation, L.L., Z.C. and P.F.; Writing-Original Draft Preparation, L.L.; Writing-Review & Editing, L.L.; Supervision, L.L.; Funding Acquisition, L.L.

Funding: This work is supported by the National Natural Science Foundation of China (No. 51605084 and U1708265) and the Fundamental Research Funds for the Central Universities of China (N170304019 and N170308028).

Conflicts of Interest: The authors declare no conflict of interest.

References

1. Garcia-Junceda, A.; Capdevila, C.; Caballero, F.G.; Garcia de Andres, C. Dependence of martensite start temperature on fine austenite grain size. *Scr. Mater.* **2008**, *58*, 134–137. [CrossRef]
2. Morito, S.; Saito, H.; Ogawa, T.; Furuhara, T.; Maki, T. Effect of austenite grain size on the morphology and crystallography of lath martensite in low carbon steels. *ISIJ Int.* **2005**, *45*, 91–94. [CrossRef]
3. Furuhara, T.; Kikumoto, K.; Saito, H.; Sekine, T.; Ogawa, T.; Morito, S.; Maki, T. Phase transformation from fine-grained austenite. *ISIJ Int.* **2008**, *48*, 1038–1045. [CrossRef]
4. Hanamura, T.; Torizuka, S.; Tamura, S.; Enokida, S.; Takechi, H. Effect of austenite grain size on transformation behavior, microstructure and mechanical properties of 0.1C-5Mn martensitic steel. *ISIJ Int.* **2013**, *53*, 2218–2225. [CrossRef]
5. Hidalgo, J.; Santofimia, M.J. Effect of prior austenite grain size refinement by thermal cycling on the microstructural features of as-quenched lath martensite. *Metall. Mater. Trans. A* **2016**, *47*, 5288–5301. [CrossRef]
6. Matsuzaki, A.; Bhadeshia, H.K.D.H. Effect of austenite grain size and bainite morphology on overall kinetics of bainite transformation in steels. *Mater. Sci. Technol.* **1999**, *15*, 518–522. [CrossRef]
7. Rees, G.I.; Bhadeshia, H.K.D.H. Bainite transformation kinetics Part 1 Modified model. *Mater. Sci. Technol.* **1992**, *8*, 985–993. [CrossRef]
8. Lan, L.Y.; Qiu, C.L.; Zhao, D.W.; Gao, X.H.; Du, L.X. Effect of austenite grain size on isothermal bainite transformation in low carbon microalloyed steel. *Mater. Sci. Technol.* **2011**, *27*, 1657–1663. [CrossRef]
9. Shome, M.; Gupta, O.P.; Mohanty, O.N. Effect of simulated thermal cycles on the microstructure of the heat-affected zone in HSLA-80 and HSLA-100 steel plates. *Metall. Mater. Trans. A* **2004**, *35*, 985–996. [CrossRef]
10. Lee, S.J.; Park, J.S.; Lee, Y.K. Effect of austenite grain size on the transformation kinetics of upper and lower bainite in a low-alloy steel. *Scr. Mater.* **2008**, *59*, 87–90. [CrossRef]
11. Lan, L.Y.; Qiu, C.L.; Zhao, D.W.; Gao, X.H.; Du, L.X. Effect of reheat temperature on continuous cooling bainite transformation behavior in low carbon microalloyed steel. *J. Mater. Sci.* **2013**, *48*, 4356–4364. [CrossRef]
12. abbasi, M.; Nelson, T.W.; Sorensen, C.D. Analysis of variant selection in friction-stir-processed high-strength low-alloy steels. *J. Appl. Crystallogr.* **2013**, *46*, 716–725. [CrossRef]
13. Lan, L.Y.; Kong, X.W.; Qiu, C.L. Characterization of coarse bainite transformation in low carbon steel during simulated welding thermal cycles. *Mater. Charact.* **2015**, *105*, 95–103. [CrossRef]
14. Zhao, H.; Wynne, B.P.; Palmiere, E.J. Effect of austenite grain size on the bainitic ferrite morphology and grain refinement of a pipeline steel after continuous cooling. *Mater. Charact.* **2017**, *123*, 128–136. [CrossRef]
15. Garcia de Andres, C.; Caballero, F.G.; Capdevila, C.; Alvarez, L.F. Application of dilatometric analysis to the study of solid-solid phase transformations in steels. *Mater. Charact.* **2002**, *48*, 101–111. [CrossRef]

16. Chen, X.W.; Qiao, G.Y.; Han, X.L.; Wang, X.; Xiao, F.R.; Liao, B. Effects of Mo, Cr and Nb on microstructure and mechanical properties of heat affected one for Nb-bearing X80 pipeline steels. *Mater. Des.* **2014**, *53*, 888–901. [CrossRef]
17. Lan, L.Y.; Yu, M.; Qiu, C.L. On the local mechanical properties of isothermally transformed bainite in low carbon steel. *Mater. Sci. Eng. A* **2019**, *742*, 442–450. [CrossRef]
18. Li, X.D.; Shang, C.J.; Ma, X.P.; Gault, B.; Subramanian, S.V.; Sun, J.B.; Misra, R.D.K. Elemental distribution in the martensite-austenite constituent in intercritically reheated coarse-grained heat-affected zone of a high-strength pipeline steel. *Scr. Mater.* **2017**, *139*, 67–70. [CrossRef]
19. Biss, V.; Cryderman, R.L. Martensite and retained austenite in hot-rolled low-carbon bainitic steel. *Metall. Mater. Trans.* **1971**, *2*, 2267–2276. [CrossRef]
20. Zhang, M.X.; Kelly, P.M. Determination of carbon content in bainitic ferrite and carbon distribution in austenite by using CBKLDP. *Mater. Charact.* **1998**, *40*, 159–168. [CrossRef]
21. Tomozawa, M.; Miyahara, Y.; Kako, K. Solute segregation onΣ3 and random grain boundaries in type 316L stainless steel. *Mater. Sci. Eng. A* **2013**, *578*, 167–173. [CrossRef]
22. Thibaux, P.; Metenier, A.; Xhoffer, C. Carbon diffusion measurement in austenite in the temperature range 500 to 900 °C. *Metall. Mater. Trans. A* **2007**, *38*, 1169–1176. [CrossRef]
23. Escobar, J.D.; Faria, G.A.; Wu, L.; Oliveira, J.P.; Mei, P.R.; Ramirez, A.J. Austenite reversion kinetics and stability during tempering of a Ti-stabilized supermartensitic stainless steel: Correlative in situ synchrotron X-ray diffraction and dilatometry. *Acta Mater.* **2017**, *138*, 92–99. [CrossRef]
24. Wan, X.L.; Wu, K.M.; Nune, K.C.; Li, Y.; Cheng, L. In situ observation of acicular ferrite formation and grain refinement in simulated heat affected zone of high strength low alloy steel. *Sci. Technol. Weld. Join.* **2015**, *20*, 254–263. [CrossRef]
25. Sainis, S.; Farahani, H.; Gamsjager, E.; van der Zwaag, S. An In-situ LSCM study on bainite formation in a Fe-0.2C-1.5Mn- 2.0Cr alloy. *Metals* **2018**, *8*, 498–512. [CrossRef]
26. Takayama, N.; Miyamoto, G.; Furuhara, T. Effects of transformation temperature on variant pairing of bainitic ferrite in low carbon steel. *Acta Mater.* **2012**, *60*, 2387–2396. [CrossRef]
27. Borgenstam, A.; Hillert, M.; Agren, J. Metallographic evidence of carbon diffusion in the growth of bainite. *Acta Mater.* **2009**, *57*, 3242–3252. [CrossRef]
28. Slane, J.A.; Wolverton, C.; Gibala, R. Experimental and theoretical evidence for carbon-vacancy binding in austenite. *Metall. Mater. Trans. A* **2004**, *35*, 2239–2245. [CrossRef]
29. Abe, T.; Tsukada, K.; Tagawa, H.; Kozasu, I. Grain boundary segregation behavior of phosphorus and carbon under equilibrium and non-equilibrium conditions in austenitic region of steels. *ISIJ Int.* **1990**, *30*, 444–450. [CrossRef]
30. Timokhina, I.B.; Liss, K.D.; Raabe, D.; Rakha, K.; Beladi, H.; Xiong, X.Y.; Hodgson, P.D. Growth of bainitic ferrite and carbon partitioning during the early stages of bainite transformation in a 2 mass% silicon steel studied by in situ neutron diffraction, TEM and APT. *J. Appl. Crystallogr.* **2016**, *49*, 399–414. [CrossRef]
31. Kempen, A.T.W.; Sommer, F.; Mittemeijer, E.J. The kinetics of the austenite-ferrite phase transformation of Fe-Mn: Differential thermal analysis during cooling. *Acta Mater.* **2002**, *50*, 3545–3555. [CrossRef]
32. Song, T.; Cooman, B.C.D. Effect of boron on the isothermal bainite transformation. *Metall. Mater. Trans. A* **2013**, *44*, 1686–1705. [CrossRef]
33. Madariaga, I.; Gutierrez, I. Role of the particle-matrix interface on the nucleation of acicular ferrite in a medium carbon microalloyed steel. *Acta Mater.* **1999**, *47*, 951–960. [CrossRef]

© 2018 by the authors. Licensee MDPI, Basel, Switzerland. This article is an open access article distributed under the terms and conditions of the Creative Commons Attribution (CC BY) license (http://creativecommons.org/licenses/by/4.0/).

Article

Microstructure Formation of Low-Carbon Ferritic Stainless Steel during High Temperature Plastic Deformation

Yi Shao, Xiaohua Li, Junjie Ma, Chenxi Liu * and Zesheng Yan

State Key Lab of Hydraulic Engineering Simulation and Safety, School of Materials Science & Engineering, Tianjin University, Tianjin 300072, China; shzyshy@163.com (Y.S.); lixh@bohaisteel.com (X.L.); 15002233349@163.com (J.M.); mouselcx@21cn.com (Z.Y.)
* Correspondence: cxliutju@163.com; Tel.: +86-22-8740-1873

Received: 11 March 2019; Accepted: 18 April 2019; Published: 20 April 2019

Abstract: In this paper, the effects of the deformation temperature, the deformation reduction and the deformation rate on the microstructural formation, ferritic and martensitic phase transformation, stress–strain behaviors and micro-hardness in low-carbon ferritic stainless steel were investigated. The increase in deformation temperature promotes the formation of the fine equiaxed dynamic strain-induced transformation ferrite and suppresses the martensitic transformation. The higher deformation temperature results in a lower starting temperature for martensitic transformation. The increase in deformation can effectively promote the transformation of DSIT ferrite, and decrease the martensitic transformation rate, which is caused by the work hardening effect on the metastable austenite. The increase in the deformation rate leads to an increase in the ferrite fraction, because a high density of dislocation remains that can provide sufficient nucleation sites for ferrite transformation. The slow deformation rate results in dynamic recovery according to the stress–strain curve.

Keywords: ferritic stainless steel; plastic deformation; dynamic strain-induced transformation

1. Introduction

Ferritic stainless steels have been widely used in railway transportation equipment, mining machinery, the auto industry and nuclear fission power plant components, because of their remarkable corrosion resistance, outstanding strength and toughness, good weld ability and high cost performance [1–5]. In particular, low-carbon 11–13% Cr ferritic stainless steels have lower material cost than austenitic stainless steels and high Cr ferritic stainless steels, due to their lower contents of Cr and Ni [6,7]. To maintain the ferritic phase at room temperature, the austenite-stabilization elements C, N, and Ni are strictly controlled at a low level [8,9].

Severe plastic deformation is a practical route for the improvement of mechanical properties in the ferritic stainless steels, since in their case it is difficult to realize the effect of phase transformation strengthening [10]. It has been recognized that ultrafine grain structure can be obtained during severe plastic deformation in ferritic stainless steels [11]. The increase in deformation temperature accelerates the kinetics of ultrafine grain evolution significantly. During hot deformation, dynamic recrystallization and dynamic recovery occur, while dynamically recovered and sub-microstructures can be obtained during warm deformation [12]. The dynamic recrystallization may occur in the coarse-grained structure during severe plastic deformation at ambient temperature [13]. However, due to their restricted ferrite-forming elements, low-carbon ferritic stainless steels may enter the austenite phase regions at high temperature. Thus, the effect of austenitic transformation and the subsequent decomposing of austenite on severe plastic deformation in low-carbon ferritic stainless steels should be considered, though it has been rarely reported until now.

This project focuses on the high temperature plastic deformation in the austenite phase region in a low-carbon ferritic stainless steel. The effects of the deformation temperature, the deformation reduction and the deformation rate on the microstructural formation, ferritic and martensitic phase transformation, stress–strain behaviors and micro-hardness in the low-carbon ferritic stainless steel were studied.

2. Experimental Details

The chemical composition of the employed low-carbon ferritic stainless steel is given in Table 1. The chemical compositions of the sample were obtained by inductively coupled plasma optical emission spectrometry (ICP-OES). The original state of the samples was the as-rolled plate, whose rolling parameters were that the rolling passes occurred six times, the total rolling reduction was 80%, and the finishing rolling temperature was about 600 °C. Due to this low finishing temperature during rolling, the dynamic recrystallization process would be incomplete, and thus the as-rolled microstructure remained at room temperature. The microstructure of the as-rolled sample is presented in Figure 1, which shows the typical rolled morphology. The grains are elongated along the rolling direction. Besides, the severe deformation would result in a high density of dislocation, although it cannot be observed in the optical micrograph, due to the low resolution.

Table 1. Chemical compositions of the employed low-carbon ferritic stainless steel (wt. %).

C	Cr	Si	Ni	Mn	Nb	Ti	Fe
0.01	11.54	0.2	0.57	1.12	0.09	0.12	Bal.

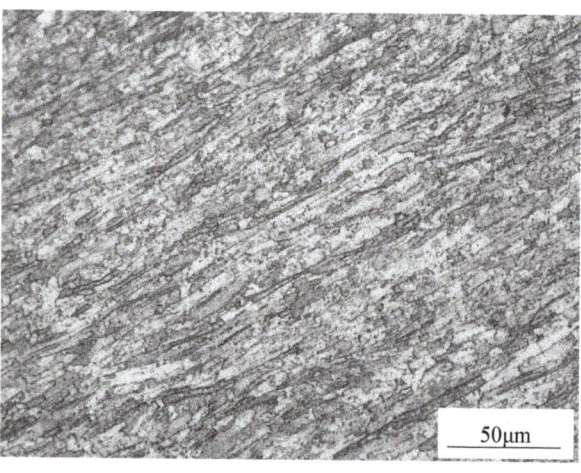

Figure 1. The original microstructure of the employed low-carbon ferritic stainless steel.

The deformation experiments were conducted on the Gleeble-3500 thermal simulated test machine (DSI, New York, NY, USA). In this paper, the effect of the deformation temperature, the deformation reduction (referring to the relative deformation in this project, the same as below) and the deformation rate were investigated. The process procedures are illustrated in Figure 2 and Table 2. For the deformation temperature experiments, the samples were heated to 1000 °C and held for 10 min, with a heating rate of 100 °C/min, followed by cooling to the deformation temperatures (600, 700, 800 and 900 °C), and then deformed with the deformation reduction of 0.9 and the deformation rate of 0.1 s^{-1}, before being finally cooled to room temperature with a cooling rate of 100 °C/min. For the deformation reduction experiments, the samples were also austenitized at 1000 °C for 10 min, followed by cooling to 900 °C and deformation with the different reductions (0.2, 0.5 and 0.9), and then cooled to room temperature. For the deformation rate experiments, the samples were also deformed at 900 °C after austenitization, with the different deformation rates (0.01, 0.1 and 1 s^{-1}).

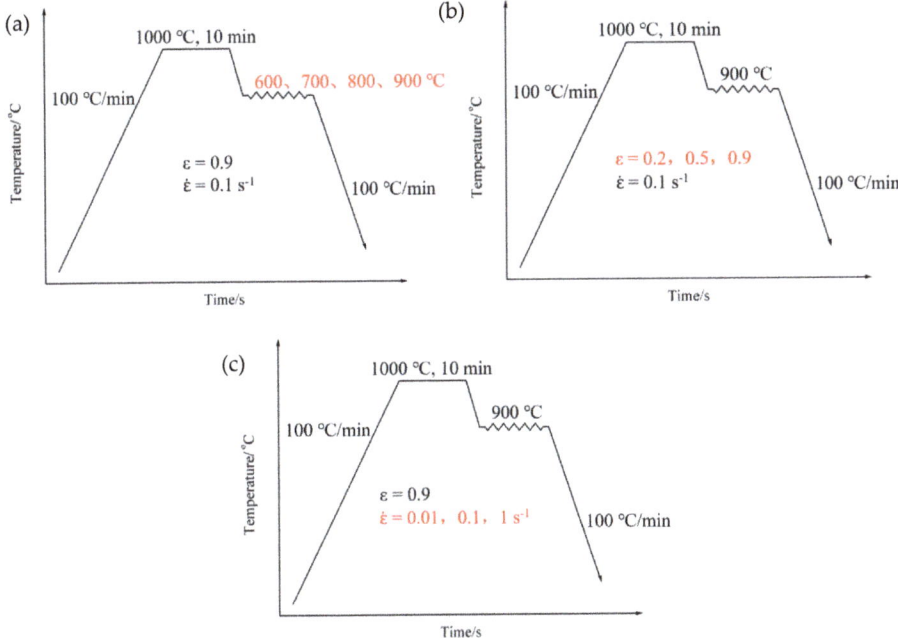

Figure 2. Schematic illustration of the different process parameters: (a) the different deformation temperatures, (b) the different deformation reductions, (c) the different deformation rates.

Table 2. The process parameters of the samples for the different conditions.

Experimental Conditions	Deformation Temperature/°C	Deformation Reduction	Deformation Rate/s^{-1}
For the different deformation temperatures	600, 700, 800, 900	0.9	0.1
For the different deformation reductions	900	0.2, 0.5, 0.9	0.1
For the different deformation rates	900	0.9	0.01, 0.1, 1

After the deformation experiments, the samples were mounted, polished, and etched in a solution of hydrochloric acid (15 mL), ethanol (150 mL), and ferric chloride (5 g). The microstructures were observed by the C-35A OLYMPUS Optical Microscope (Tokyo, Japan). The stress–strain curves and the phase transformation points were captured by the accessory dilatometer of the Gleeble-3500 thermal simulated equipment (DSI, New York, NY, USA). The Vickers micro-hardness tests were carried out by the Duramin-A300 Vickers hardness tester (Struers, Ohio, OH, USA), with a load of 200 g and a pressure time of 10 s.

3. Results and Discussion

3.1. Effect of the Deformation Temperature

As mentioned above, the original sample is in an as-rolled state, so the isothermal treatment at high temperature (annealing or normalizing) is necessary before deformation to remove the deformation texture and the residual stress caused by rolling. Figure 3 shows the equilibrium phase diagram of the employed low-carbon ferritic stainless steel calculated by JMatPro 7.0 software (Sente Software, Guildford, UK). It is found that, in the temperature range from 854 to 1068 °C, the steel is in the single-phase austenite region. Figure 4 gives the proof for the occurrence of austenitic transformation during heating with a rate of 100 °C/min. It can be seen that the inflection points resulted from austenitic transformation on the thermal expansion curve captured by Gleeble-3500. The A_{c1} and A_{c3} points can be determined as 837 and 991 °C. The transformation temperatures were determined from the dilatometer curves using the tangent line method [14].

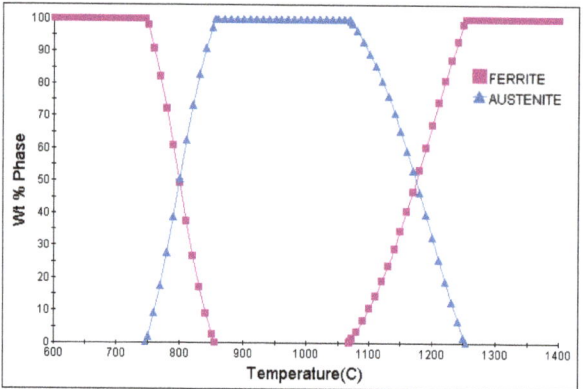

Figure 3. The equilibrium phase diagram of the employed low-carbon ferritic stainless steel calculated by JMatPro software.

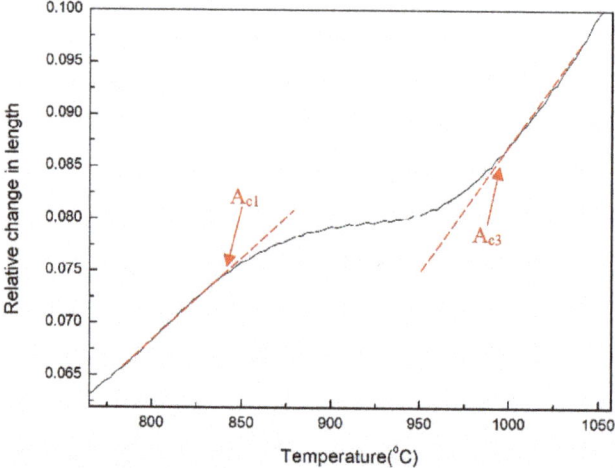

Figure 4. Thermal expansion curve of the low-carbon ferritic stainless steel sample during heating with a rate of 100 °C/min.

Figure 5 gives a continuous cooling transformation (CCT) diagram for the explored low-carbon ferritic stainless steel in this work, calculated by JMatPro software. It is found from Figure 5 that martensitic transformation occurs under the cooling rate between 3 °C/min (0.05 °C/s) and 6000 °C/min (100 °C/s). When the cooling rate is higher than 18 °C/min (0.3 °C/s), the ferritic transformation would not happen, and the austenite would be decomposed to martensite completely. According to the experimental CCT diagram for 3Cr12 steel [15], a typical low-carbon 12% Cr ferritic stainless steel, martensitic transformation still occurs when the cooling rate is as low as 0.084 °C/min. Figure 6 presents the microstructure of the low-carbon ferritic stainless steel sample after austenitization, whose treatment parameter is heating to 1000 °C with a heating rate of 100 °C/min, holding for 10 min, and cooling to room temperature with a cooling rate of 100 °C/min. Due to the moderate cooling rate after austenitization (air cooling), the bainitic and pearlite transformation would be avoided, and only martensite would form during cooling. As a result, it can be confirmed that the microstructure of the sample is composed of martensite (denoted as "M" in Figure 6), and δ-ferrite distributed along the prior austenitic boundaries (denoted as "F" in Figure 6).

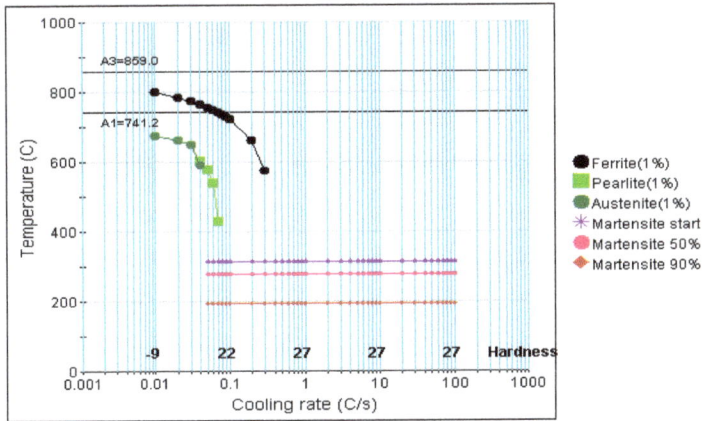

Figure 5. Continuous cooling transformation (CCT) diagram of the explored low-carbon ferritic stainless steel in this work calculated by JMatPro software.

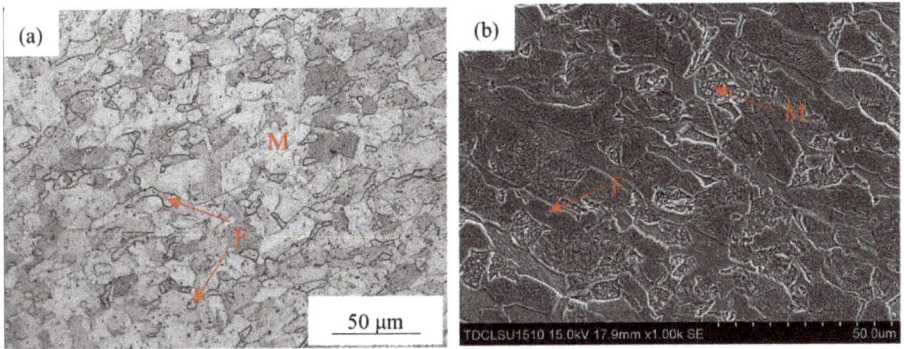

Figure 6. Microstructure of the low-carbon ferritic stainless steel sample after austenitization: (a) optical micrograph, (b) SEM micrograph.

A schematic illustration of the different deformation temperatures is given in Figure 2a. The deformation temperatures are 600, 700, 800 and 900 °C, respectively. Figure 7 presents the optical micrographs of the low-carbon ferritic stainless steel samples deformed at the different temperatures. It is found that the microstructures of all samples are composed of ferrite and lath martensite. The sample with the deformation temperature of 600 °C shows the typical rolled microstructure, with the elongated grain morphology. This suggests that the deformation temperature of 600 °C is too low to cause dynamic recrystallization. On the other hand, the deformation texture with the elongated grain is not found in the microstructures of the samples deformed at 700, 800 and 900 °C, which are composed of the martensitic laths, and the fine equiaxed ferrite grain (with the diameter of about 5 μm) distributed among the martensitic laths. With the increase in deformation temperature, the amount of the fine equiaxed ferrite is increased, which results in the microstructures appearing finer in the samples deformed at the higher temperature.

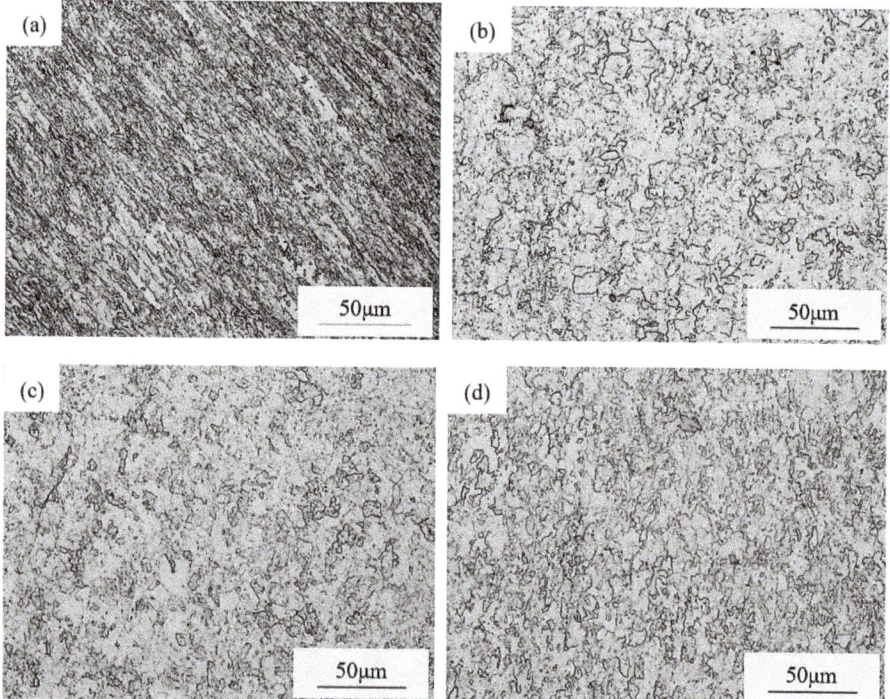

Figure 7. Optical micrographs of the low-carbon ferritic stainless steel samples deformed at the different temperatures: (**a**) 700 °C, (**b**) 800 °C, (**c**) 900 °C, (**d**) 1000 °C.

As discussed above, austenitic transformation occurs when heating up to 837 °C (namely A_{c1}), and austinite transforms to lath martensite when the cooling rate is higher than 18 °C/min. Hence, it is recognized that the meta-stable austenite remains intact before deformation. During or after deformation, the austenite would be transformed to lath martensite or ferrite. Furthermore, it also found that the higher deformation temperature is favorable to the formation of fine equiaxed ferrite, while the lower deformation temperature promotes the formation of lath martensite. Generally speaking, a high temperature would result in a coarse grain [16]. However, this phenomenon was not observed in this project. This may be due to the dynamic transformation and recrystallization of ferrite during the high temperature deformation.

Figure 8 shows the phase transformation temperatures, determined by Gleeble-3500, for the low-carbon ferritic stainless steel samples deformed at the different temperatures. It can be confirmed as martensitic transformation, in consideration of the low transformation temperature. With the increase in deformation temperature, the starting temperature for martensite transformation, M_s, is decreased, while the martensitic transformation finishing point remains almost unchanged, regardless of the deformation temperature. Besides, the M_s of the samples after deformation is higher than that of the samples without deformation (determined as 528 °C). This is because a large number of dislocations and other defects form during the deformation process, which provides more nucleation sites for martensitic transformation, and thus promotes martensitic transformation [17]. The higher the deformation temperature is, the more favorable the dynamic recovery of defects is. Hence, the number density of the defects is decreased accordingly, which results in a lower M_s in the sample with the higher deformation temperature.

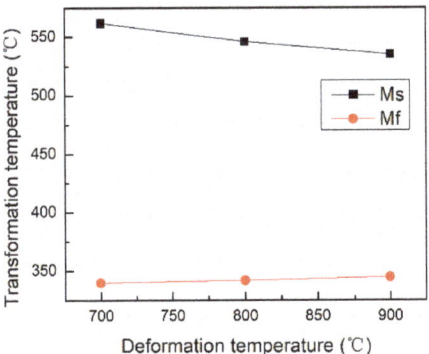

Figure 8. Phase transformation temperatures for the low-carbon ferritic stainless steel samples deformed at the different temperatures.

The formation of fine equiaxed ferrite was not detected in the thermal expansion curve determined by Gleeble-3500. This implies that ferritic transformation is likely to occur during the deformation process, rather than during continuous cooling after deformation. In general, this dynamic formation of fine ferrite during deformation is considered to be a dynamic strain-induced transformation (DSIT) [18]. When deformation is in the meta-stable austenite phase region, the dynamic strain-induced ferrite nucleates at the defects [19]. The movement of austenite-ferrite phase boundaries is hindered, due to the enhanced strength of the matrix caused by strain hardening, accompanied by dynamic recrystallization of ferrite grains. As a result, the DSIT ferrite grain size is very small.

The stress–strain curves of the samples with the different deformation temperatures are presented in Figure 9. A decrease in deformation temperature leads to an increase in loading stress for reaching the same strain amount, since the yield strength of the steel is decreased by the increase in temperature. The stress–strain curves of the samples with the deformation temperatures of 600, 700 and 800 °C are monotonically increased, showing the typical characteristic of work hardening. On the other hand, in the sample deformed at 900 °C, when the strain is more than 0.3, the stress–strain curve is horizontal, representing the dynamic softening process [20]. Firstly, during deformation at high temperature, dynamic recovery and recrystallization occurs, offsetting the work hardening effect resulting from dislocation multiplication [21,22]. Secondly, according to the microstructural observation, the higher deformation temperature promotes the DSIT ferrite formation, accompanied by the movement and annihilation of dislocation, which also results in the loss of work hardening effect [19].

Figure 10 shows the values for Vickers hardness of the samples deformed at the different temperatures. It can be seen that the results of the hardness tests agree with the stress–strain curves. With the increase in deformation temperature, the hardness is decreased. This is mainly due to the fact that the rate of dynamic recovery and recrystallization will be faster when the sample is deformed at the higher temperature, which offsets the effect of work hardening. The difference in the phase ratio of martensite to ferrite may also affect the hardness results. The size of the DSIT ferrite is higher up to about 5 μm in this project, so the hardness of ferrite would be lower than that of lath martensite [23]. As mentioned above, the lower deformation temperature is more favorable to the formation of martensite, leading to the relatively high hardness. Besides, the residual stress remains at the lower deformation temperature, which may also lead to an increase in hardness.

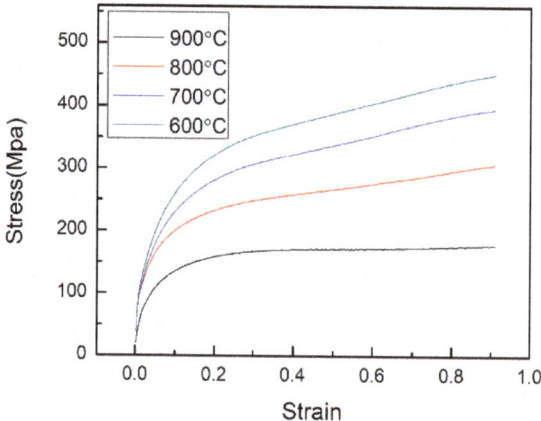

Figure 9. The true stress–strain curves for the low-carbon ferritic stainless steel samples deformed at the different temperatures.

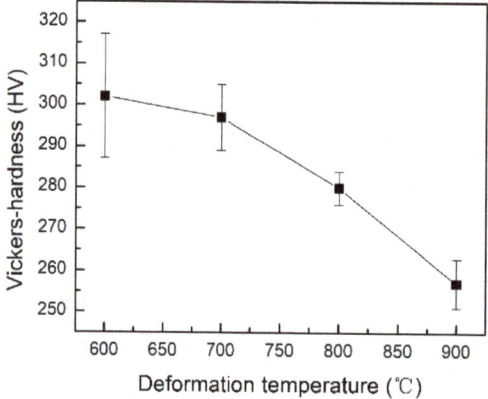

Figure 10. Vickers hardness of the low-carbon ferritic stainless steel samples deformed at the different temperatures.

3.2. Effect of Deformation Reduction

It can be recognized that the stress–strain curve for the sample deformed at 900 °C shows the feature of dynamic softening. Hence, the effect of deformation reduction was carried out at this temperature. A schematic illustration of the different deformation reductions is given in Figure 2b. Figure 11 presents the optical micrographs of the low-carbon ferritic stainless steel samples with the different deformation reductions. With the increase in deformation reduction, the fine equiaxed DSIT ferrite fraction is also increased. Obviously, a larger deformation reduction gives rise to more defects, and thus provides more nucleation sites for DSIT ferrite transformation.

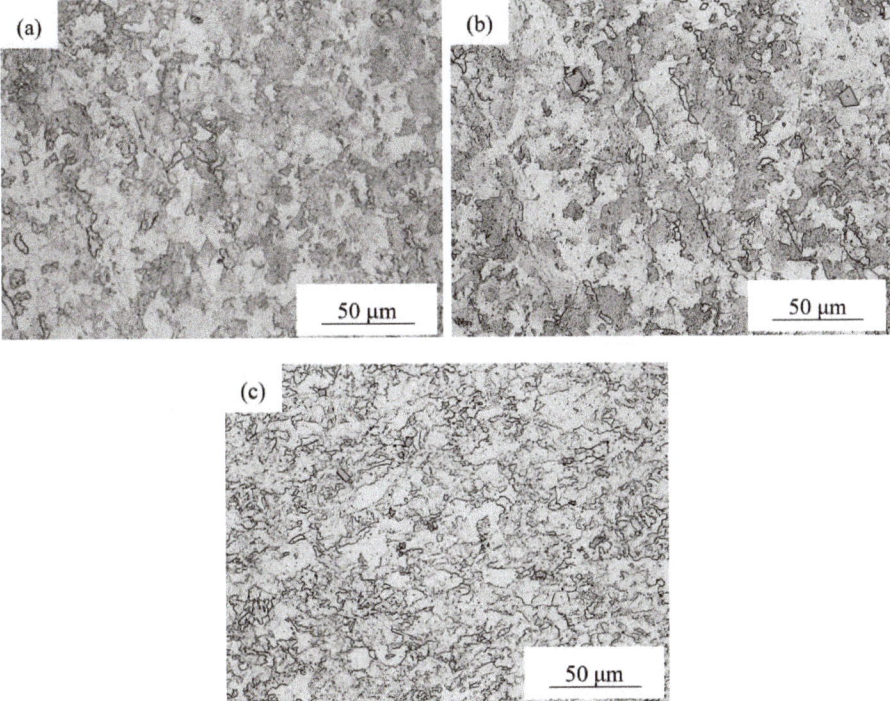

Figure 11. Optical micrographs of the low-carbon ferritic stainless steel samples with the different deformation reductions: (**a**) 0.2, (**b**) 0.5, (**c**) 0.9.

According to the stress–strain curve for the sample with the deformation reduction of 0.9 (Figure 12), before the deformation reduction reaches 0.2, the stress–strain curve belongs to the typical work hardening stage. In this stage, dislocation multiplication occurs with the increase in deformation. When the deformation reduction is between 0.2 and 0.5, the stress–strain curve steps into the stage of dynamic softening, including the dynamic recovery of dislocation and formation of DSIT ferrite. After the deformation reduction reaches 0.5, annihilation of dislocation and recrystallization of ferrite are dominant, resulting in the horizontal stress–strain curve.

Figure 13 presents the martensitic starting and finishing temperatures for the samples with the different deformation reductions. It is found that the M_s points are not significantly affected by the deformation reduction, but only slightly increased. Due to the high deformation temperature, the occurrence of recovery during cooling after deformation decreases the defect density, resulting in the decrease in nucleation sites for martensitic transformation accordingly. However, the M_f points are brought down by the increase in deformation reduction. This may be due to the fact that the movement of austenite/martensite phase boundaries is hindered during martensitic transformation. The martensitic transformation mechanism belongs to the non-diffusion type, with the shear-controlled phase transformation characteristics [24]. The yield strength of the parent phase (namely austenite) is improved by the increase in deformation reduction, impeding the migration of the martensite/austenite interface. Thus, the rate of martensitic transformation is decreased, reflected as the decrease in M_f points.

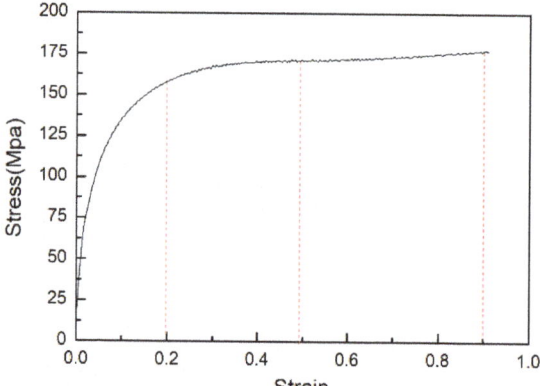

Figure 12. The true stress–strain curves for the low-carbon ferritic stainless steel samples with a deformation reduction of 0.9.

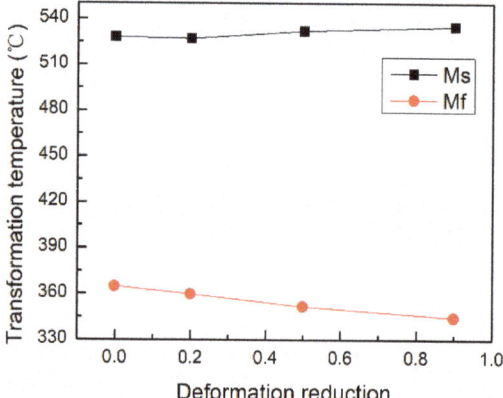

Figure 13. Phase transformation temperatures for the low-carbon ferritic stainless steel samples with the different deformation reductions.

Figure 14 shows the Vickers hardness of the samples with the different deformation reductions. The hardness values do not change greatly, though their general trend is to firstly increase and then decrease with the increase in deformation reduction. On the one hand, as discussed above, the hardness of lath martensite may be higher than that of ferrite. Therefore, with the increase in deformation reduction, the martensite fraction decreases, resulting in the decrease in hardness value. On the other hand, the increase in deformation increases the defect density, leading to the increase in strength and hardness. The combined effects of these two aspects result in a hardness value that does not monotonously increase or decrease with the increase in deformation reduction.

3.3. Effect of the Deformation Rate

A schematic illustration of the different deformation rates is given in Figure 2c. Figure 15 presents the optical micrographs of the samples with the different strain rates. It can be seen that the microstructures of all samples are composed of martensitic laths and fine ferrite grains. The increase in the strain rate leads to the increase in the phase fraction of ferrite and the decrease in lath martensite fraction. There is not enough time for the recovery of dislocations and other defects at the high deformation rate, which provides more nucleation positions for DSIT ferrite and thus greatly promotes the ferrite transformation.

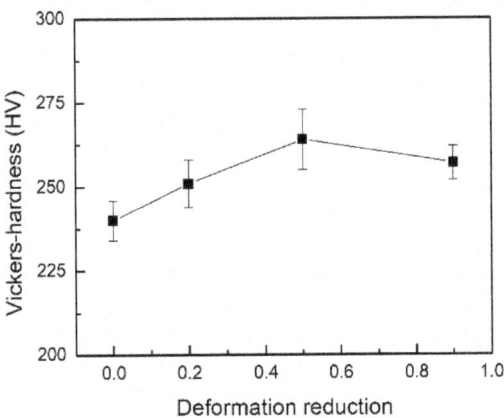

Figure 14. Vickers hardness of the low-carbon ferritic stainless steel samples with the different deformation reductions.

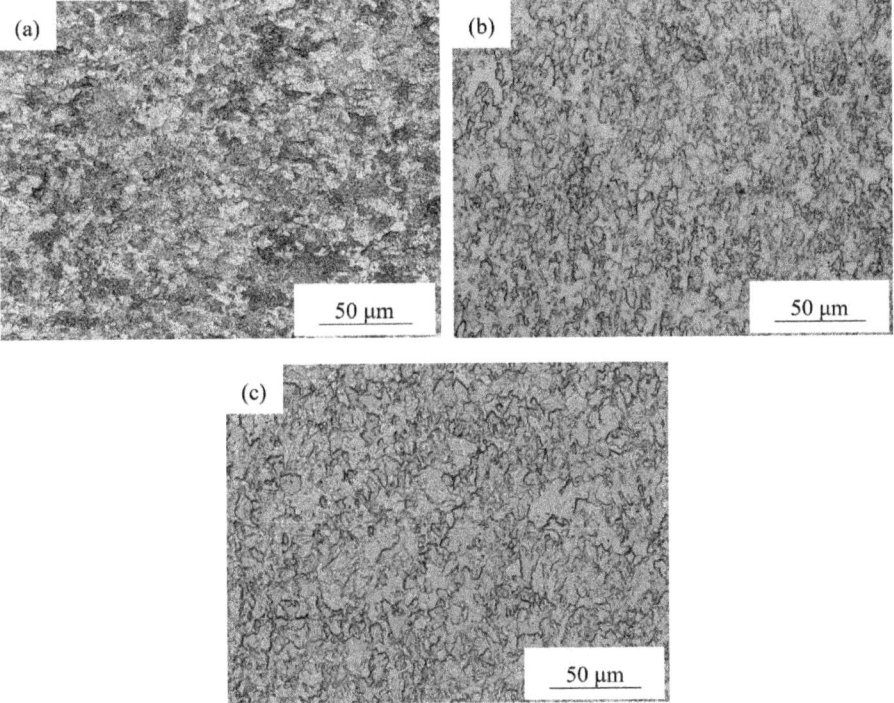

Figure 15. Optical micrographs of the low-carbon ferritic stainless steel samples with the different strain rates: (**a**) 0.01 s^{-1}, (**b**) 0.1 s^{-1}, (**c**) 1 s^{-1}.

The stress–strain curves for the different strain rates are shown in Figure 16. The stress–strain curve shows a more obvious softening with the decrease in the strain rate. The stress–strain curve of the sample undergoing the highest deformation rate (1 s^{-1}) reflects the typical work hardening characteristics. When the strain rate is the slowest (0.01 s^{-1}), the phase fraction of ferrite is the lowest. Thus, it is considered that the main origin of the softening phenomenon would be the dynamic recovery of dislocations, and the dynamic recrystallization of a small amount of ferrite grains.

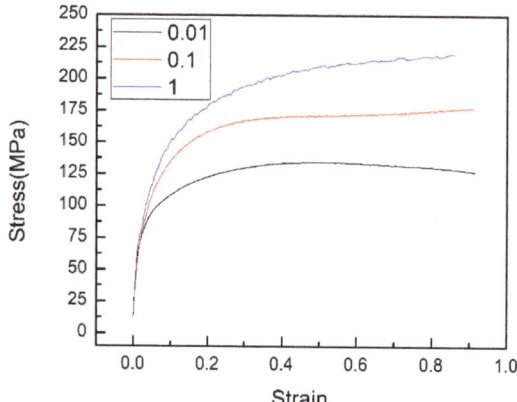

Figure 16. The true stress–strain curves for the low-carbon ferritic stainless steel samples with the different strain rates.

According to [25], the critical stress for recrystallization, σ_c, can be determined in the curves for the work-hardening rate, θ, versus flow stress, σ, where θ is expressed as $d\sigma/d\varepsilon$. As seen in Figure 17, the value for σ_c is associated with the point at which the second derivative of the work-hardening rate θ with respect to stress, i.e., $d^2\theta/d\sigma^2$, is zero. The simplest equation for fitting the θ-σ curve can be given by [26]:

$$\theta = A\sigma^3 + B\sigma^2 + C\sigma + D \tag{1}$$

where A, B, C and D are constants depending on the deformation parameters. Obviously, differentiation of the above equation with respect to σ is expressed as:

$$\frac{d\theta}{d\sigma} = 3A\sigma^2 + 2B\sigma + C \tag{2}$$

Thus, it can be derived that

$$\frac{d^2\theta}{d\sigma^2} = 6A\sigma + 2B \tag{3}$$

When $d^2\theta/d\sigma^2 = 0$, σ_c can be obtained by

$$\sigma_c = -\frac{B}{3A} \tag{4}$$

For the sample deformed at 900 °C, with a deformation rate of 0.01 s^{-1} the fitted polynomial is $\theta = -0.01122\sigma^3 + 3.83996\sigma^2 - 443.62646\sigma + 17452.03963D$ (see Figure 1 below). Hence, the critical stress for recrystallization, σ_c, is calculated as 114.08 MPa. However, σ_c for all other samples is higher than the loading stresses. It should be noted that the sample in Figure 1 undergoes the highest deformation temperature and the slowest deformation rate. As the deformation temperature or the deformation rate is decreased, the dynamic recrystallization is blocked.

The martensitic starting and finishing temperatures for the samples with the different deformation rates are presented in Figure 18. It is found that an increase in the strain rate results in an increase in

M_s. Under the condition of high strain rate, the high density of dislocation remains and provides more nucleation sites for the subsequent martensitic transformation, reflected as the increase in martensitic starting transformation temperature. Furthermore, the higher deformation rate results in a higher yield strength of the metastable austenite, which would hinder the movement of austenite/martensite boundaries during martensite. Hence, the increase in the deformation rate results in the decrease in the martensitic transformation rate, and thus the decrease in M_f.

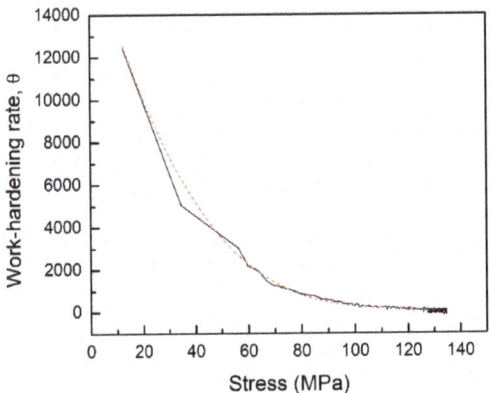

Figure 17. The experimental (black solid line) and fitted (red dashed line) θ–σ curves for the sample deformed at 900 °C, with a deformation rate of 0.01 s^{-1}.

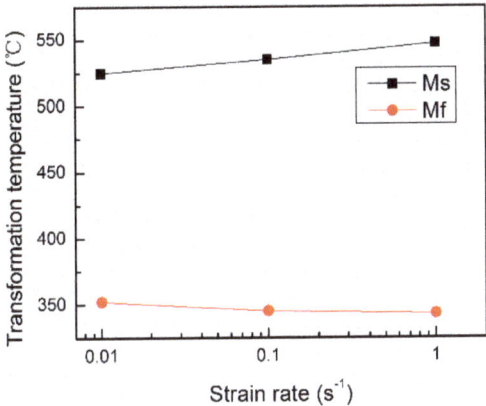

Figure 18. Phase transformation temperatures for the low-carbon ferritic stainless steel samples with the different strain rates.

Figure 19 shows the Vickers hardness of the samples with the different strain rates. With the increase in the strain rate, the hardness increases monotonically. This is attributed to the fact that the microstructure has sufficient time for recovery and recrystallization under the condition of slow strain rate, which leads to the reduction of hardness.

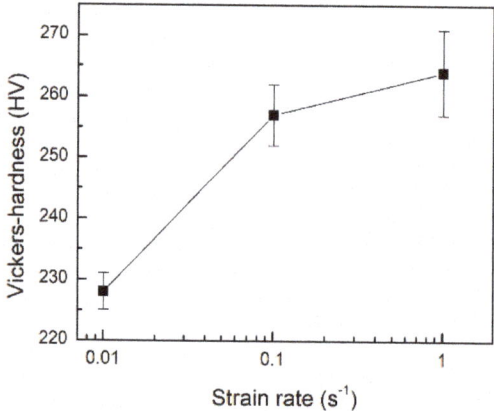

Figure 19. Vickers hardness of the low-carbon ferritic stainless steel samples with the different strain rates.

4. Conclusions

In this paper, the effects of the deformation temperature, the deformation reduction and the deformation rate on the microstructures, phase transformation behaviors, stress–strain curves and mechanical properties were investigated. The main conclusions are as follows:

(1) The increase in deformation temperature promotes the formation of the fine equiaxed DSIT ferrite and suppresses the martensitic transformation. The higher deformation temperature results in a lower starting temperature for martensitic transformation. When the deformation temperature reaches 900 °C, the stress–strain curve shows an obvious softening phenomenon, and the hardness decreases to minimum.

(2) The increase in deformation can effectively promote the transformation of DSIT ferrite, and decrease the martensitic transformation rate, which is caused by the work hardening effect on the metastable austenite.

(3) The increase in the deformation rate leads to an increase in the ferrite fraction, because a high density of dislocation remains that can provide sufficient nucleation sites for ferrite transformation. The slow deformation rate results in dynamic recovery according to the stress–strain curve.

Author Contributions: Conceptualization, C.L.; methodology, Y.S. and J.M.; software, C.L.; validation, X.L.; formal analysis, Y.S.; investigation, Y.S. and J.M.; resources, C.L. and Z.Y.; data curation, X.L.; writing—original draft preparation, Y.S.; writing—review and editing, C.L.; supervision, C.L. and Z.Y.; project administration, X.L.; funding acquisition, C.L.

Funding: This research was funded by National Magnetic Confinement Fusion Energy Research Project, grant number 2015GB119001 and the Project of Natural Science Foundation of Tianjin, grant number 18JCQNJC03300 and 18YFZCGX00050.

Conflicts of Interest: The authors declare no conflict of interest.

References

1. Huh, M.Y.; Engler, O. Effect of intermediate annealing on texture, formability and ridging of 17%Cr ferritic stainless steel sheet. *Mater. Sci. Eng. A* **2001**, *308*, 74–87. [CrossRef]
2. Yan, H.; Bi, H.; Li, X.; Xu, Z. Microstructure and texture of Nb + Ti stabilized ferritic stainless steel. *Mater. Charact.* **2008**, *59*, 1741–1746. [CrossRef]

3. Seo, M.; Hultquist, G.; Leygraf, C.; Sato, N. The influence of minor alloying elements (Nb, Ti and Cu) on the corrosion resistivity of ferritic stainless steel in sulfuric acid solution. *Corros. Sci.* **1986**, *26*, 949–960. [CrossRef]
4. Mohandas, T.; Reddy, G.M.; Naveed, M. A comparative evaluation of gas tungsten and shielded metal arc welds of a "ferritic" stainless steel. *J. Mater. Process. Technol.* **1999**, *94*, 133–140. [CrossRef]
5. Wang, J.; Qian, S.; Li, Y.; Macdonald, D.D.; Jiang, Y.; Li, J. Passivity breakdown on 436 ferritic stainless steel in solutions containing chloride. *J. Mater. Sci. Technol.* **2019**, *35*, 637–643. [CrossRef]
6. Song, C.; Guo, Y.; Li, K.; Sun, F.; Han, Q.; Zhai, Q. In Situ Observation of Phase Transformation and Structure Evolution of a 12 pct Cr Ferritic Stainless Steel. *Metall. Mater. Trans. B* **2012**, *43*, 1127–1137. [CrossRef]
7. Shao, Y.; Liu, C.; Yue, T.; Liu, Y.; Yan, Z.; Li, H. Effects of Static Recrystallization and Precipitation on Mechanical Properties of 00Cr12 Ferritic Stainless Steel. *Metall. Mater. Trans. B* **2018**, *49*, 1560–1567. [CrossRef]
8. Fujita, N.; Ohmura, K.; Yamamoto, A. Changes of microstructures and high temperature properties during high temperature service of Niobium added ferritic stainless steels. *Mater. Sci. Eng. A* **2003**, *351*, 272–281. [CrossRef]
9. Hu, X.; Du, Y.; Yan, D.; Rong, L. Effect of Cu content on microstructure and properties of Fe-16Cr-2.5Mo damping alloy. *J. Mater. Sci. Technol.* **2018**, *34*, 774–781. [CrossRef]
10. Maki, T. Stainless steel: Progress in thermomechanical treatment. *Curr. Opin. Solid State Mater. Sci.* **1997**, *2*, 290–295. [CrossRef]
11. Sakai, T.; Belyakov, A.; Miura, H. Ultrafine Grain Formation in Ferritic Stainless Steel during Severe Plastic Deformation. *Metall. Mater. Trans. A* **2008**, *39*, 2206. [CrossRef]
12. Dobatkin, S.V. Severe Plastic Deformation of Steels: Structure, Properties and Techniques. In *Investigations and Applications of Severe Plastic Deformation*; Lowe, T.C., Valiev, R.Z., Eds.; Springer: Dordrecht, The Netherlands, 2000; pp. 13–22.
13. Belyakov, A.; Kaibyshev, R. Structural changes of ferritic stainless steel during severe plastic deformation. *Nanostruct. Mater.* **1995**, *6*, 893–896. [CrossRef]
14. Liu, C.; Liu, Y.; Zhang, D.; Yan, Z. Kinetics of isochronal austenization in modified high Cr ferritic heat-resistant steel. *Appl. Phys. A* **2011**, *105*, 949–957. [CrossRef]
15. van Warmelo, M.; Nolan, D.; Norrish, J. Mitigation of sensitisation effects in unstabilised 12%Cr ferritic stainless steel welds. *Mater. Sci. Eng. A* **2007**, *464*, 157–169. [CrossRef]
16. Wu, Y.; Liu, Y.; Li, C.; Xia, X.; Wu, J.; Li, H. Coarsening behavior of γ' precipitates in the $\gamma' + \gamma$ area of a Ni3Al-based alloy. *J. Alloys Compd.* **2019**, *771*, 526–533. [CrossRef]
17. Liu, C.; Zhao, Q.; Liu, Y.; Wei, C.; Li, H. Microstructural evolution of high Cr ferrite/martensite steel after deformation in metastable austenite zone. *Fusion Eng. Des.* **2017**, *125*, 367–371. [CrossRef]
18. Zheng, C.; Xiao, N.; Hao, L.; Li, D.; Li, Y. Numerical simulation of dynamic strain-induced austenite–ferrite transformation in a low carbon steel. *Acta Mater.* **2009**, *57*, 2956–2968. [CrossRef]
19. Liu, Y.; Shao, Y.; Liu, C.; Chen, Y.; Zhang, D. Microstructure Evolution of HSLA Pipeline Steels after Hot Uniaxial Compression. *Materials* **2016**, *9*, 721. [CrossRef]
20. Dong, J.; Li, C.; Liu, C.; Huang, Y.; Yu, L.; Li, H.; Liu, Y. Hot deformation behavior and microstructural evolution of Nb–V–Ti microalloyed ultra-high strength steel. *J. Mater. Res.* **2017**, *32*, 3777–3787. [CrossRef]
21. Chen, J.; Liu, Y.; Liu, C.; Zhou, X.; Li, H. Study on microstructural evolution and constitutive modeling for hot deformation behavior of a low-carbon RAFM steel. *J. Mater. Res.* **2017**, *32*, 1376–1385. [CrossRef]
22. Zhou, Y.; Liu, Y.; Zhou, X.; Liu, C.; Yu, L.; Li, C.; Ning, B. Processing maps and microstructural evolution of the type 347H austenitic heat-resistant stainless steel. *J. Mater. Res.* **2015**, *30*, 2090–2100. [CrossRef]
23. Mao, C.; Liu, C.; Yu, L.; Li, H.; Liu, Y. The correlation among microstructural parameter and dynamic strain aging (DSA) in influencing the mechanical properties of a reduced activated ferritic-martensitic (RAFM) steel. *Mater. Sci. Eng. A* **2019**, *739*, 90–98. [CrossRef]
24. Tamura, I. Deformation-induced martensitic transformation and transformation-induced plasticity in steels. *Met. Sci.* **1982**, *16*, 245–253. [CrossRef]

25. Jonas, J.J.; Quelennec, X.; Jiang, L.; Martin, É. The Avrami kinetics of dynamic recrystallization. *Acta Mater.* **2009**, *57*, 2748–2756. [CrossRef]
26. Najafizadeh, A.; Jonas, J.J. Predicting the Critical Stress for Initiation of Dynamic Recrystallization. *ISIJ Int.* **2006**, *46*, 1679–1684. [CrossRef]

© 2019 by the authors. Licensee MDPI, Basel, Switzerland. This article is an open access article distributed under the terms and conditions of the Creative Commons Attribution (CC BY) license (http://creativecommons.org/licenses/by/4.0/).

Article

Interaction between Microalloying Additions and Phase Transformation during Intercritical Deformation in Low Carbon Steels

Unai Mayo [1,2], Nerea Isasti [1,2], Jose M. Rodriguez-Ibabe [1,2] and Pello Uranga [1,2,*]

1. Materials and Manufacturing Division, CEIT, 20018 San Sebastian, Basque Country, Spain; umayo@ceit.es (U.M.); nisasti@ceit.es (N.I.); jmribabe@ceit.es (J.M.R.-I.)
2. Mechanical and Materials Engineering Department, Universidad de Navarra, Tecnun, 20018 San Sebastian, Basque Country, Spain
* Correspondence: puranga@ceit.es; Tel.: +34-943-212-800

Received: 29 August 2019; Accepted: 25 September 2019; Published: 27 September 2019

Abstract: Heavy gauge line pipe and structural steel plate materials are often rolled in the two-phase region for strength reasons. However, strength and toughness show opposite trends, and the exact effect of each rolling process parameter remains unclear. Even though intercritical rolling has been widely studied, the specific mechanisms that act when different microalloying elements are added remain unclear. To investigate this further, laboratory thermomechanical simulations reproducing intercritical rolling conditions were performed in plain low carbon and NbV-microalloyed steels. Based on a previously developed procedure using electron backscattered diffraction (EBSD), the discretization between intercritically deformed ferrite and new ferrite grains formed after deformation was extended to microalloyed steels. The austenite conditioning before intercritical deformation in the Nb-bearing steel affects the balance of final precipitates by modifying the size distributions and origin of the Nb (C, N). This fact could modify the substructure in the intercritically deformed grains. A simple transformation model is proposed to predict average grain sizes under intercritical deformation conditions.

Keywords: intercritical rolling; microalloying; microstructure; EBSD

---

1. Introduction

Intercritical rolling is extensively employed in the production of heavy gauge structural plates, with the aim of meeting the increasing material demands of a variety of structural applications. Rolling in the austenite/ferrite two-phase region has already been explored for plain carbon steels [1]. However, the effect of intercritical rolling for microalloyed steels is less investigated [2]. Therefore, a deeper understanding of the microstructural evolution under intercritical conditions and the influence of different austenite–ferrite balances at high temperature is required for microalloyed steels in order to define stable processing windows. It is well established that the addition of Nb as an alloying element can retard or inhibit the recrystallization of austenite and ferrite due to two mechanisms: the solute drag effect related to Nb in solid solution and the pinning effect caused by strain-induced precipitation [3].

In intercritical rolling, several microstructural mechanisms could be activated, such as restoration and recrystallization. The recovery and recrystallization phenomena occurring during deformation in the two-phase region has been extensively analyzed for CMn steels [4,5]. It is well known that the restoration process taking place during or after the intercritical deformation could be affected by the available niobium during austenite to ferrite transformation. However, the interaction between Nb in solution and softening kinetics is less explored in the intercritical region [2,6]. Therefore, a deeper understanding is needed regarding this issue.

In a recently published work [7], the microstructural evolution during intercritical deformation was explored for low carbon steels, and a methodology capable of differentiating different ferrite populations (intercritically deformed and non-deformed ferrite formed during the final cooling) using EBSD was developed. This methodology will provide a better understanding of the exact effect of the rolling process parameters on each ferrite population. For this purpose, intercritical deformation simulations were carried out via dilatometry tests using CMn steels with different C content, and an exhaustive EBSD characterization procedure was developed to classify and quantify the different phases obtained after air cooling [7]. The procedure can be summarized in two steps. First, pearlite has to be removed from the calculations, and to that end, the grain average image quality (IQ) parameter is used [7,8]. Taking into account that pearlite is of a lower quality than ferrite, the lowest IQ value points are removed. The removed fraction from EBSD scans are close to the pearlite contents measured by optical microscopy. Then, using the grain orientation spread (GOS) parameter, which is the average deviation between the orientation of each point in the grain and the average orientation of the grain, the remaining ferrite grains are separated in two populations: DF (deformed ferrite) and NDF (non-deformed ferrite). This differentiation will allow for a better understanding of the effect of the different parameters and processes, such as restoration, precipitation etc., that occur during intercritical deformation for each NDF and DF family. It is assumed that during deformation, a distortion in the crystal lattice is introduced, leading to higher GOS parameter values [7,9]. Therefore, the GOS parameter distribution is strongly affected by the ferrite content prior to intercritical deformation. In the recently published work, a GOS value of 2° was set to differentiate NDF from DF, yielding reasonable results. This procedure was developed for polygonal ferritic microstructures transformed from fine austenitic structures. However, in the current work, slightly different intercritically deformed microstructures were formed due to the addition of microalloying elements. Depending on the austenite condition and chemical composition, different ferrite morphology could be achieved. Coarser austenite grains, as well as addition of microalloying elements, promote the delaying of phase transformation, leading to the formation of more non-polygonal transformation products [10,11]. In this study, given that more bainitic phases, such as quasi-polygonal ferrite, are observed, differentiating non-deformed and deformed ferrite grains becomes more complex. Therefore, the threshold able to distinguish non-deformed ferrite from deformed ferrite (previously shown in [7]) has to be redesigned.

The current work shows the complex interaction between austenite–ferrite content prior to deformation, microalloying elements, austenite condition (recrystallized and deformed austenite) and microstructural evolution during intercritical rolling. To that end, intercritical deformation simulations were performed in a deformation dilatometer for CMn and NbV-microalloyed steels. In addition to obtaining microstructural characterization by means of conventional characterization techniques (optical and electron microscopy), a precipitation analysis was also performed for the NbV-microalloyed steel.

2. Materials and Methods

The chemical compositions of the steels are listed in Table 1. The materials were laboratory cast and hot-rolled to 16 mm-thick plates/slabs.

Table 1. Chemical composition of the studied steels (weight percent).

Steels	C	Mn	Si	Cr	V	Ti	Al	Nb	N
CMn	0.063	1.53	0.25	0.012	0.005	0.002	0.035	0.002	0.003
NbV	0.062	1.52	0.25	0.012	0.034	0.002	0.038	0.056	0.004

Uniaxial compression tests, depicted schematically in Figure 1, were performed using a Bähr DIL805D deformation dilatometer (BÄHR Thermoanalyse GmbH, Hüllhorst, Germany). Solid cylinders of 5 mm in diameter and 10 mm in length were used. Two different thermomechanical schedules were defined with the purpose of obtaining different austenite conditions (recrystallized and deformed austenite, in Cycle A and Cycle B, respectively) prior to transformation. As shown in Figure 1, both schedules include a solubilization treatment at 1250 °C for 15 min to ensure the total dissolution of Nb and V precipitates. Afterwards, a multipass deformation sequence was designed. Both cycles include a deformation of 0.4 at 1050 °C, in order to ensure a fine recrystallized austenite. In Cycle B, a second deformation pass is applied at 900 °C, below the non-recrystallized temperature, in order to obtain a deformed austenite prior to transformation. Cycle A was applied to the CMn steel, whereas both schedules were applied to the NbV microalloyed steel. The samples were cooled down slowly (1 °C/s) to three different deformation temperatures (Tdef25, Tdef50, and Tdef75), in order to obtain three ferrite fractions (25%, 50%, and 75%) before the intercritical deformation. Finally, a deformation of 0.4 was applied in the intercritical region. After that, the specimens were cooled down to room temperature (1 °C/s) in both steels.

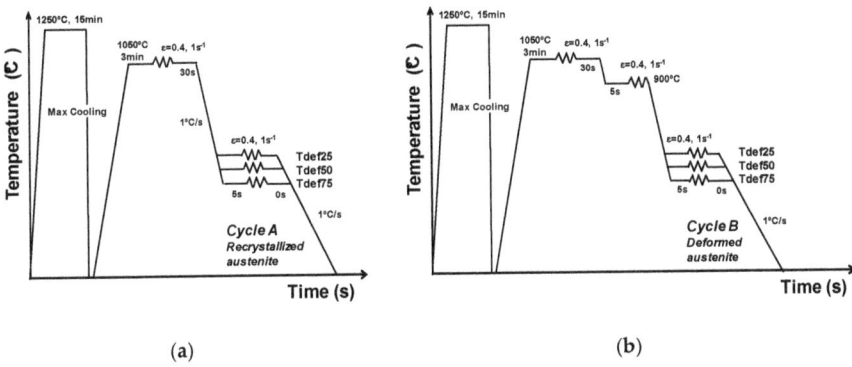

Figure 1. Thermomechanical schedule performed in the dilatometer: (a) CMn steel and NbV-microalloyed steel; (b) NbV-microalloyed steel only.

The deformation temperatures for achieving the predefined 25%, 50%, and 75% ferrite fractions were determined based on the dilatometry curves. The lever rule was considered in defining the evolution of transformed ferrite fraction and proportionality between transformed fraction and measured sample length change was assumed. The lever rule is based on extrapolating the linear expansion behavior from the temperature regions where no transformation occurs and subsequently assuming proportionality between the fraction of decomposed austenite and the observed length change [12]. The ferrite fractions prior to deformation were measured in each deformation temperature by systematic manual point count [13].

The dilatometry samples for the corresponding microstructural characterization were prepared according to the following procedure. First, the selected samples were cut along their longitudinal axis, at the region corresponding to a maximum area fraction of nominal strain and reduced strain gradient [14]. After that, the samples were mechanically grounded with SiC abrasive papers and polished with different diamond paste grades (6, 3, and 1 µm) to get mirror surfaces. Finally, the samples were etched with a 2% Nital solution. The microstructures were analyzed using different characterization techniques like optical microscopy (OM, LEICA DMI5000 M, Leica Microsystems, Wetzlar, Germany) and field-emission gun scanning electron microscopy (FEG-SEM, JEOL JSM-7000F, JEOL Ltd., Tokyo, Japan).

In order to evaluate the crystallographic features of the dilatometry tests in more detail, EBSD scans were performed for all cases. For that purpose, samples were polished with a colloidal silica suspension. Orientation imaging microscopy was performed on the Philips XL 30CP SEM with W-filament using TSL (TexSEM Laboratories, UT, USA) equipment. The total scanned area of the

EBSD mappings was 350 × 350 µm², using a step size of 0.5 µm and an accelerating voltage of 20 kV. The scans were analyzed using TSL OIM™ Analysis 5.31 software (EDAX, Mahwah, NJ, USA). During the post-processing, a clean-up procedure was applied to the raw data obtained from the scans to assimilate any non-indexed points into the surrounding neighborhood grains. A single iteration dilation clean-up routine with a tolerance angle of 5° and a minimum size of 3 pixels was defined. In addition, the neighbor CI correlation procedure was taken into account, defining 0.1 as the minimum accepted value. If there was any point with a CI value lower than 0.1, the orientation and CI of the particular point were reassigned to match the orientation and CI of the neighbor with the maximum CI.

The size and morphology of the precipitates were studied by TEM on a scanning transmission electron microscope JEOL JEM 2100 (TEM, JEOL 2100, JEOL Ltd., Tokyo, Japan) operated at 200 kV. Carbon extraction replicas were obtained from dilatometry samples. Copper grids were employed to support the carbon replicas.

3. Results and Discussion

3.1. Definition of Deformation Temperatures

In Figure 2a, the dilation curves obtained during air cooling are plotted for both steels and austenite conditions, represented as ΔL/L0 as a function of temperature. Figure 2b illustrates the evolution of the transformed ferrite fraction for CMn, NbV-recrystallized austenite, and NbV-deformed austenite (these curves are obtained from the dilatometry curves shown in Figure 2a). In Figure 2b, the desired fractions of ferrite (25%, 50%, and 75%) are also drawn. As Figure 2 shows, depending on the chemical composition, the transformation temperature varies. Given the aim of generating different ferrite contents of approximately 25%, 50%, and 75%, deformation temperatures of 750, 740, and 730 °C were selected for the CMn steel. Meanwhile, for the NbV steel, the formation of different fractions of ferrite were achieved at deformation temperatures of 750, 730, and 720 °C in Cycle A and 770, 750, and 740 °C in Cycle B (Tdef25, Tdef50, and Tdef75, respectively).

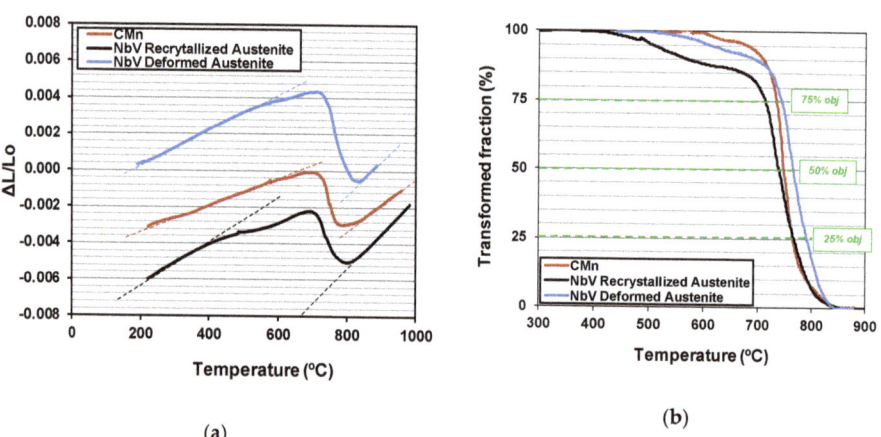

Figure 2. (a) Dilation curves and (b) evolution of transformed fraction in both steels and both austenite conditions for the NbV steel.

Using the deformation temperatures predicted in the dilatometry study as a reference, several interrupted quenching tests were performed to check that the ferrite content formed at those temperatures agreed with the objective ferrite fractions. Figure 3 shows the quenched microstructures obtained before different intercritical deformation temperatures for CMn steel (no deformation is applied in the austenite–ferrite domain). In Figure 4, the microstructures corresponding to NbV quenched specimens before intercritical deformation are shown for both austenite conditions and the entire range of deformation temperatures. The microstructures shown in Figures 3 and 4 illustrate the variation of the ferrite-martensite balance caused by the modification of deformation temperature, where more ferrite is formed as the temperature decreases. Together with the microstructure, the measured ferrite fraction is also noted in all cases. All these data are summarized in Table 2.

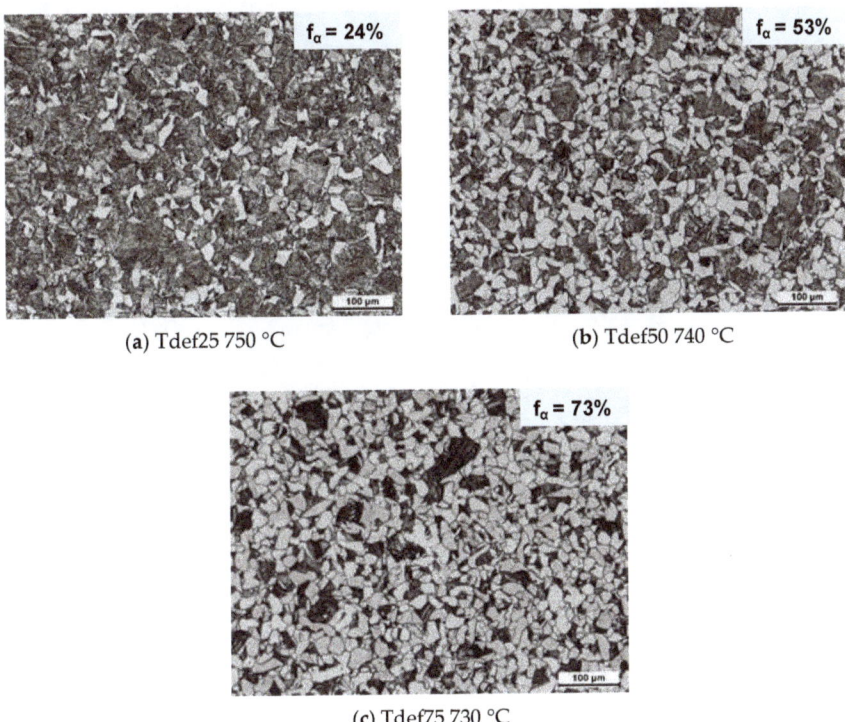

Figure 3. Optical micrographs obtained after a quenching at different deformation temperatures and CMn steel (before intercritical deformation): (**a**) Tdef25, (**b**) Tdef50, and (**c**) Tdef75.

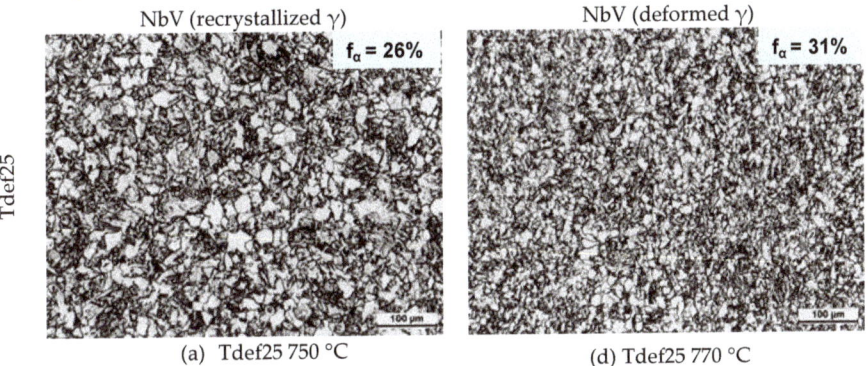

Figure 4. Cont.

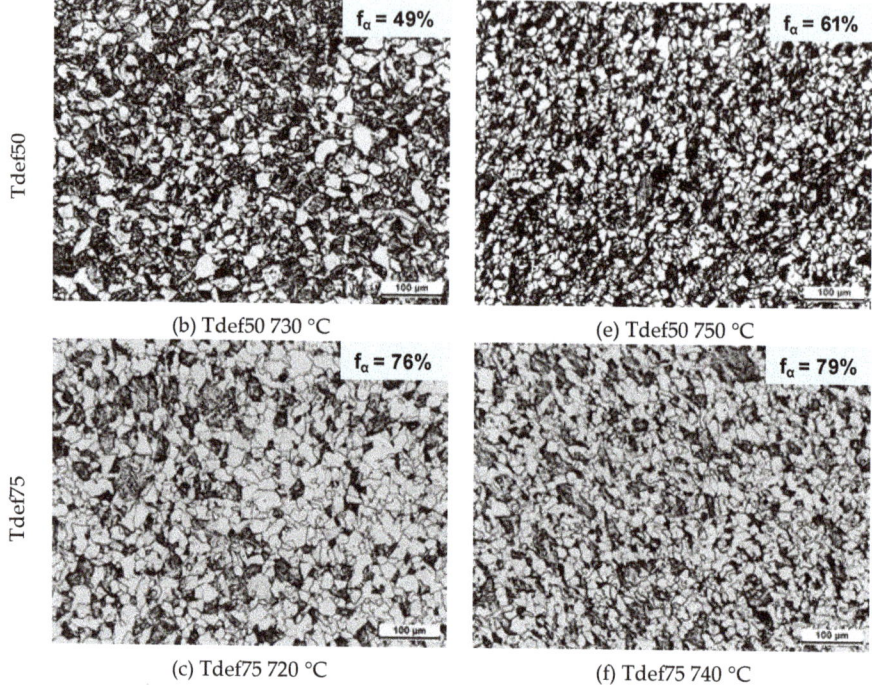

Figure 4. Optical micrographs obtained after a quenching at different deformation temperatures (before intercritical deformation) and both austenite conditions for NbV-microalloyed steel (recrystallized and deformed): (a,d) Tdef25, (b,e) Tdef50, and (c,f) Tdef75.

Table 2. Measured ferrite fraction in each deformation temperature, both chemical compositions, and austenite conditions after quenching.

Steel and Austenite Condition		25% Ferrite		50% Ferrite		75% Ferrite	
		Tdef25 (°C)	Measured $f\alpha$	Tdef50 (°C)	Measured $f\alpha$	Tdef75 (°C)	Measured $f\alpha$
NbV	Cycle A recrystallized γ	750	26%	730	49%	720	76%
	Cycle B deformed γ	770	31%	750	61%	740	79%
CMn	Cycle A recrystallized γ	750	24.3%	740	53.3%	730	73.1%

For NbV steel (see Figure 4), a microstructural refinement (finer ferrite grains) is observed when the transformation occurs from deformed austenite (Cycle B). This refinement is related to the increase in the density of nucleation sites introduced by deformation in the austenite [11]. The ferrite fractions prior to deformation were measured in each deformation temperature and the results are presented in Table 2. As shown in Table 2, the measured ferrite fractions are very close to the objective ones.

3.2. Microstructures at Room Temperature after Intercritical Deformation

In Figure 5, optical micrographs corresponding to room temperature after intercritical deformation for both steels (CMn and NbV) and both austenite conditions (recrystallized in Cycle A and deformed austenite in Cycle B) are presented. Combinations of non-deformed ferrite (NDF) and deformed ferrite (DF) are clearly distinguished in all the cases. Moreover, the formation of pearlite is observed in the resulting microstructures. Pearlite fractions of 13.7%, 6.1%, and 6.2% have also been measured for CMn, NbV recrystallized austenite, and NbV deformed austenite Tdef25 samples, respectively. In both chemical compositions (see Figure 5), the fraction of deformed ferrite increases as the deformation temperature decreases, due to a higher amount of ferrite formed prior to the intercritical deformation. This deformed ferrite is characterized by a significant presence of substructure (see Figure 5c), reflecting the fact that ferrite is restored after deformation in the intercritical region [6]. In addition, the morphology of non-deformed ferrite changes from a polygonal ferrite (see Figure 5d) to a quasi-polygonal ferrite (non-equiaxed) as deformation temperature decreases (see Figure 5f).

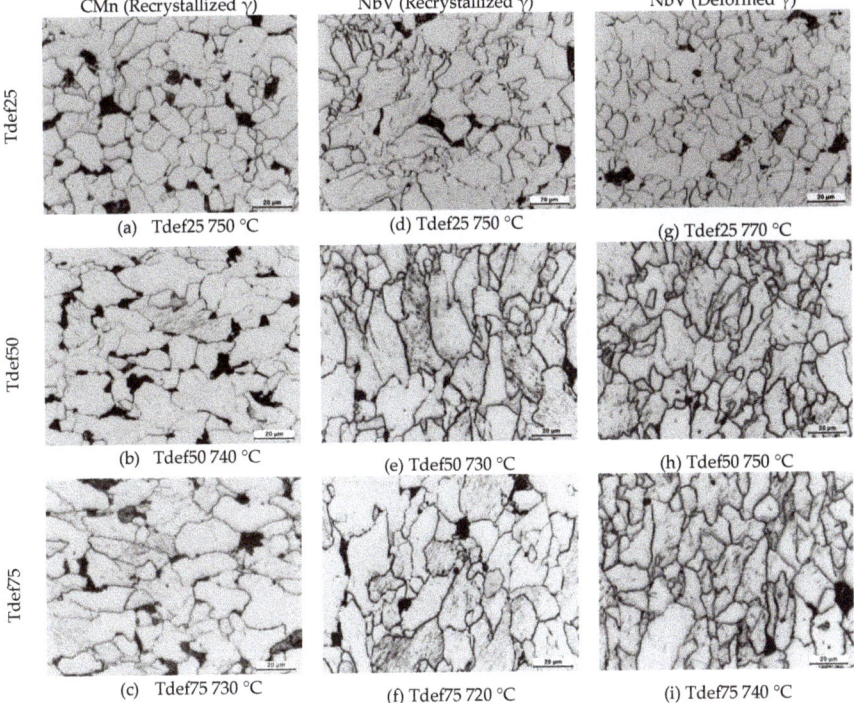

Figure 5. Optical micrographs obtained after air cooling at different deformation temperatures ((a,d,g) Tdef25, (b,e,h) Tdef50, and (c,f,i) Tdef75) and different conditions: (a,b,c) CMn recrystallized austenite, (d,e,f) NbV recrystallized austenite, and (g,h,i) NbV deformed austenite.

With the aim of evaluating the effect of deformation temperature and ferrite fraction before deformation on the final microstructure more precisely, the microstructural analysis was extended via EBSD. The EBSD technique was used to analyze the microstructural features of the samples deformed in the intercritical region, which cannot be correctly quantified by any other standard microstructural characterization techniques such as optical microscopy and/or FEG-SEM. In Figure 6, the grain boundary maps related to the NbV recrystallized austenite and CMn steels can be compared for Tdef25 (a,c) and Tdef75 (b,d). The boundaries between $2° < \vartheta < 15°$ are considered the low angle boundary, while those with $\vartheta > 15°$ are assumed to be high angle boundaries [14]. In Figure 6, the low and high angle boundaries are drawn in red and black, respectively. With regard to the effect of chemical composition, it is observed that the addition of Nb and V promotes a more intense substructure within the non-deformed ferrite grains (higher fraction of low angle boundaries drawn in red) within the ferrite grains, mainly in the highest ferrite content (see Figure 6b,d). This could be related to the effect of microalloying elements (mainly Nb) on the delaying of transformation start temperatures. It is known that the addition of Nb retards the austenite to ferrite transformation [15,16], which implies that ferrite will be formed at lower transformation temperatures, leading to the formation of more bainitic phases. Quasipolygonal ferrite grains are characterized by irregular grain boundaries and often show etching evidence of substructure [17].

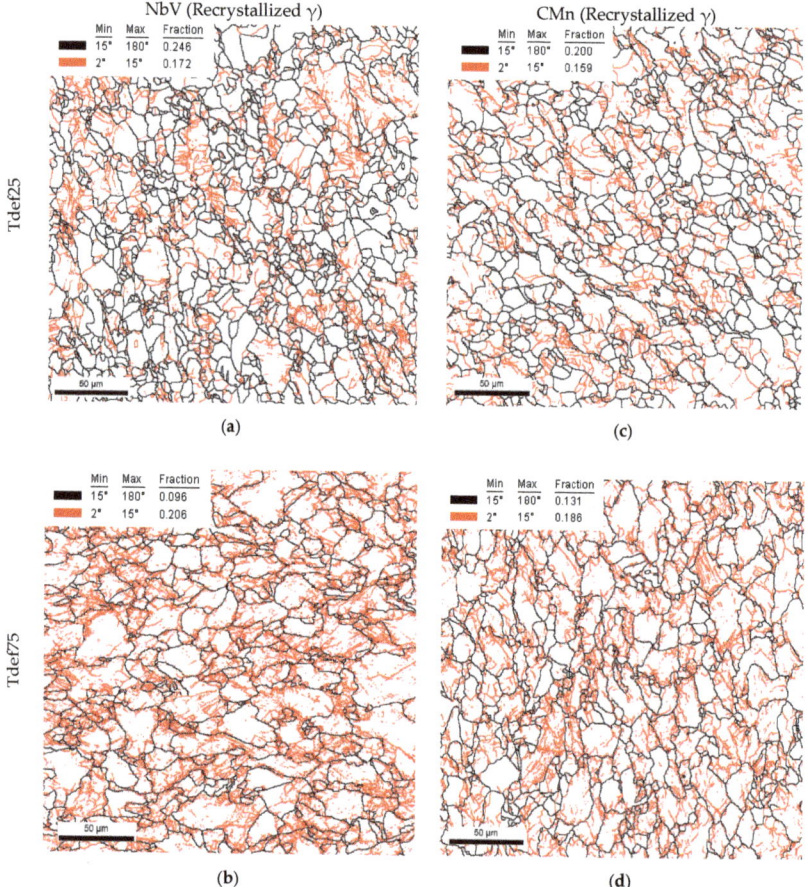

Figure 6. Grain boundary maps (low and high angle boundaries, in red and black color, respectively) corresponding to (**a**,**b**) NbV recrystallized austenite and (**c**,**d**) CMn. Different ferrite levels prior to deformation have been included: (**a**,**c**) Tdef25 and (**b**,**d**) Tdef75.

In addition, the EBSD maps shown in Figure 6 suggest that depending on the deformation temperature and, consequently, the fraction of ferrite prior to intercritical deformation, completely different microstructures are formed. When the microstructure contains a low fraction of ferrite prior to intercritical deformation, the presence of relatively polygonal ferrite grains is clearly observed (see Figure 6a,c). As the deformation temperature decreases, more elongated and coarser grains can be observed. As mentioned before, the applied deformation promotes the modification of the formed ferrite, reflected in the presence of a higher substructure. Figure 6 shows that the deformed ferrite is characterized by a significant presence of substructure, reflected in a higher fraction of low angle boundaries drawn in red. This could be clearly observed for both chemical compositions. In both steels, for Tdef75 condition, a lower deformation temperature than in Tdef25 is applied, leading to the formation of a more intense substructure as the deformation temperature decreases.

The differences between different ferrite morphologies can be appreciated more properly in the FEG-SEM micrographs shown in Figure 7, which presents the micrographs corresponding to the NbV steel and the lowest deformation temperatures (Tdef75). Figure 7a corresponds to the microstructure obtained after transformation from recrystallized austenite (Cycle A) and intercritically deformed at 720 °C, whereas in Figure 7b, the micrograph is related to the sample transformed from deformed austenite (Cycle B) with an intercritical deformation at 740 °C. The addition of microalloying elements, especially Nb, promotes the formation of more bainitic phases and reduces the presence of polygonal phases, leading to non-equiaxed grains being predominant in the microstructure [12]. This quasi-polygonal ferrite is observed in both austenite conditions of the NbV-microalloyed steel. However, the differences between the microstructures formed from recrystallized austenite and deformed austenite are clear. Furthermore, when the transformation occurs from recrystallized austenite (Figure 7a), deformation bands are identified inside the deformed ferrite grains, reflecting a lack of ferrite restoration. The addition of Nb delays or suppresses the restoration of ferrite when the transformation occurs from Cycle A (recrystallized austenite). The drag effect at low temperatures (ranging between 740 °C and 720 °C), delays and suppresses the restoration of ferrite during intercritical deformation [2,3]. However, in the sample transformed from deformed austenite (Figure 7b), there is a substructure composed of subgrains, which is associated with the activation of restoration during the deformation pass [13]. The deformation of austenite below the non-recrystallization temperature (T_{nr}) promotes strain-induced precipitates which are effective for austenite pancaking but reduce the Nb available during and after transformation to interact with ferrite restoration or recrystallization phenomena [3].

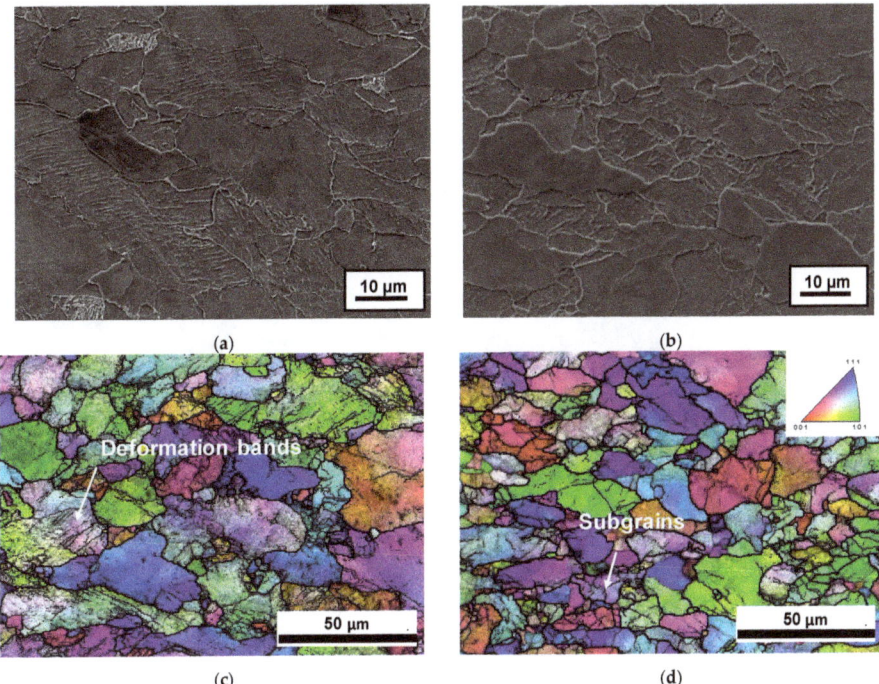

Figure 7. (**a**,**b**) FEG-SEM micrographs and (**c**,**d**) inverse pole figure (IPF) + image quality (IQ) maps corresponding to the NbV and Tdef75: (**a**,**c**) Cycle A (transformation from recrystallized austenite) and intercritical deformation temperature of 720 °C and (**b**,**d**) Cycle B (transformation from deformed austenite) and deformation temperature of 740 °C.

In Figure 7c,d, inverse pole figures (IPFs) and image quality maps are superimposed, as are high and low angle boundaries (coarse and fine black lines, respectively). As mentioned previously, a completely different substructure is observed inside the deformed ferrite grains for each austenite condition (see the low angle boundaries drawn in the IPF maps). When the ferrite comes from a deformed austenite, well-defined subgrains are observed in the deformed ferrite (see Figure 7d). By contrast, when the transformation occurs from a recrystallized austenite, the deformed ferrite presents microbands (see Figure 7c). The addition of Nb delays or suppresses the restoration of ferrite when transformation takes place from a recrystallized austenite, promoting the formation of microbands. In the sample transformed from deformed austenite, a clear substructure associated with the activation of restoration during the deformation pass is noticed [2–5,13].

3.3. Effect of Austenite Conditioning and Addition of Microalloying Elements

As shown in Figure 5, the microstructures generated after intercritical deformation are composed of different balances of NDF and DF, with some pearlite (P) islands dispersed in the ferritic matrix. In order to analyze the microstructural features of the deformed (DF) and non-deformed ferrite (NDF) populations separately, a recently developed discretization methodology was employed [7].

In the microstructures shown above, three different ferrite populations are identified: polygonal ferrite (PF), quasi-polygonal ferrite (QF), and deformed ferrite. GOS distributions were analyzed in each case, defining a grain tolerance angle of 5° and a minimum grain size of 0.91 µm. For each GOS distribution, a threshold GOS value was defined for differentiating deformed and non-deformed ferrite grains and optimum GOS values were defined for obtaining the desired ferrite fraction (25%, 50%, and 75%). From this analysis, a common value of 4° was estimated as the average of all the threshold angles defined in each scan. This threshold GOS value was able to distinguish both non-deformed and deformed ferrite in the final microstructure for both compositions, the ferrite content before deformation, and austenite condition (recrystallized and deformed ferrite). In this case, the ferrite population with GOS values lower than 4° is considered non-deformed ferrite, where PF and QF are included. GOS values higher than 4° correspond to deformed ferrite grains. As an example, in Figure 8, the grain boundary maps corresponding to both non-deformed (polygonal and quasi-polygonal ferrite in Figure 8a) and deformed ferrite families (in Figure 8b) are shown. The differentiation of both ferrite populations is shown for CMn steel and a deformation temperature of 740 °C (Tdef50).

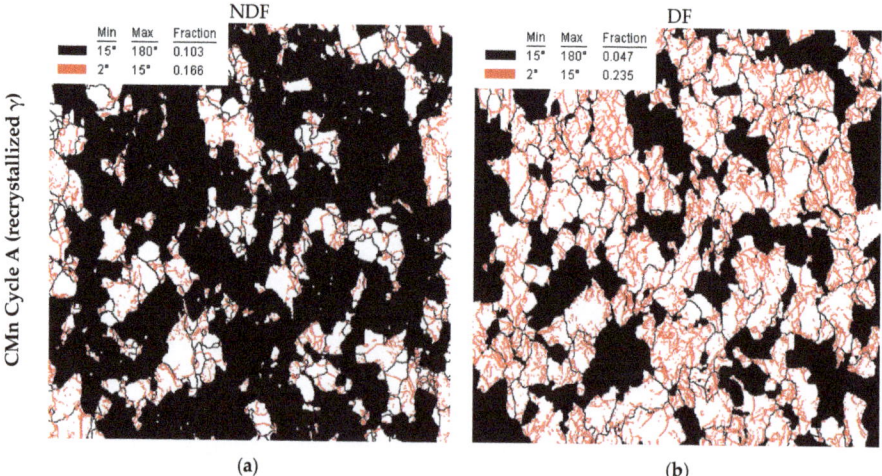

Figure 8. Grain boundary maps (low and high angle boundaries, in red and black color, respectively) corresponding to Tdef50 and CMn steel: (**a**) non-deformed ferrite (NDF) and (**b**) deformed ferrite (DF) family.

Based on the discretization methodology described above, high angle misorientation unit sizes for each ferrite type were quantified, considering the 15° misorientation criterion ($D_{15°}$). The effective grain size was calculated as the equivalent circle diameter corresponding to the individual grain area. The considered minimum grain size is equal to 3 pixels (equivalent to 0.9 µm for the 0.5 µm step size employed in the current EBSD analysis). In Figure 9a,b, unit size distributions are presented for both non-deformed and deformed ferrite families corresponding to Tdef50. Comparing the grain size distributions obtained in each ferrite population, significantly finer grains were measured for non-deformed ferrite (see Figure 9a) compared to deformed ferrite (Figure 9b). Additionally, in Figure 9c,d, mean grain size values are plotted for both ferrite types and different ferrite levels, as well as both chemistries and austenite conditions. As mentioned previously, in the analysis of the results shown in Figure 9c,d, considerably finer microstructures were obtained in the non-deformed ferrite formed during the final air cooling step. With regard to the mean grain size trends corresponding to non-deformed ferrite population, no significant effect of ferrite fraction before deformation and addition of microalloying elements on grain size is observed. Similar mean unit size values were measured in entire ferrite content for both steels and different austenite conditions. Nevertheless, a different behavior could be detected regarding the evolution of mean grain size of the deformed ferrite family. The addition of Nb and V promotes microstructural refinement when the transformation occurs from deformed austenite. In summary, a microstructural refinement is ensured when deformation is accumulated in the austenite prior to transformation. The benefit of the accumulation of deformation in the austenitic range is associated with the increase of the specific grain boundary, which leads to a significant increase in the density of ferrite nucleation sites introduced by deformation [18–21].

Different equations have been proposed in the literature for predicting the ferrite grain size after austenite-to-ferrite transformation for different steels [19,22–24]. All of the mentioned expressions take into account the initial austenite grain size (d_γ), the cooling rate under continuous cooling conditions ($\dot{T}$), and the accumulated strain in the austenite prior to transformation (ε_{acc}). However, depending on the equation, significant differences can be observed in the predicted ferrite grain size. After comparing the predicted ferrite grain size with the experimental grain size values, the approach proposed by Bengochea et al. [19] (see Equation (1)) was selected.

$$d_\alpha = \left(1 - 0.5\varepsilon_{acc}^{0.47}\right)\left(4.5 + 3\dot{T}^{-\frac{1}{2}} + 13.4(1 - \exp(-0.015d_\gamma))\right), \tag{1}$$

where d_γ, ε_{acc}, and $\dot{T}$ are the austenite mean grain size (in µm), the accumulated strain in the austenite prior to transformation, and the cooling rate (°C/s), respectively.

Even though Equation (1) [19] was initially developed for continuous cooling conditions (austenite–ferrite phase transformation), in the present work, the applicability of this approach for intercritically deformed microstructure was evaluated. Considering this equation, the intercritically deformed ferrite grain size, as well as the non-deformed ferrite grain size were predicted. In Table 3, besides the predicted mean ferrite grain sizes, the variables considered in each condition are summarized for both ferrite families. For the deformed ferrite population (DF), the mean austenite grain size at 1050 °C is considered (after the deformation pass at this temperature) (73 and 55 µm, for CMn and NbV-microalloyed steel, respectively). In the NbV deformed austenite condition, an accumulated strain (ε_{acc}) of 0.4 was considered in order to take into account the accumulation of deformation in the austenite prior to transformation. In analyzing the predicted mean ferrite grain sizes, it is observed that finer ferrite grains are estimated for NbV deformed austenite.

Regarding the non-deformed ferrite grain size (transformed after intercritical deformation and during the final air cooling step), the ε_{acc} term is considered to be 0.4 for CMn and NbV recrystallized austenite, whereas for NbV deformed austenite, a ε_{acc} of 0.8 is taken into account (the sum of the deformation applied below T_{nr} in austenite and deformation applied in the intercritical region). To estimate mean austenite grain size prior to transformation, the ferrite content prior to intercritical deformation must be taken into account [25,26]. It is well known that as ferrite content increases before deformation, the remaining austenite grain size decreases. Therefore, the remaining austenite sizes were measured in selected quenched samples prior to intercritical deformation (by measuring the martensitic regions in the quenched samples) in order to find the relation between ferrite fraction before intercritical deformation and the remaining austenite size. For example, in the NbV recrystallized austenite condition, the mean austenite size decreased from 47 to 26 µm, increasing the intercritically deformed ferrite fraction from 24% to 71%. As the content of ferrite prior to intercritical deformation increased, the predicted mean non-deformed ferrite size decreased, due to the reduction of the mean austenite grain size. For example, for NbV recrystallized austenite, calculated dα decreased from 9.6 to 8.0 µm when the ferrite fraction increased from 24% to 71%.

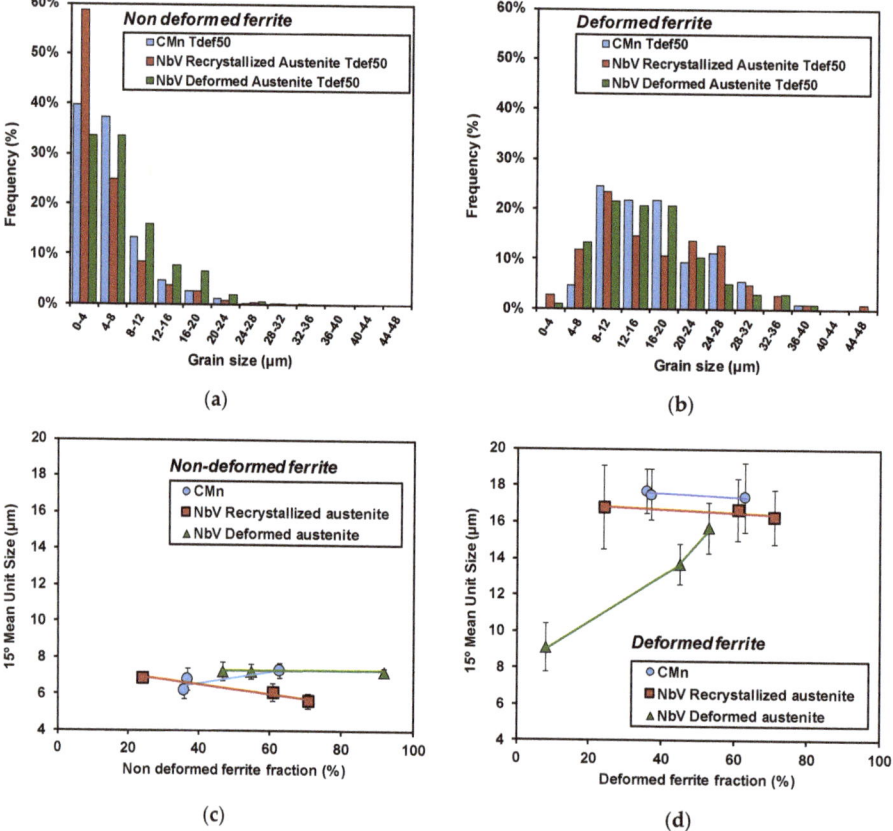

Figure 9. Grain size distributions corresponding to Tdef50 and different chemical composition (CMn, NbV recrystallized austenite, and NbV deformed austenite): (**a**) Non-deformed ferrite and (**b**) deformed ferrite population. Influence of austenite/ferrite balance, chemical composition, and austenite condition on 15° mean grain size of both ferrite populations: (**c**) non-deformed ferrite and (**d**) deformed ferrite population.

Table 3. Predicted mean ferrite grain sizes for both ferrite families considering Equation (1) [19] and the variables considered in each condition.

Ferrite Family	Condition	Ferrite Content Prior to Deformation (%)	Mean Austenite Size, $d\gamma$ (μm)	ε_{acc}	$\dot{T}$ (°C/s)	$d\alpha$ Predicted (μm) by Equation (1)
DF	CMn	37	73	0	1	16.4
		64	73	0	1	16.4
		63	73	0	1	16.4
	NbV recrystallized austenite	24	55	0	1	15
		61	55	0	1	15
		71	55	0	1	15
	NbV deformed austenite	8	55	0.4	1	10.1
		53	55	0.4	1	10.1
		45	55	0.4	1	10.1
NDF	CMn	37	36	0.4	1	8.9
		64	16	0.4	1	7.0
		63	16	0.4	1	7.0
	NbV recrystallized austenite	24	47	0.4	1	9.6
		61	31	0.4	1	8.4
		71	26	0.4	1	8.0
	NbV deformed austenite	8	52	0.8	1	8.1
		53	35	0.8	1	7.1
		45	38	0.8	1	7.3

In Figure 10, the predicted ferrite sizes are plotted as a function of ferrite grain size measured by the EBSD technique for the different compositions, austenite conditions, and ferrite families (DF and NDF, deformed and non-deformed ferrite, respectively). With regard to the non-deformed ferrite, a reasonable fitting can be observed for all the conditions. However, for the NDF population, the experimentally measured mean grain sizes were slightly larger than the predicted ferrite grain sizes, principally for the NbV deformed austenite sample. This deviation is more significant for Tdef50 Tdef75 conditions. For NbV grade, further analysis is required in order to understand the effect of deformation temperature on ferrite grain size when the transformation takes place from a deformed austenite.

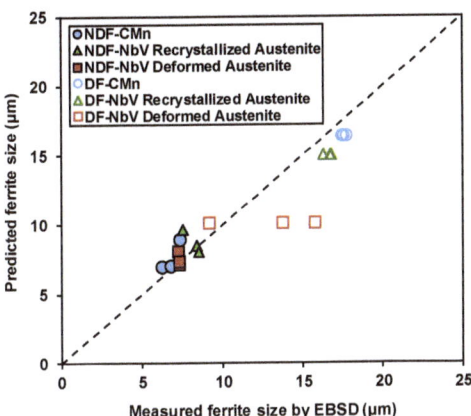

Figure 10. Predicted ferrite grain sizes considering Bengochea's [19] equation as a function of ferrite size measured by EBSD for both chemical composition, austenite conditions, and both ferrite populations (NDF and DF).

3.4. Interaction between Precipitation, Nb in Solution, and Intercritically Deformed Ferrite

In order to evaluate the role of Nb and V precipitates in the different processes (restoration, recrystallization) occurring during the intercritical rolling, a study of fine precipitates was carried out on carbon extraction replicas, and the average precipitate size was measured in selected conditions. Figure 11 shows differences in precipitation for both recrystallized and deformed austenite for the

Tdef75 condition. Regarding the effect of austenite conditioning, different precipitation populations were formed depending on the applied rolling strategy (recrystallized or deformed austenite). When the transformation occurs from recrystallized austenite, a high density of fine precipitates can be detected. These precipitates are considered to be formed in ferrite during or after the intercritical deformation. Nevertheless, in the sample corresponding to deformed austenite, strain-induced precipitates formed in austenite are also observed. In all cases, most of the particles are Nb-rich precipitates, as well as NbV and NbTiV-rich precipitates (see microanalysis shown in Figure 11e).

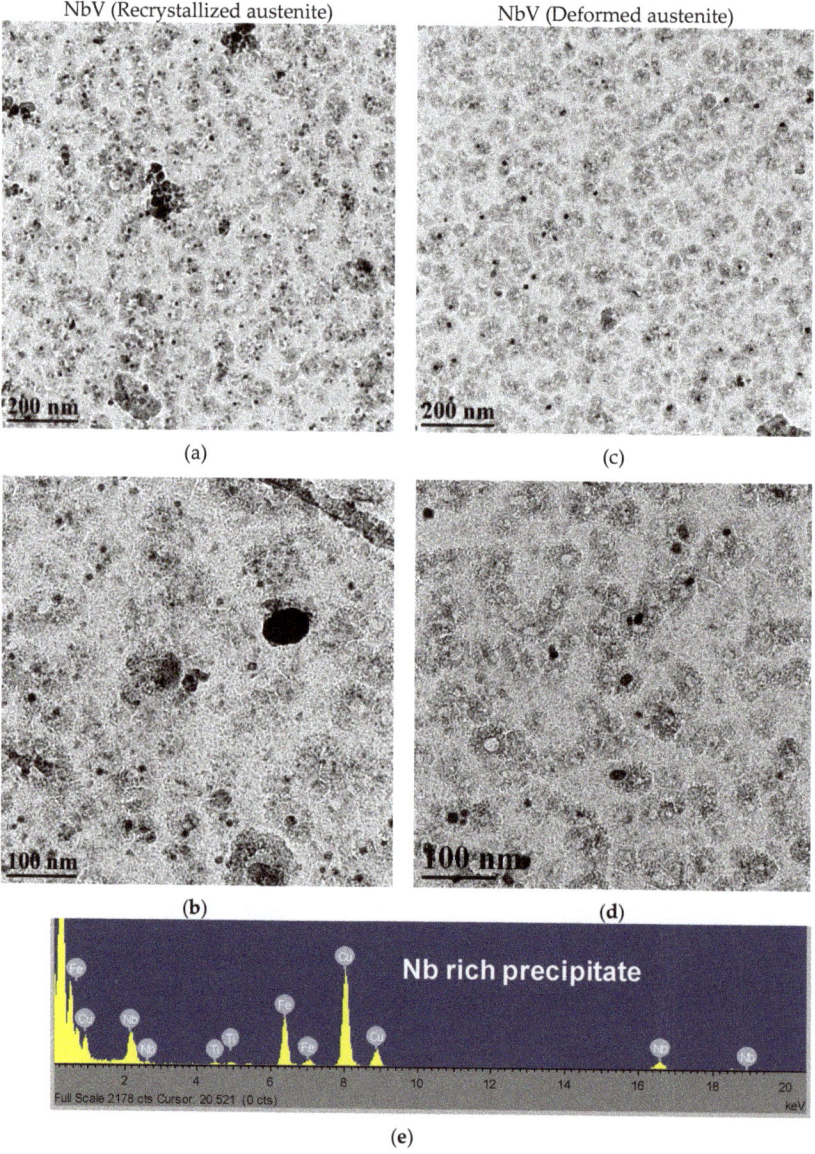

Figure 11. TEM images obtained in (**a**,**b**) NbV recrystallized austenite and (**c**,**d**) NbV deformed austenite at different magnifications for Tdef75. (**e**) Microanalysis of the precipitates.

In Figure 12a, the precipitate size distributions are plotted. The precipitate diameter distributions corresponding to recrystallized austenite and deformed austenite conditions (Tdef75 case) can be compared. The results indicate that noticeably finer precipitates were measured when transformation occurs from recrystallized austenite (mean precipitate size is decreased from 15.9 to 10.2 nm). This could be related to the fact that a lower intercritical deformation temperature is applied in Cycle A (recrystallized austenite), which leads to the formation of finer precipitates in ferrite. In addition, as previously mentioned, a higher density of fine precipitates was observed.

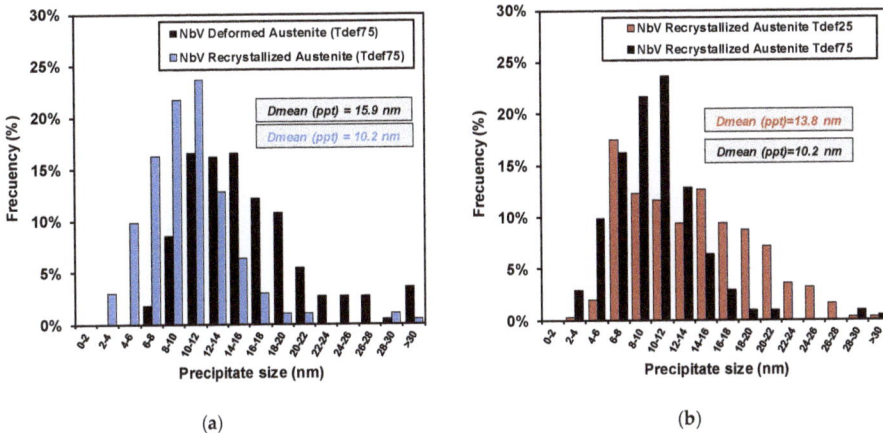

Figure 12. Precipitate size distribution measured for (a) NbV recrystallized austenite and NbV deformed austenite for Tdef75. (b) Tdef25 and Tdef75 for NbV recrystallized austenite.

The results indicate that the nanometric precipitates could suppress the restoration of ferrite, forming microbands when transformation occurs from recrystallized austenite. In the case of Cycle B (deformed austenite), strain-induced precipitates formed in the austenitic region—which are bigger but fewer in number—were not able to stop the restoration process that happens during or after the intercritical deformation. Differences in the niobium available during transformation are a key factor in maintaining the deformation bands or in developing a substructure within the intercritically deformed ferrite grains.

Besides the analysis of the influence of the accumulation of deformation in the austenite prior to transformation, the effect of ferrite fraction before intercritical deformation was also analyzed. An example is given in Figure 13, which shows the comparison between TEM images corresponding to Tdef25 and Tdef75. Furthermore, in Figure 12b, the precipitate size distributions are also presented. Quantified mean precipitate sizes are also included in the graphs. A significant precipitate size refinement is clearly noticeable as ferrite content prior to deformation increases and intercritical deformation temperature decreases. Moreover, as shown in Figure 13, a higher fraction of fine precipitates is observed for Tdef75. Mean precipitate size decreases from 13.8 to 10.2 nm when the fraction of ferrite before intercritical deformation increases from 25% to 75%. Finer precipitates are formed when lower intercritical deformation is applied, confirming that the precipitation occurs during or after intercritical deformation.

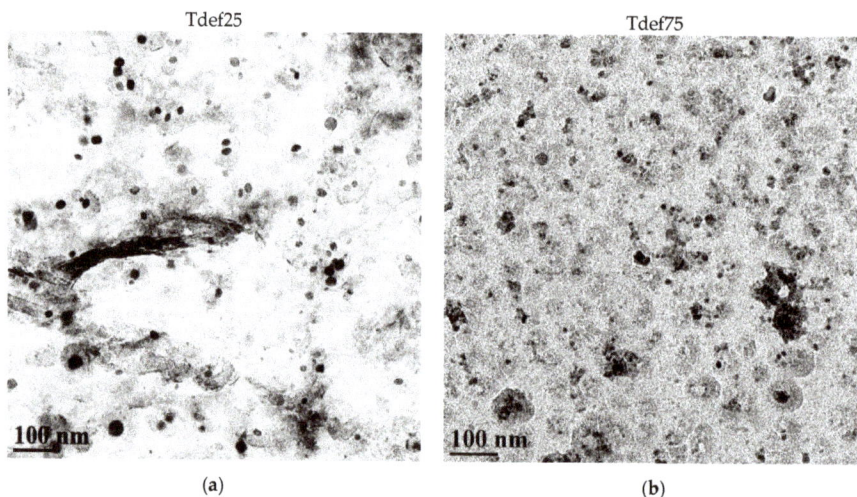

Figure 13. TEM images obtained in NbV recrystallized austenite and different ferrite contents prior to intercritical deformation: (**a**) Tdef25 and (**b**) Tdef75.

4. Conclusions

The analysis of intercritically deformed microstructures has been validated and extended to NbV-microalloyed steels following an EBSD characterization procedure. The threshold values for differentiating between deformed and non-deformed ferrite depend on the ferritic structure formed. Non-deformed equiaxed ferrite grains can be distinguished more easily from deformed ferrite grains, while in situations where non-polygonal and more bainitic structures are formed, the differentiation between deformed and non-polygonal grains is more diffuse due to a higher dislocation density in the latter.

The transformation of ferrite from recrystallized and deformed austenite implies modifications in the intercritically deformed ferrite. Differences in the niobium available during transformation are a key factor in maintaining the deformation bands and in developing a substructure within the intercritically deformed grains.

Author Contributions: U.M. carried out the experiments and wrote the manuscript; N.I. analyzed the data and wrote the manuscript; J.M.R.-I. supervised the results and edited the manuscript; P.U. managed the project and edited the manuscript.

Funding: This research was funded by Spanish Ministry of Economy and Competitiveness (MAT2015-69752) and by the European Commission Research Fund for Coal and Steel (RFSR-CT-2015-00014).

Conflicts of Interest: The authors declare no conflict of interest.

References

1. Humphreys, A.O.; Liu, D.; Toroghinejad, M.R.; Essadiqi, E.; Jonas, J.J. Warm rolling behaviour of low carbon steels. *Mater. Sci. Technol.* **2003**, *19*, 709–714. [CrossRef]
2. Simielli, E.A.; Yue, S.; Jonas, J.J. Recrystallization kinetics of microalloyed steels deformed in the intercritical region. *Met. Trans. A* **1992**, *23*, 597–608. [CrossRef]
3. Kwon, O.; De Ardo, A.J. Interactions between recrystallization and precipitation in hot-deformed microalloyed steels. *Acta Mater.* **1991**, *39*, 529–538. [CrossRef]
4. Dunne, D.P.; Feng, B.; Chandra, T. The effect of Ti and Ti-Nb additions on α formation and restoration during intercritical rolling and holding of C-Mn structural steels. *ISIJ Int.* **1991**, *31*, 1354–1361. [CrossRef]
5. Smith, A.; Luo, H.; Hanlon, D.N.; Sietsma, J.; Van Der Zwaag, S. Recovery processes in the ferrite phase in C–Mn steel. *ISIJ Int.* **2004**, *44*, 1188–1194. [CrossRef]

6. Eghbali, B. Microstructural development in a low carbon Ti microalloyed steel during deformation within the ferrite region. *Mat. Sci. Eng. A* **2008**, *480*, 84–88. [CrossRef]
7. Mayo, U.; Isasti, N.; Jorge-Badiola, D.; Rodriguez-Ibabe, J.M.; Uranga, P. An EBSD-based methodology for the characterization of intercritically deformed low carbon steel. *Mat. Charact.* **2019**, *147*, 31–42. [CrossRef]
8. Wright, S.I.; Nowell, M.M. EBSD image quality mapping. *Microsc. Microanal.* **2006**, *12*, 72–84. [CrossRef] [PubMed]
9. Llanos, L.; Pereda, B.; Jorge-Badiola, D.; Rodriguez-Ibabe, J.M.; López, B. Study of recrystallization in high manganese steels by means of the EBSD technique. *Mat. Sci. Forum* **2013**, *753*, 443–448. [CrossRef]
10. Olasolo, M.; Uranga, P.; Rodriguez-Ibabe, J.M.; López, B. Effect of austenite microstructure and cooling rate on transformation characteristics in a low carbon Nb–V microalloyed Steel. *Mater. Sci. Eng. A* **2011**, *528*, 2559–2569. [CrossRef]
11. Bengochea, R.; Lopez, B.; Gutierrez, I. Microstructural evolution during the austenite-to-ferrite transformation from deformed austenite. *Metall. Mater. Trans. A* **1998**, *29*, 417–426. [CrossRef]
12. Petrov, R.; Kestens, L.; Houbaert, Y. Characterization of the microstructure and transformation behaviour of strained and nonstrained austenite in Nb–V-alloyed C–Mn steel. *Mater. Charact.* **2004**, *53*, 51–61. [CrossRef]
13. American Society for Testing and Materials (ASTM). *ASTM E-562 Standard Test Method for Determining Volume Fraction by Systematic Manual Point Count*; ASTM International: West Conshohocken, PA, USA, 2019.
14. Isasti, N.; Jorge-Badiola, D.; Taheri, M.L.; López, B.; Uranga, P. Effect of composition and deformation on coarse-grained austenite transformation in nb-mo microalloyed steels. *Met. Trans. A* **2011**, *42*, 3729–3742. [CrossRef]
15. Isasti, N.; Jorge-Badiola, D.; Taheri, M.L.; Uranga, P. Phase transformation study in Nb-Mo microalloyed steels using dilatometry and EBSD quantification. *Metall. Mater. Trans. A* **2013**, *44*, 3552–3563. [CrossRef]
16. Cizek, P.; Wynne, B.P.; Davies, C.H.J.; Muddle, B.C.; Hodgson, P.D. Effect of composition and austenite deformation on the transformation characteristics of low-carbon and ultralow-carbon microalloyed steels. *Met. Trans. A* **2002**, *33*, 1331–1349. [CrossRef]
17. Krauss, G.; Thompson, S.W. Ferritic microstructures in continuously cooled low- and ultralow-carbon steels. *ISIJ Int.* **1995**, *35*, 937–945. [CrossRef]
18. Bengochea, R.; López, B.; Gutierrez, I. Influence of the prior austenite microstructure on the transformation products obtained for C-Mn-Nb steels after continuous cooling. *ISIJ Int.* **1999**, *39*, 583–591. [CrossRef]
19. Cizek, P.; Wynne, B.P.; Davies, C.H.J.; Hodgson, P.D. The effect of simulated thermomechanical processing on the transformation behavior and microstructure of a low-carbon mo-nb linepipe steel. *Met. Trans. A* **2015**, *46*, 407–425. [CrossRef]
20. Tamura, I.; Sekine, H.; Tanaka, T.; Ouchi, C. *Thermomechanical Processing of High-Strength Low-Alloy Steels*; Butterworth and Company: London, UK, 1988.
21. Speich, G.R.; Cuddy, L.J.; Gordon, C.R.; DeArdo, A.J. *Phase Transformations in Ferrous Alloys*; Marder, A.R., Goldstein, J.I., Eds.; TMS-AIME: Warrendale, PA, USA, 1983; pp. 341–389.
22. Sellars, C.M.; Beynon, J.H. *Proceedings of the Conference on High Strength Low Alloy Steels*; Dunee, D., Chandra, T., Eds.; South Coast Printers: Wollongong, Australia, 1984; p. 142.
23. Gibbs, R.K.; Parker, B.A.; Hodgson, P. Low-carbon steels for the 90's. In *Proceedings of the International Symposium on Low-Carbon Steels for the 90s*; Asfahani, R., Tiher, G., Eds.; The Minerals, Metals and Materials Society: Pittsburgh, PA, USA, 1993; p. 173.
24. García-Riesco, P.M.; Uranga, P.; López, B.; Rodriguez-Ibabe, J.M. Modelling the austenite to ferrite phase transformation in low carbon microalloyed steels in terms of grain size distributions. In Proceedings of the International Conference on Solid-Solid Phase Transformations in Inorganic Materials 2015, PTM'2015, Whistler, BC, Canada, 28 June–3 July 2015; pp. 917–924.
25. Hernandez, D.; López, B.; Rodriguez-Ibabe, J.M. Ferrite grain size refinement in vanadium microalloyed structural steels. *Mater. Sci. Forum* **2015**, *500*, 411–418. [CrossRef]
26. Vandermerr, R.A.; Juul Jensen, D. Microstructural path and temperature dependence of recrystallization in commercial aluminum. *Acta Mater.* **2001**, *49*, 2083–2094. [CrossRef]

© 2019 by the authors. Licensee MDPI, Basel, Switzerland. This article is an open access article distributed under the terms and conditions of the Creative Commons Attribution (CC BY) license (http://creativecommons.org/licenses/by/4.0/).

Article

Study on ς Phase in Fe–Al–Cr Alloys

Jintao Wang [1,*], Shouping Liu [1] and Xiaoyu Han [2]

1. College of Materials Science and Engineering, Chongqing University, Chongqing 400044, China; LSP@cqu.edu.cn
2. Chongqing Materials Research Institute, Chongqing 400707, China; xiaoyuhan@cqu.edu.cn
* Correspondence: 201709131171@cqu.edu.cn; Tel.: +86-133-6406-1236

Received: 12 September 2019; Accepted: 9 October 2019; Published: 11 October 2019

Abstract: In this paper, a method of using the second phase to control the grain growth in Fe–Al–Cr alloys was proposed, in order to obtain better mechanical properties. In Fe–Al–Cr alloys, austenitic transformation occurs by adding austenitizing elements, leading to the formation of the second phase and segregation at the grain boundaries, which hinders grain growth. FeCr(σ) phase was obtained in the Fe–Al–Cr alloys, which had grains of several microns and was coherent and coplanar with the matrix (Fe$_2$AlCr). The nucleation of σ phase in Fe–Al–Cr alloy was controlled by the ratio of nickel to chromium. When the Ni/Cr (eq) ratio of alloys was more than 0.19, σ phase could nucleate in Fe–Al–Cr alloy. The relationship between austenitizing and nucleation of FeCr(σ) phase was given by thermodynamic calculation.

Keywords: high-aluminum steel; second phase; phase transition; thermodynamic calculation

1. Introduction

Heat-resistant ferritic steels have better heat capacity and a lower thermal expansion rate than nickel-based alloys. The growth rate and adhesion of Al-rich oxide film on Fe–Al–Cr alloy are not affected by water vapor because of its good thermal cycling resistance. Depending on the protective thermal growth oxides (TGO) to prolong the service life of alloys under harsh working conditions and maintain the stability of material systems and isolate pollutants, many researchers have proposed that Fe–Al–Cr alloys can be regarded as a new generation of nuclear fuel cladding materials [1,2]. The coarse grains are the main factor hindering the development of Fe–Al–Cr alloys. In recent years, researchers generally intend to use the second phase in Fe–Al–Cr alloys to refine the grains and strengthen the matrix. The (Ni, Fe) Al precipitates have been studied most [3–5]. C. Stallybrass et al. have suggested that the high temperature strength of nickel-based superalloys is attributed to the dispersion strengthening of Ni3Al. Similar microstructures could be obtained in the Fe–Al–Ni–Cr system with B2 ordered (Ni, Fe) Al precipitates in the ferritic matrix superalloys.

In the Fe–Cr–Al ternary system, there is a miscible zone between the disordered A2 phase and the ordered B2 phase [6]. These two phases have similar lattice parameters, and they are coherent and coplanar, which is a characteristic shared by alloys consisting of miscible phases and nickel-based superalloys. In general, the volume fraction of precipitates strengthened by coherent B2 (Ni, Fe) Al precipitates is 13% and the average precipitation radius is 62 nm [7]. The creep mechanism is the repulsive elastic interaction between the general dislocation climb, the coherent precipitate, and the matrix dislocation [8–12].

However, the Ni–Al phase plays an important role in precipitation strengthening and does not fundamentally refine the grain size of Fe–Al–Cr alloys. The aim of this study was to limit the grain growth of Fe–Al–Cr alloys by searching for a continuous and inhomogeneous micron-sized second phase in order to achieve good mechanical properties. FeCr has entered our field of vision, because it has a body-centered cubic lattice similar to ferrite, its crystal group is Im3m (229), and it is easy to nucleate at the grain boundary. As a common second phase in nickel-based superalloys, we tried to study its effects on iron-based superalloys.

2. Experimental Procedures

Four alloys with different chemical compositions (Table 1) were melted in a vacuum induction furnace. The raw melting materials included DT4C pure iron, pure aluminum, pure nickel, and GCr15 bearing steel. It is noteworthy that the alloy was stirred uniformly under the action of the magnetic field. After the shell was formed by air cooling, the alloy, the billet of which had just become solidified, was impacted by a large amount of water (of about 20 °C temperature), and the metal was cooled to room temperature in a few minutes.

Table 1. Alloy chemical composition (in wt.%) determined by XRF (X-ray Fluorescence, 20°/min).

No.	Fe	Al	Cr	Ni	C	V	Ti
#1	72.27	7.91	13.17	6.65	-	-	-
#2	81.97	8.46	8.76	-	0.81	-	-
#3	75.15	12.25	10.29	2.11	0.20	-	-
#4	75.54	11.73	11.48	8.68	-	-	1.21
#5	82.57	8.80	4.86	2.19	-	1.36	-
#6	86.91	8.38	4.71	-	-	-	-

The grain morphology of the alloys was observed under a metallographic microscope, and their phase composition was analyzed by X-ray diffraction (XRD, D/mAX 2500, Japanese Neo-Confucianism Corporation, Tokyo, Japan). The surface morphology of the alloys was observed by field emission scanning electron microscopy (FESEM, JOEL JSM-7800F, Japan Electronics Corporation, Tokyo, Japan), and the main chemical components of the phases were analyzed by energy dispersive spectroscopy (EDS, JOEL JSM-7800F, Japan Electronics Corporation, Tokyo, Japan) coupled with FESEM. Subsequently, the as-cast alloys were cut into the specimens of 15 × 15 × 5 mm in size.

The #1, #2 and #3 alloys were heated from 20 to 1000 °C for 30 min, cooled by water (20 °C), re-heated to 1000 °C for 30 min, and then rolled by a rolling mill with a pair of work rolls under a rolling pressure of 25 T. The thickness of the samples was decreased from 5 to 2.5 mm during the rolling process. After rolling, the alloys were placed into a DHG-9053 thermal oven and dried at 200 °C for 72 h.

3. Results and Discussion

3.1. Second Phase of Alloys

From the metallographic observations, the matrix structure was divided into smaller grains by the second phase in #1, #2 and #3 alloys, and the matrix grains encountered the hindrance of the second phase during the growth process (Figures 1 and 2). Combining the EDS results (Table 2, Figures 3 and 4), it can be observed that the matrix structure was α-Fe solid solution of Fe_2AlCr. However, the second phases of the alloys with different compositions varied from each other. In particular, alloy #1 formed FeCr phase (σ phase), alloy #2 formed a mixed phase of FeCr phase and a great deal of carbide precipitation, and alloy #3 formed a mixed phase of FeCr and a little amount of carbide precipitation. In contrast, samples #4, #5, and #6 alloy did not have FeCr phase; they had a large grain size, which can reach several millimeters in diameter. The average rain diameter of each alloy was calculated by Image-pro software, #1 alloy's grain diameter was 86.7 μm, #2 alloy was 93.2 μm, #3 alloy was 95.5 μm, #4 alloy was 1035.3 μm, #5 was 4369.7 μm, and #6 was 3180.9 μm.

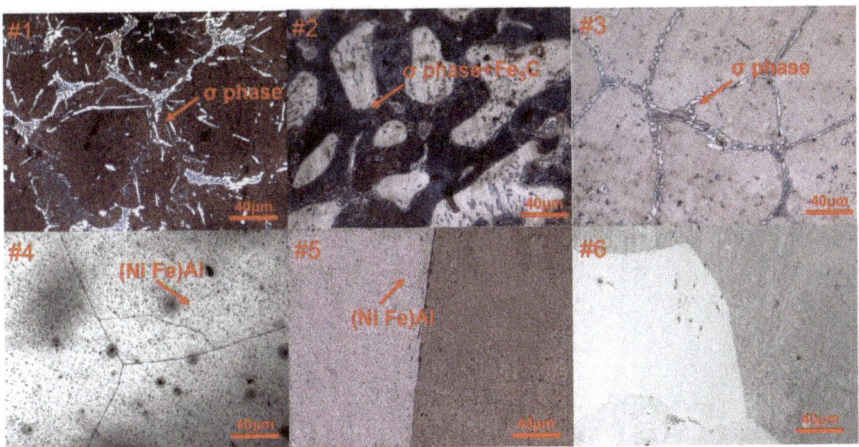

Figure 1. Metallography structure of alloys. FeCr phase appeared in #1, 2, and 3 alloys, (Ni Fe)Al appeared in #4 and #5 alloys. FeCr and (Ni Fe)Al's components are presented in Table 2.

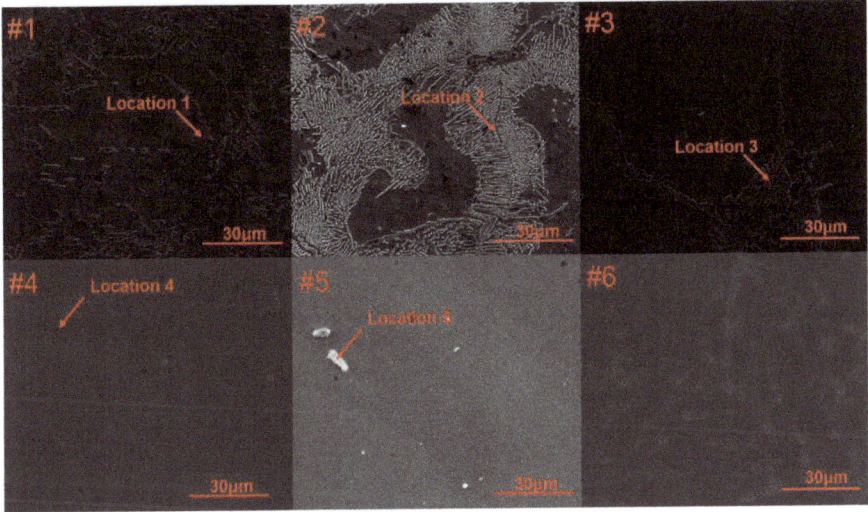

Figure 2. SEM image of alloys grain. A large number of FeCr phases accumulated at the grain boundaries in #1 alloy. In #2 alloy, FeCr phase nucleated on the carbide. There was distribution of FeCr along the grain boundary in 3# alloy. (Ni Fe)Al phase was precipitated in #4 and #5 alloys, which played the role of precipitation strengthening.

Table 2. Chemical compositions of the respective locations shown in Figure 2. There were both FeCr and carbides (Locations 1, 2 and 3), and there were (Ni Fe)Al in Locations 4 and 5.

Element	Location 1 (at. %)	Location 2 (at. %)	Location 3 (at. %)	Location 4 (at. %)	Location 5 (at. %)
Fe	53.64	57.68	59.72	73.19	75.86
Al	1.46	7.31	9.96	18.31	16.72
Cr	42.45	14.07	34.48	-	-
C	-	20.94	3.76	-	-
Ni	2.45	-	2.04	8.60	4.32

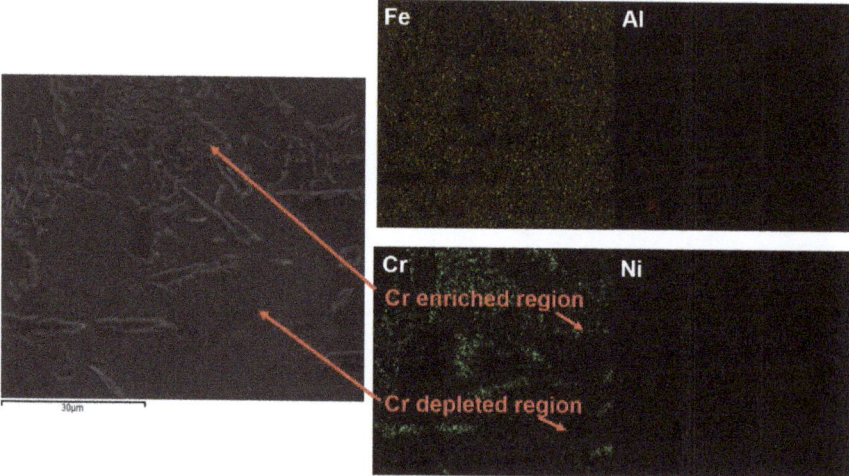

Figure 3. FESEM image of alloy #1, showing a sample grain and element distribution at a grain boundary. The distribution of elemental Cr is particularly important, as the distribution of elements coincides with that of phases.

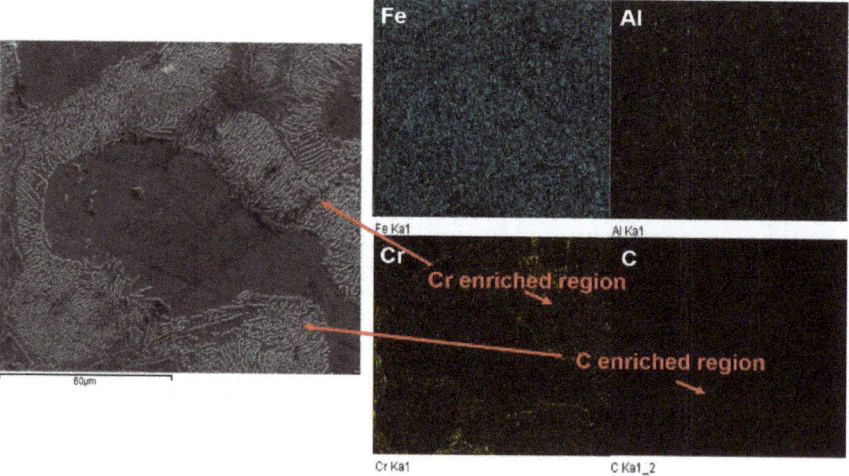

Figure 4. FESEM image of alloy #2 sample grain and element distribution at a grain boundary. The distribution of Cr elements coincides with that of the phases in the grain boundary.

There was no obvious grain boundary defect in the matrix in #1, #2 and #3 alloys' grain, whilst the second phase had an obvious grain boundary. The grain size of the matrix could reach tens of microns, whilst the grain size of the second phase was only a few microns. The second-phase grains were dendritic, a small portion of which were scattered in the matrix grains. The grain sizes of #4, #5 and #6 alloys reached hundreds of microns, and the alloys' grain boundaries were clean, without the second phase at the grain boundaries.

3.2. Thermodynamic Calculation

The matrix phase of Fe–Al–Cr alloys is ferrite structure, which is a body-centered cubic lattice. In our conception, we need to find a second phase of body-centered cubic crystal to meet two requirements. The second phase is needed to prevent the growth of ferrite grains, and to form a semi-eutectic lattice structure. σ phase is a destructive second phase in nickel-based superalloys, which is body-centered cubic crystal (its space group is Im3m (229)). The destructive effect of σ phase in nickel-based superalloys is that σ phase nucleates easily at grain boundaries and the lattice of nickel-based superalloys is usually a face-centered cubic structure [13].

σ phase is a kind of topological close-packed phase (TCP), which is easy to nucleate on grain boundaries and carbides. Thermodynamic calculation by FactSage6.2 (Equilib Module) shows that adding Ni element to the alloy is beneficial for the nucleation of σ phase (Figure 5) [14].

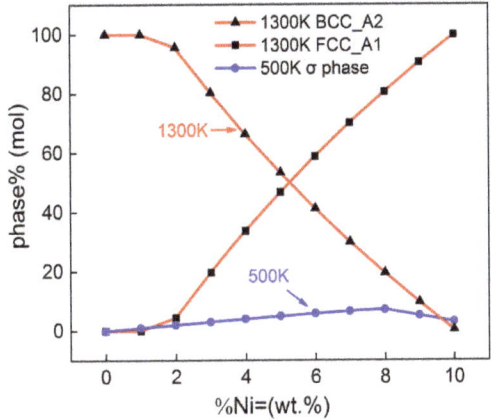

Figure 5. Relationship between Ni content in alloy and σ phase volume content in Fe-8 wt.%Cr-4 wt.% Al.

The nucleation of σ phase is related to the electron vacancy concentration in the lattice [15,16].

$$\Delta N_v = \overline{N_v^c} - \overline{N_V} \tag{1}$$

When ΔN_v is less than zero, σ phase will nucleate. The smaller ΔN_v is, the more σ phase will precipitate. $\overline{N_v^c}$ means critical electron vacancy and $\overline{N_V}$ means electrons vacancy density. Austenite's primary cell has higher electron vacancy density than ferrite. Thus, σ phase is easy to nucleate in austenite solid solution. Ni is a strong austenitic stabilization element, and a small amount of Ni can achieve stability austenitic phase (Figure 6). This explains the positive effect of Ni on σ phase nucleation.

Another reason is that the diffusion of Cr atoms in ferrite is easier than that in austenite. After alloy austenitizing, Cr atoms are confined to those regions and mixed with some free Fe atoms to form σ phase, which hinders the growth of Fe–Al–Cr alloy grains. Other austenitizing elements have similar effects with Ni elements, such as C (Figure 7). The effect of austenitizing on the second phase of the alloy is shown in Figure 5. Nickel equivalence has an important effect on austenitizing of alloys, and austenitizing affects the nucleation of FeCr.

The equilibrium multicomponent system has the smallest Gibbs free energy (Equation (1)). The thermodynamic description of a system requires that each phase has its corresponding thermodynamic function (Equation (2)), which is used to describe the relationship between temperature, pressure, concentration and various free energy functions. In the calculation of phase diagrams, Gibbs free energy of two-element phase in a multicomponent system can be decomposed into three independent parts (Equation (3)). The sub-lattice model is used for the ordered phase with ordered/disordered phase transition. The lattice position fraction occupied by atoms in the primary cell is used to replace the stoichiometric ratio in the solution model (Equations (4) and (5)).

$$G_{eq} = \min\left(\sum_{i=1}^{p} n_i G_i^\phi\right) \quad (2)$$

$$G^\phi = G_T^\phi(T, x) + G_p^\phi(p, T, x) + G_m^\phi(T_c, \beta_0, T, x) \quad (3)$$

$$G^\phi = x_A^0 G_A^0 + x_B^0 G_B^0 + \Delta G^f \quad (4)$$

$$x_A = a^1 y_A^1 + a^2 y_A^2 \quad (5)$$

$$x_B = a^1 y_B^1 + a^2 y_B^2 \quad (6)$$

The results of phase diagram calculation are as follows (Figures 6–8).

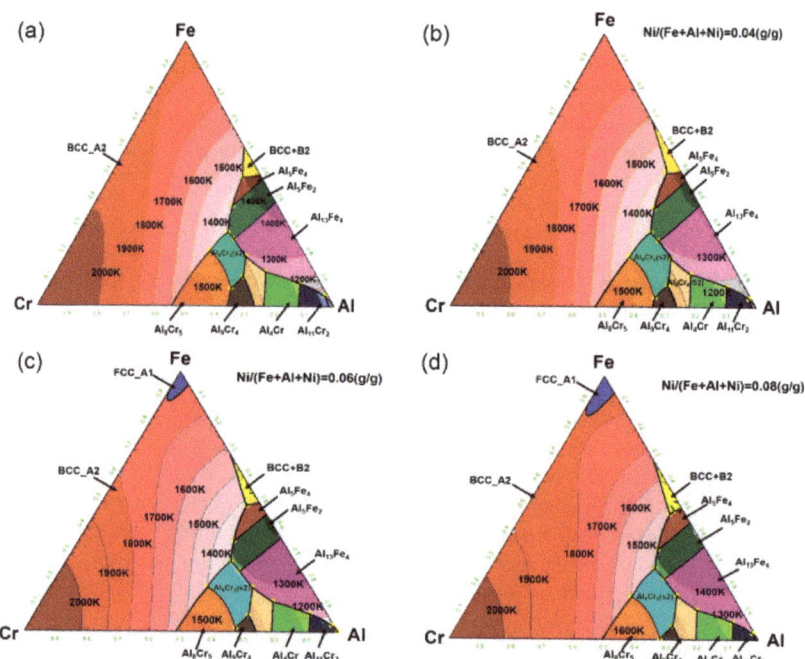

Figure 6. Effect of Ni addition on the Fe–Al–Cr phase diagram (1000–2000 K). (a) Ni/(Fe + Al + Cr) = 0 wt.%; (b) Ni/(Fe + Al + Cr) = 4 wt.%; (c) Ni/(Fe + Al + Cr) = 6 wt.%; and (d) Ni/(Fe + Al + Cr) = 8 wt.%.

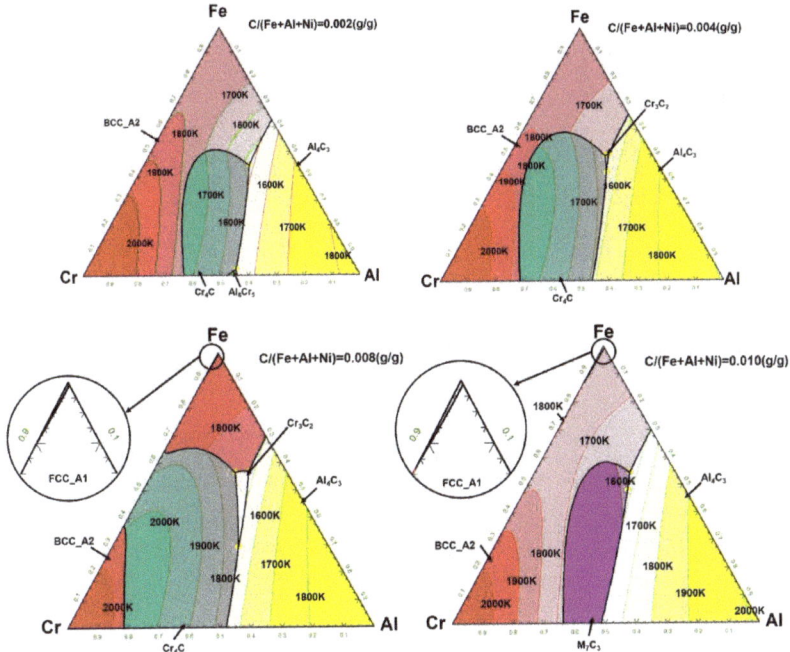

Figure 7. Effect of C addition on the Fe–Al–Cr phase diagram (1000–2000 K). (**a**) C/(Fe + Al + Cr) = 0.2 wt.%; (**b**) C/(Fe + Al + Cr) = 0.4 wt.%; (**c**) C/(Fe + Al + Cr) = 0.8 wt.%; and (**d**) C/(Fe + Al + Cr) = 1 wt.%.

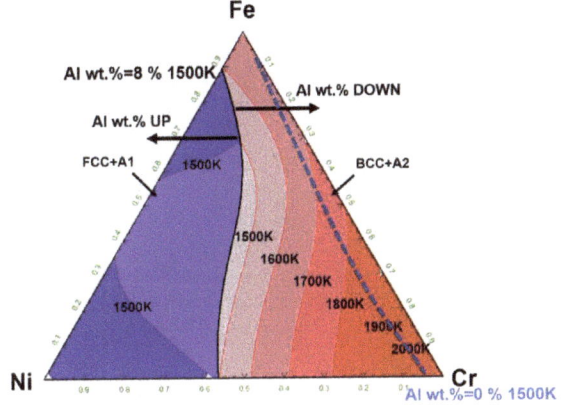

Figure 8. Effect of Al addition on the Fe–Ni–Cr phase diagram (1000–2000 K).

With the increase of aluminium content, the boundary between the austenite phase and the ferrite phase diagram shifts to the left until the austenite phase disappears completely. From Figure 9, austenitizing of Fe–Al–Cr alloys does not depend on the content of Ni, but on the ratio of Ni equivalent to Cr equivalent. This conclusion is confirmed by experiments (Figure 9). When the Ni/Cr ratio of alloys is more than 0.19, σ phase can nucleate in Fe–Al–Cr alloys.

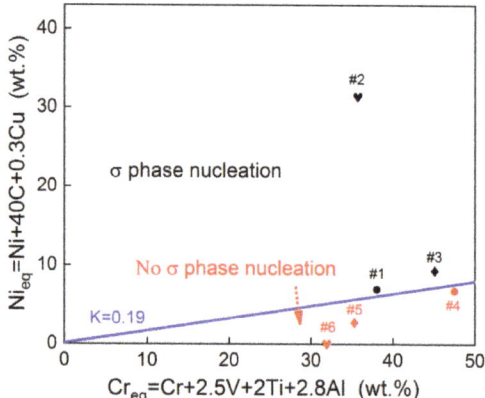

Figure 9. The different locations indicate the Cr/Ni ratio of the alloys.

3.3. Effect of High Temperature Environment on σ Phase

The matrix phase (Fe$_2$AlCr) of the alloys was coherent and coplanar with the second phase (FeCr). It was impossible to observe the behavior of the second phase at high temperature by XRD (Figure 10), because the matrix phase causes confusion. The effect of high temperature on phase A was studied using high temperature experiments.

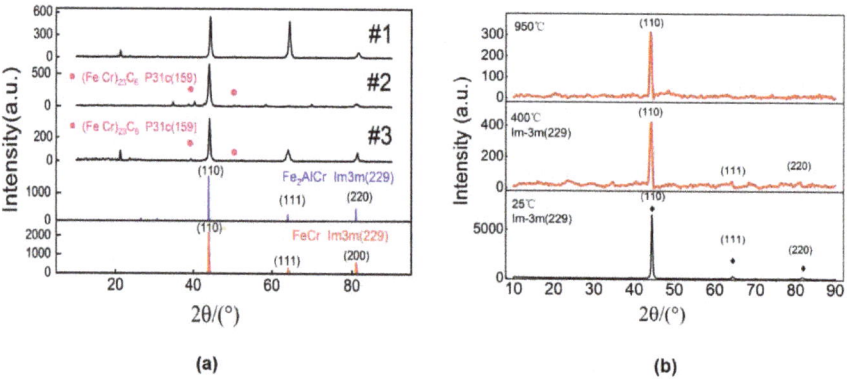

Figure 10. XRD results of the #1, #2 and #3 alloys. (a) Comparison of XRD images of Alloys with XRD of Standard Substances. The Fe$_2$AlCr and FeCr phases were coherent and coplanar. (b) XRD results of #1 alloy at different temperatures.

The phase distribution and grain size around the grain boundary of the alloys was hardly changed by quenching. Thermomechanical treatment results (rolling) in obvious grain orientation: tangential direction of the deformation direction of the alloys. The grain size of the alloys tended to grow after aging at 200 °C for 72 h after deformation. In addition, the distribution of the second phase was significantly changed after thermomechanical treatment. The distribution became more dense and the grains became more chaotic whilst more second phase particles entered into the grains (Figure 11). Unfortunately, these heat treatments cannot refine the grain size of these alloys from the results. Moreover, after thermomechanical treatment, the toughness will be further decreased due to the formation of texture (Figure 12). In summary, thermomechanical treatment did not achieve the goal of grain refinement, which again indicated that the two phases are coherent and coplanar. σ phase will not decompose rapidly in a 1000 °C high temperature working environment, and it will not dissolve into the matrix in a short time heating process.

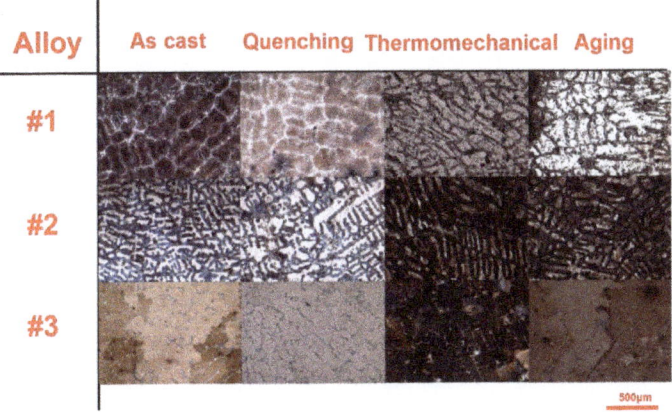

Figure 11. Metallographic structure changes in the four alloys during heat treatment. Alloys were heated from 20 to 1000 °C for 30 min, cooled by water (20 °C) re-heated to 1000 °C for 30 min, and after rolling, aging 72 h in 200 °C.

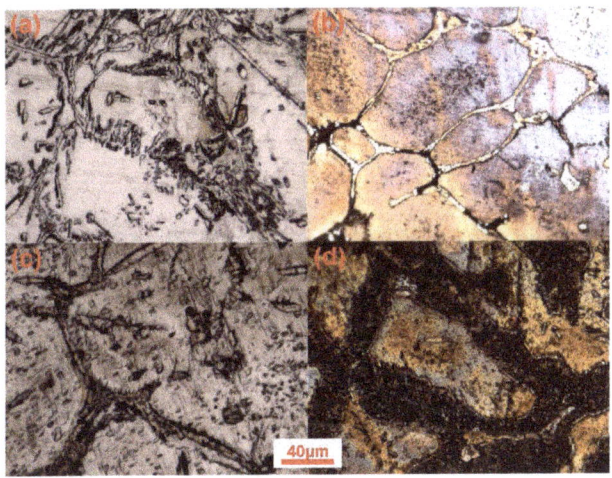

Figure 12. Alloy microstructure after a series of heat treatments. (a) Alloy #1; (b) alloy #2; (c) alloy #3; and (d) alloy #4.

4. Conclusions

(1) This study provides a new way for grain refinement of Fe–Al–Cr alloys and a new idea for commercial application on a large scale. In this study, FeCr(σ) phase was obtained in the Fe–Al–Cr alloys, which had grains of several microns and was coherent and coplanar with the matrix (Fe_2AlCr).

(2) σ phase nucleated in austenite. The nucleation of σ phase in Fe–Al–Cr alloys is controlled by the ratio of nickel to chromium. When the Ni/Cr (eq) ratio of alloys is more than 0.19, σ phase can nucleate in Fe–Al–Cr alloys.

(3) σ phase will not decompose rapidly in a 1000 °C high temperature working environment, and it will not dissolve into the matrix in a short time heating process.

Author Contributions: conceptualization, J.W.; methodology, S.L.; software, J.W.; validation, X.H.; formal analysis, J.W.; investigation, X.H.; resources, S.L.; data curation, J.W.; writing—original draft preparation, J.W.; writing—review and editing, J.W.; visualization, J.W.; supervision, J.W.; project administration, S.L.

Conflicts of Interest: The authors declare no conflict of interest.

References

1. Gregory, O.J.; Busch, E.; Fralick, G.C.; Chen, X. Preparation and characterization of ceramic thin film thermocouples. *Thin Solid Film* **2010**, *518*, 6093–6098. [CrossRef]
2. Deodeshmukh, V.P.; Matthews, S.J.; Klarstrom, D.L. High-temperature oxidation performance of a new alumina-forming Ni–Fe–Cr–Al alloy in flowing air. *Int. J. Hydrogen Energy* **2011**, *36*, 4580–4587. [CrossRef]
3. Stallybrass, C.; Schneider, A.; Sauthoff, G. The strengthening effect of (Ni, Fe)al precipitates on the mechanical properties at high temperatures of ferritic Fe–Al–Ni–Cr alloys. *Intermetallics* **2005**, *13*, 1263–1268. [CrossRef]
4. Stallybrass, C.; Sauthoff, G. Ferritic Fe–Al–Ni–Cr alloys with coherent precipitates for high-temperature applications. *Mater. Sci. Eng. A (Struct. Mater. Propert. Microstruct. Process.)* **2004**, *387–389*, 985–990. [CrossRef]
5. Sarkar, S.; Bansal, C. Atomic disorder–order phase transformation in nanocrystalline Fe–Al. *J. Alloys Compd.* **2002**, *334*, 135–142. [CrossRef]
6. Vo, N.Q.; Liebscher, C.H.; Rawlings, M.J.S.; Asta, M.; Dunand, D.C. Creep properties and microstructure of a precipitation-strengthened ferritic Fe–Al–Ni–Cr alloy. *Acta Mater.* **2014**, *71*, 89–99. [CrossRef]
7. Teng, Z.K.; Zhang, F.; Miller, M.K.; Liu, C.T.; Huang, S.; Chou, Y.T. Thermodynamic modeling and experimental validation of the Fe-Al-Ni-Cr-Mo alloy system. *Mater. Lett.* **2012**, *71*, 36–40. [CrossRef]
8. Liebscher, C.H.; Radmilovic, V.R.; Dahmen, U.; Vo, N.Q.; Dunand, D.C.; Asta, M.; Ghosh, G. A hierarchical microstructure due to chemical ordering in the bcc lattice: early stages of formation in a ferritic Fe–Al–Cr–Ni–Ti alloy. *Acta Mater.* **2015**, *92*, 220–232. [CrossRef]
9. Janda, D.; Ghassemiarmaki, H.; Bruder, E.; Hockauf, M.; Heilmaier, M.; Kumar, K.S. Effect of strain-rate on the deformation response of DO3-ordered Fe3Al. *Acta Mater.* **2016**, *103*, 909–918. [CrossRef]
10. Wang, J.; Han, X.; Liu, S.; Hou, W. Effects of Si and V on high temperature oxidation resistance of Fe-Al-Cr alloys. *J. Chongqing Univ.* **2019**, *42*, 86–97.
11. Minamino, Y.; Koizum, Y.; Tsuji, N.; Hirohata, N.; Mizuuchi, K.; Ohkanda, Y. Microstructures and mechanical properties of bulk nanocrystalline Fe–Al–C alloys made by mechanically alloying with subsequent spark plasma sintering. *Sci. Technol. Adv. Mater.* **2004**, *5*, 133–143. [CrossRef]
12. Risanti, D.D.; Sauthoff, G. Strengthening of iron aluminide alloys by atomic ordering and Laves phase precipitation for high-temperature applications. *Intermetallics* **2005**, *13*, 1313–1321. [CrossRef]
13. Palm, M.; Sauthoff, G. Deformation behaviour and oxidation resistance of single-phase and two-phase L21-ordered Fe–Al–Ti alloys. *Intermetallics* **2004**, *12*, 1345–1359. [CrossRef]
14. Ha, M.C.; Koo, J.M.; Lee, J.K.; Hwang, S.W.; Park, K.T. Tensile deformation of a low density Fe–27Mn–12Al–0.8C duplex steel in association with ordered phases at ambient temperature. *Mater. Sci. Eng. A* **2013**, *586*, 276–283. [CrossRef]
15. Zhao, X.B.; Dang, Y.Y.; Yin, H.F.; Lu, J.T.; Yuan, Y.; Cui, C.Y.; Gu, Y.F. Super-supercritical power stations with nickel-iron-based high-temperature alloy TCP phase and carbide precipitation thermodynamic calculations. *Mater. Eng.* **2015**, *43*, 38–43.
16. Baik, S.I.; Rawlings, M.J.S.; Dunand, D.C. Atom probe tomography study of Fe-Ni-Al-Cr-Ti ferritic steels with hierarchically-structured precipitates. *Acta Mater.* **2018**, *144*, 707–715. [CrossRef]

© 2019 by the authors. Licensee MDPI, Basel, Switzerland. This article is an open access article distributed under the terms and conditions of the Creative Commons Attribution (CC BY) license (http://creativecommons.org/licenses/by/4.0/).

Article

Tensile Properties and Microstructural Evolution of an Al-Bearing Ferritic Stainless Steel at Elevated Temperatures

Ying Han [1],*, Jiaqi Sun [1], Yu Sun [2], Jiapeng Sun [3] and Xu Ran [1],*

1. Key Laboratory of Advanced Structural Materials, Ministry of Education, Changchun University of Technology, Changchun 130012, China; 2201702015@stu.ccut.edu.cn
2. National Key Laboratory for Precision Hot Processing of Metals, Harbin Institute of Technology, Harbin 150001, China; yusun@hit.edu.cn
3. College of Mechanics and Materials, Hohai University, Nanjing 211100, China; Sun.jiap@gmail.com
* Correspondence: hanying_118@sina.com (Y.H.); ranxu@ccut.edu.cn (X.R.); Tel.: +86-133-4157-6601 (Y.H.)

Received: 29 November 2019; Accepted: 27 December 2019; Published: 4 January 2020

Abstract: The influence of temperature and strain rate on the hot tensile properties of 0Cr18AlSi ferritic stainless steel, a potential structural material in the ultra-supercritical generation industry, was investigated at temperatures ranging from 873 to 1123 K and strain rates of 1.7×10^{-4}–1.7×10^{-2} s^{-1}. The microstructural evolution linked to the hot deformation mechanism was characterized by electron backscatter diffraction (EBSD). At the same strain rate, the yield strength and ultimate tensile strength decrease rapidly from 873 K to 1023 K and then gradually to 1123 K. Meanwhile, both yield strength and ultimate tensile strength increase with the increase in strain rate. At high temperatures and low strain rates, the prolonged necking deformation can be observed, which determines the ductility of the steel to some extent. The maximum elongation is obtained at 1023 K for the strain rates of 1.7×10^{-3} and 1.7×10^{-2} s^{-1}, while this temperature is postponed to 1073 K once decreasing the strain rate to 1.7×10^{-4} s^{-1}. Dynamic recovery (DRV) and continuous dynamic recrystallization (CDRX) are found to be the main softening mechanisms during the hot tensile deformation. With the increase of temperature and the decrease of strain rate (i.e., 1123 K and 1.7×10^{-4} s^{-1}), the sub-grain coalescence becomes the main mode of CDRX that evolved from the sub-grain rotation. The gradual decrease in strength above 1023 K is related to the limited increase of dynamic recrystallization and the sufficient DRV. The area around the new small recrystallized grains on the coarse grain boundaries provides the nucleation site for cavity, which generally results in a reduction in ductility. Constitutive analysis shows that the stress exponent and the deformation activation energy are 5.9 and 355 kJ·mol^{-1} respectively, indicating that the dominant deformation mechanism is the dislocations motion controlled by climb. This work makes a deeply understanding of the hot deformation behavior and its mechanism of the Al-bearing ferritic stainless steel and thus provides a basal design consideration for its extensive application.

Keywords: ferritic heat resistant stainless steel; hot tensile deformation; tensile property; dynamic recrystallization; flow behavior

1. Introduction

Ferritic stainless steels are widely used in automobile, furnace part, construction and environmental protecting industries owing to their higher thermal conductivities, smaller thermal expansions, better resistance to atmospheric corrosion and stress corrosion cracking, and lower cost in comparison with austenitic stainless steels [1,2]. However, low strength and poor resistance to oxidation at high temperatures generally limit their extensive application.

Hot tensile property is an important performance index of heat resistant steel. Generally, the dynamic recovery (DRV) and dynamic recrystallization (DRX) are the most softening mechanisms during the deformation [3,4]. The microstructural reconstitution including grain refinement and structure homogenization can be achieved, which will be beneficial to improve the high temperature strength and toughness of the materials [5]. Of course, the microstructural evolution during the deformation depends strongly on the deformation parameters, such as temperature and strain rate, and thus affecting the mechanical property. Chiu et al. [6] investigated the hot tensile property of Crofer 22 APU ferritic stainless steel at the temperatures of 873–1073 K and found a remarkable drop of yield strength between 973 and 1023 K due to the obvious DRV. In addition, the micro-alloying technique as an efficient approach has been applied in ferritic stainless steels to improve the high temperature property [7,8]. The solute atoms, such as Nb, Mo, Ti, and Zr, are used to enhance the high temperature strength of ferritic stainless steels [9–11]. Moreover, W and Ce can improve their high temperature oxidation resistance [12,13]. However, these alloying elements are commonly expensive, and it is no doubt that they would increase the manufacturing cost of the products of ferritic stainless steels.

Recently, with the development of metallurgical technologies, it is proved that Al as a cheap element becomes an important alternative of the alloying design for many steels. For instance, adding a small amount of Al (~1 wt.%) in transformation induced plasticity steels (TRIPs) facilitated the suppression of cementite precipitation, the refinement of the bainite laths and the retention of austenite [14–16], all which would improve the strength and toughness. For high Mn steels, the addition of Al could increase the stacking fault energy and produce short-range ordering and/or κ′-carbide precipitation [17]. In addition, Al can lead to a specific weight reduction, and a 1.3% reduction in density was obtained per 1 wt% addition of Al [17]. Due to the addition of the large amount of Al (~10 wt.%), a new type of steel, namely low-density steel, has been formed [18]. The strength of oxide dispersion strengthened steels (ODSs) also increased with Al addition due to the back stress strengthening combined with Orowan strengthening [19]. However, for ferritic stainless steels, the high temperature properties [20], particularly in oxidation resistance [21,22], could be improved considerably by alloying them with Al. However, the addition of Al could decrease the recrystallization temperature of ferritic stainless steels [23], and thus resulted in untimely softening during applications at high temperatures. Hence, the research on the hot deformation behavior of Al-bearing ferritic stainless steels is significant. However, the studies on this aspect are limited.

0Cr18AlSi ferritic stainless steel is a potential structural material in the ultra-supercritical generation industry. The authors' previous results showed that this steel exhibited excellent high temperature oxidation resistance at 1073 and 1173 K due to the formation of continuous, compact, and well-adherent multicomponent oxide films containing Al_2O_3 [21]. The purpose of the present study is to investigate the hot tensile deformation behavior and fracture of 0Cr18AlSi ferritic stainless steel. Effects of temperature and strain rate on the tensile properties and microstructural evolution are analyzed in detail. The results provide a reasonable evidence for the materials design and applications of Al-bearing ferritic stainless steels.

2. Materials and Methods

The 0Cr18AlSi ferritic stainless steel used in this study was prepared by melting high-purity elements in a vacuum induction furnace with an argon atmosphere. Its composition is as follows (wt.%): C 0.09, Cr 18.4, Al 1.05, Si 1.01, Mn 0.75, P 0.017, S 0.001. The ingots were first forged at 1423 K, and then hot-rolled at 1323 K to 8 mm thick strips in a laboratory hot-rolling mill. The strips were annealed at 1123 K for 40 min in a resistance furnace, followed by quenching in water.

The tensile specimens with a gage length of 10 mm and cross section of 2 mm × 2 mm were machined from the annealed plates parallel to rolling direction. An electronic universal testing machine (WDW-200, JILIN GUANTENG AUTOMATION TECHNONOGY CO., LTD., Changchun, China) was used to carry out the tensile tests. Temperatures ranging from 873 to 1123 K and strain rates ranging from 1.7×10^{-4} to 1.7×10^{-2} s^{-1} were employed. The specimens were loaded at the target temperature for 20 min before testing in order to eliminate the temperature gradient. After the tensile tests, all specimens were immediately quenched to room temperature by water.

The fracture surfaces of the tensile specimens were observed by scanning electron microscope (SEM, JSM-5600, JEOL, Tokyo, Japan). The microstructural morphology close to the fracture parallel to tensile direction was analyzed by the optical microscope (OM, DM13000M, Leica, Vizula, Germany). After the standard metallographic procedures including grinding, polishing and etching for the specimens, a mixed solution of 5 vol.% ferric chloride (FeCl$_3$) and 5 vol.% hydrochloric acid (HCl) in distilled water was used to display the microstructures. In order to obtain the detailed information on microstructural evolution during hot deformation, electron backscattered diffraction (EBSD) technology was performed. The specimens for EBSD were ground mechanically followed by argon ion polishing. The HKL CHANNEL 5 software (Company Oxford Instruments, Oxfordshire, UK) was utilized to post-process the data obtained from the EBSD measurements.

3. Results and Discussion

3.1. Initial Microstructure

The initial microstructure of the studied steel is shown in Figure 1a, which reveals a relatively uniform equiaxed grain structure with an average grain size of 27 μm, as measured by the linear-intercept method (ASTM E112). Corresponding X-ray diffraction result is seen in Figure 1b. Only body-centered cubic (bcc) structured phase can be detected, confirming that the existence of the complete ferritic microstructure after annealing at 1123 K for 40 min.

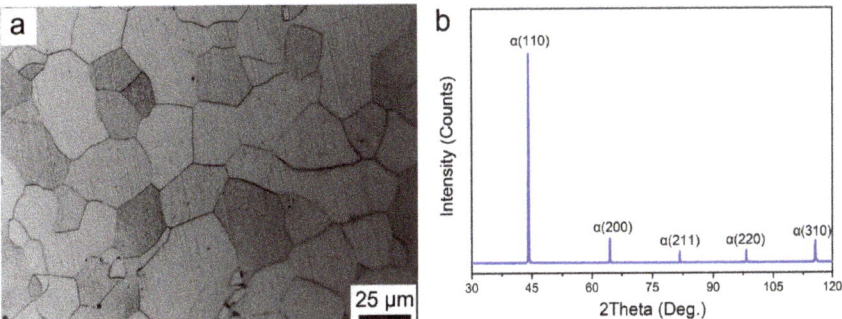

Figure 1. (a) Optical microstructure of the annealed 0Cr18AlSi steel and (b) corresponding X-ray diffraction pattern.

3.2. Tensile Properties

Figure 2 shows the tensile engineering stress–strain curves of the specimens tested at different deformation conditions. It can be seen that the flow behavior is sensitive to deformation temperature and strain rate. At low temperature and high strain rate, i.e., 873 K and 1.7×10^{-2} s^{-1}, the variation of flow stress is parabolic. It means that the flow stress increases to a peak value at a reduced increase rate as the strain proceeds, and then declines quickly until the specimen ruptures. This is related to the microstructural evolution during hot tensile deformation. As illustrated in Figure 1b, the studied steel has a bcc structure. The dislocations can cross-slip and climb easily during hot deformation because of a relatively high stacking fault energy of the bcc structure [24], and thus dynamic recovery (DRV) generally dominates the softening process of this kind of steel [24,25]. Dynamic recrystallization (DRX) will take place only when DRV is not enough to offset the strain hardening effect [26]. As can be seen, the slow increase in flow stress at the early deformation stage indicates that the flow softening, mainly DRV, occurs. This partially offsets the initial strain hardening caused by dislocation proliferation. After necking, the cavities and cracks will be formed easily. Their rapid propagation at the localized necking positions can be responsible for the sharp decrease in flow stress. In stark contrast to parabolic stress–strain curve, the stress–strain curves are characterized by the rapidly increased flow stress to a peak value followed by gradually decreasing at a steady rate as increasing the temperature or decreasing the strain rate, as shown in Figure 2b,c. This indicates the extended necking deformation though the necking has occurred. The similar phenomenon has also been reported in other alloys [27,28]. It is inferred that the flow softening associated with DRV and possible DRX can promote the progressive development of necking deformation and thus delay the necking rupture.

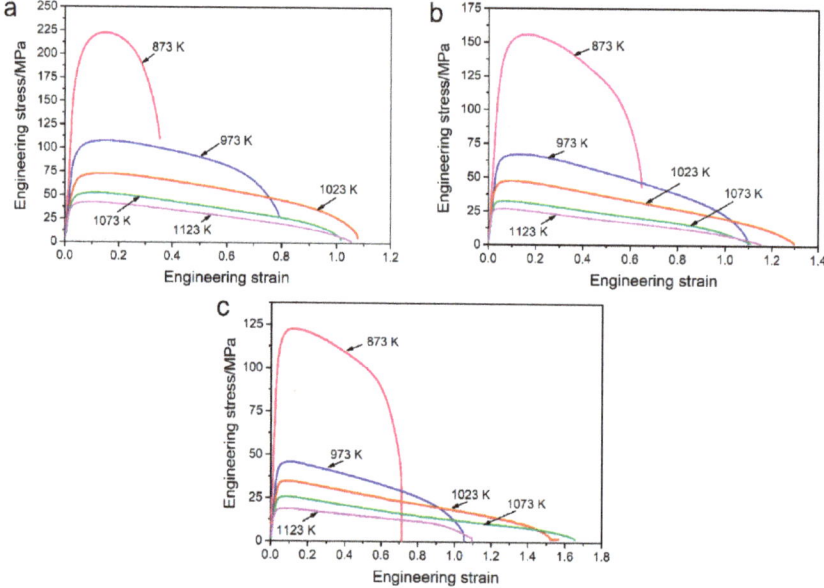

Figure 2. Tensile engineering stress–strain curves of the 0Cr18AlSi steel tested at various temperatures with strain rates of (a) 1.7×10^{-2} s^{-1}; (b) 1.7×10^{-3} s^{-1}; and (c) 1.7×10^{-4} s^{-1}.

The tensile properties, namely, yield strength (YS), ultimate tensile strength (UTS), elongation to fracture (EL), uniform elongation (UE), and reduction of area (RA) are summarized in detail, and the results are presented in Figure 3. From Figure 3a,b, it can be seen that, overall, YS and UTS drop rapidly from 873 K to 1023 K for all strain rates followed by a gradual decrease until 1123 K. Meanwhile, both YS and UTS increase with the increase of the strain rate for all temperatures. The maximum YS and UTS are obtained at 873 K and 1.7×10^{-2} s^{-1} (166.49 MPa and 222.28 MPa, respectively). It is known that the grain boundary and dislocation are easy to move at high temperatures, because the average kinetic energy of atoms increases and the critical shear stress for dislocation activation is reduced [29]. Therefore, the flow softening can be promoted by reducing the dislocation density at high temperatures, so as to decrease the strength. In addition, the increase of strain rate can enhance deformation storage energy and restrict atoms motion, which can cause an obvious work hardening, and thus results in the enhancement of strength. The influence of temperature on the EL is complex (Figure 3c). At high strain rates of 1.7×10^{-2} s^{-1} and 1.7×10^{-3} s^{-1}, EL increases with increasing temperature up to 1023 K, after which it exhibits a gradual decrease and then remains relatively stable. At the low strain rate of 1.7×10^{-4} s^{-1}, a peak value in EL (~163%) can be found at 1073 K, indicating the pleasant plasticity. However, at temperatures above 1073 K, EL decreases to ~107%. Generally, such remarkable decrease in EL at high temperatures is ascribed to carbide precipitation and grain boundary migration [30]. The detailed discussion will be demonstrated later. Furthermore, it is observed from Figure 2 that the studied steel shows a limited strain hardening. The variations of UE with temperature at different strain rates are shown in Figure 3d. It is well known that the higher the value of UE, the more the strain hardening capacity [27]. From Figure 4d, UE decreases with increasing the temperatures regardless of the strain rates. UE presents a high level at the strain rate of 1.7×10^{-2} s^{-1} in comparison to other employed strain rates, where it reaches a maximum of 12.5%. This infers that the flow softening is accelerated at low strain rate and high temperature, which thereby leads to an invisible strain hardening during the initial tensile deformation. For the RA, as shown in Figure 3e, it increases with increasing the temperature at all strain rates, which exhibits a distinct difference with the variation of EL at temperatures above 1023 K. A high RA and low EL illustrates that the local necking resistance is reduced at high temperatures. RA increases with decreasing the strain rate as expect, but it changes little when the strain rate is less than 1.7×10^{-3} s^{-1}.

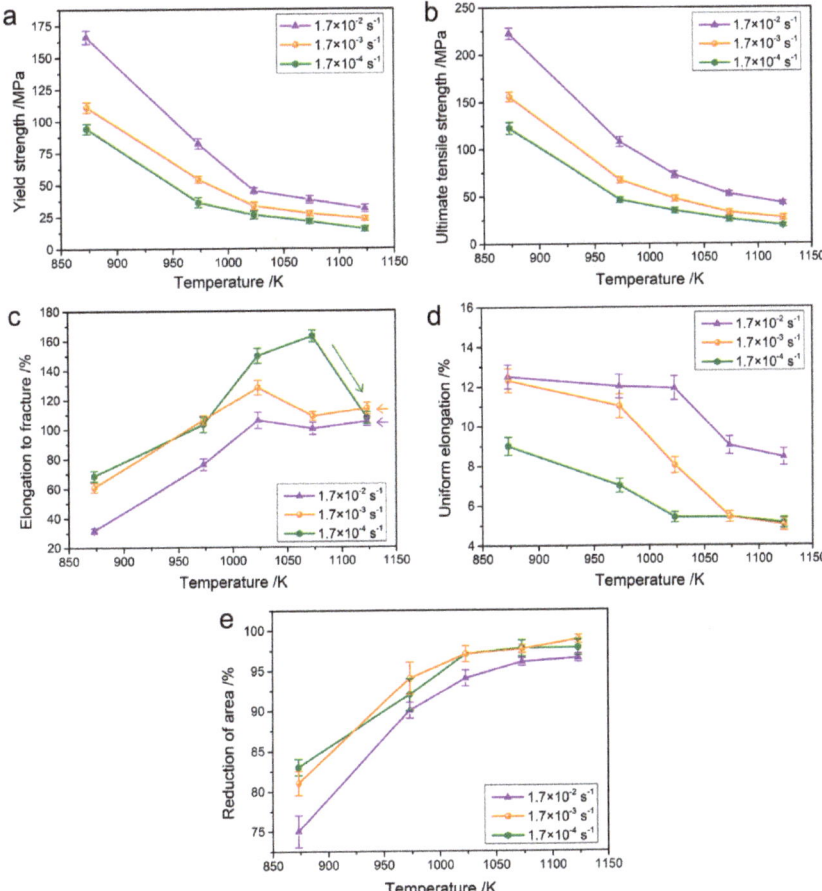

Figure 3. Tensile properties of the 0Cr18AlSi steel tested at different conditions: (**a**) yield strength, YS; (**b**) ultimate tensile strength (UTS); (**c**) elongation to fracture (EL); (**d**) uniform elongation (UE); (**e**) reduction of area (RA).

The fracture surfaces of the specimens tested at different conditions are shown in Figure 4, which reveals the rupture features. The existence of tearing ridges and dimples indicate that the fracture failure mode is ductile in nature. At the high strain rate of 1.7×10^{-2} s^{-1}, it can be seen that the small equiaxed dimples are uniformly distributed on the fracture surface of the specimen deformed at 973 K (Figure 4a), and the inner walls of these dimples are smooth. The dimples become larger and deeper with increasing the temperature up to 1023 K (Figure 4b). This indicates that the studied steel undergoes large deformation before rupture, and thus facilitates the void growth. This is also evidenced from the macro-view of the tested specimen along the gauge length (Figure 4b inset). The macro-appearance of the fracture changes from square to strip, and obvious rumpling of the side walls can be observed. Moreover, the trans-granular shear becomes apparent as the increase of test temperature, as shown in Figure 4c. When the strain rate decreases to 1.7×10^{-4} s^{-1}, there exhibits significant change on the fracture morphology. The dimples coalescence and the cracks on tearing ridges can be easily observed (Figure 4d,e). More micro-voids are also activated at low strain rates. This infers that the degree of the deformation is increased. Similar with the fracture at 1123 K and 1.7×10^{-2} s^{-1}, the trans-granular shear failure with plastic flow is the main feature of the fracture surface as increasing the temperature to 1073 K at 1.7×10^{-4} s^{-1}, as shown in Figure 4f. The severe deformation featured by oblate dimples and cracks with a large size give the reason for the achievement of best elongation.

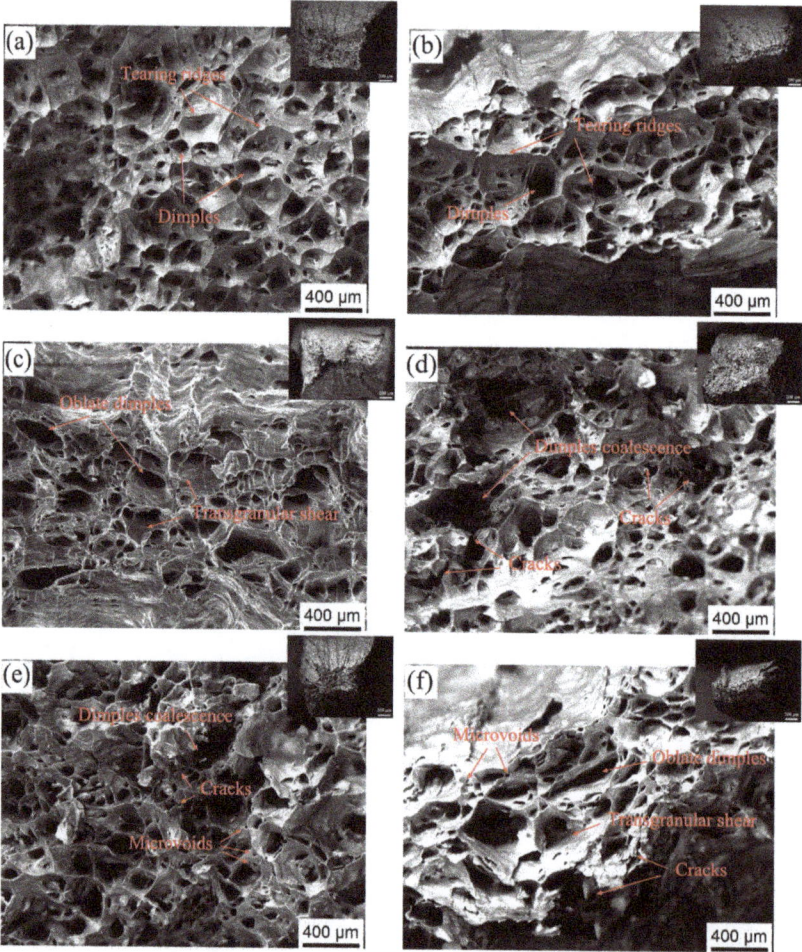

Figure 4. SEM fractographs of the 0Cr18AlSi steel tested at different conditions: (**a**) 973 K and 1.7×10^{-2} s^{-1}; (**b**) 1023 K and 1.7×10^{-2} s^{-1}; (**c**) 1123 K and 1.7×10^{-2} s^{-1}; (**d**) 873 K and 1.7×10^{-4} s^{-1}; (**e**) 973 K and 1.7×10^{-4} s^{-1}; (**f**) 1073 K and 1.7×10^{-4} s^{-1}. The insets indicated the fracture surfaces at low magnifications.

3.3. Microstructures Analyses

3.3.1. Microstructures Evolution Observed by OM

The tensile behaviors are strongly connected to the microstructures. Figure 5 shows the optical micrographs of the cross-sectional area beneath the fracture surface for the specimen fractured under the typical conditions. The characteristics of microstructures and cavities can be observed. It is seen that at 873 K and 1.7×10^{-4} s^{-1} (Figure 5a), most grains are elongated parallel to the tensile axis due to the deformation of original grains, and a large number of small spherical cavities appear near the grain boundaries, particularly in the boundary junctions. These small cavities become coarser in the vicinity of the fracture due to their coalescence, eventually causing the material to break. As can be seen, the fracture experiences cavity formation, cavity coalescence, and crack formation and growth. As the temperature goes up to 1023 K (Figure 5b), the equiaxed grains can be found in the microstructure near the fracture apart from the deformed grains, indicating the formation of DRX. Moreover, the number of cavities is significantly reduced. However, these cavities preferentially nucleated at the boundaries of the new grains, and they are then elongated and expanded from the grain boundaries to interiors. This process can consume a large amount of deformation energy. Therefore, the necking deformation ability is improved when deformed at 1023 K. At the high temperature of 1123 K (Figure 5c), the grains are obviously coarsened due to the activation of boundaries migration, which indicates the dynamic softening is accelerated. However, the number of elongated cavities decreases. Plenty of spherical cavities and their coalescence in the vicinity of the fracture can be observed once again, which also indicates that the studied steel undergoes low necking deformation before fracture. With increasing the strain rate up to 1.7×10^{-2} s^{-1} at 1123 K (Figure 5d), the microstructure is refined obviously. It is noted that most elongated cavities distribute intensely near the upper and lower surfaces (marked the blue arrows). This because that a strong three-dimensional stress has formed at the necking area during the tensile test at high strain rates, which can promote the initiation and propagation of cavities.

3.3.2. Microstructures Evolution Observed by EBSD

In order to analyze the mechanism of microstructures evolution during hot tensile deformation, EBSD measurements were carried out on the deformed specimens. Figure 6 shows the grain morphologies near the fracture tips of the specimens tested at 1.7×10^{-4} s^{-1} with different temperatures. At 873 K (Figure 6a), numerous pancake-like deformed grains can be easily observed along the tensile direction. Different areas show various colors in these deformed grains, indicating the generation of sub-structures. To reveal the development of the sub-structures within the deformed grains, the local (point-to-point) and cumulative (point-to-origin) misorientations were calculated in a special grain marked the line L1 from left to right, and the results are shown in Figure 7a. Several sharp peaks with misorientation angle of 4–7° can be seen in the point-to-point misorientation curve, while the misorientation angle between these sharp peaks is less than 2°. This indicates that the orientation of different areas within this grain changes, and the grain has been divided into several sub-grains surrounded by low-angle grain boundaries (LAGBs, 2–15°). The large amount of wavy shape peaks less than 2° represent that the high density of dislocations forms in the sub-grains. Belyakov et al. [31] pointed out that the dislocation density between the dislocation walls could exceed 10^{14} m^{-2} during warm deformation for ferritic stainless steel. Moreover, the point-to-origin misorientation can easily exceed 10° on the distances of 16–20 µm, 38–93 µm, and 122–157 µm. The large strain gradient will promote the sub-grain rotation and eventually lead to the formation of high-angle grain boundaries (HAGBs, >15°). Some new segments of HAGBs have been detected in the deformed grain interiors, marked the black arrows in Figure 6a. The variation in misorientation across a typical HAGB segment is plotted in Figure 7b, which provides an evidence for the occurrence of sub-grain boundary rotation induced HAGB. Obviously, these new segments of HAGBs will promote the formation of DRX nuclei through their closed loop. Such DRX formation mechanism caused by the progressive sub-grain rotation within the deformed grain is generally referred as continuous DRX (CDRX) [32–34]. Therefore, the emergence of HAGBs segments means the initial stage of CDRX nucleation. In addition, some fine equiaxed grains can be observed around the original grain boundaries, except for a few clusters of small grains around the particles within the grains (white arrows in Figure 6a), thus forming the typical necklace structure. These fine grains are formed dynamically during the hot tensile deformation, indicating the formation of DRX. However, the DRX region accounts for only ~7%. The average size of the new grains is about 3.7 μm. It is referred that the growth of DRX grains is restricted due to the low deformation temperature in spite of at a very low strain rate. The serrated original grain boundaries can be easily observed in Figure 6a, and fine grains exist along these boundaries. This phenomenon is generally associated with discontinuous DRX (DDRX) characterized by grain boundary bulging [35]. The nucleation sites of DDRX can be provided by the grain boundary bulging through strain-induced grain boundary migration. Subsequently, the sub-boundaries with low misorientations caused by dislocation rearrangement can be formed at the bottom of the bulging area, and then they continuously absorb dislocations to increase their misorientations. Once HAGBs form from these LAGBs, the fine DDRX grains will develop. In fact, DDRX will spread toward the center of the prior grain with increasing the strain, and eventually swallow up the whole grain. However, under the current state, DDRX is insufficient because of the low degree of deformation. Therefore, DRV characterized by sub-grain formation and a small amount of DDRX play an important role in the microstructural evolution.

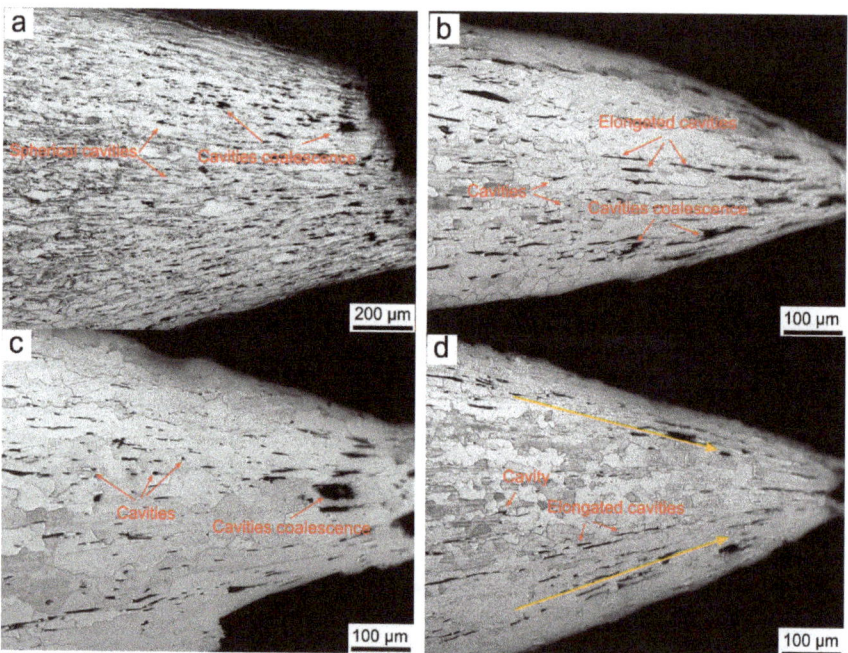

Figure 5. Cross-sectional microstructures of the fracture tips of the specimens after tensile tests at different conditions: (**a**) 873 K and 1.7×10^{-4} s^{-1}; (**b**) 1023 K and 1.7×10^{-4} s^{-1}; (**c**) 1123 K and 1.7×10^{-4} s^{-1}; (**d**) 1123 K and 1.7×10^{-2} s^{-1}.

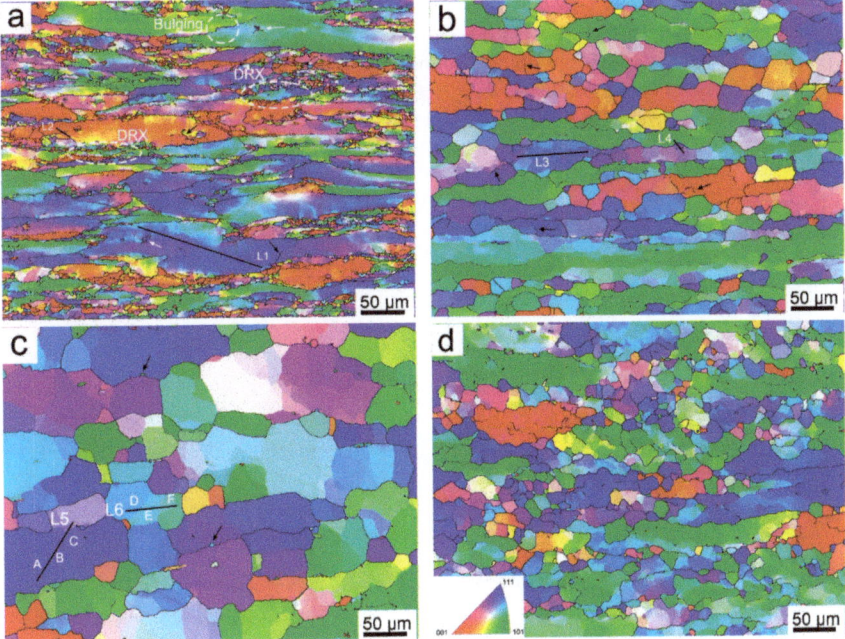

Figure 6. Inverse pole figure (IPF)-coloring orientation map for the 0Cr18AlSi steel tested under different deformation conditions: (**a**) 1.7×10^{-4} s^{-1} and 873 K; (**b**) 1.7×10^{-4} s^{-1} and 1023 K; (**c**) 1.7×10^{-4} s^{-1} and 1123 K; (**d**) 1.7×10^{-2} s^{-1} and 1123 K.

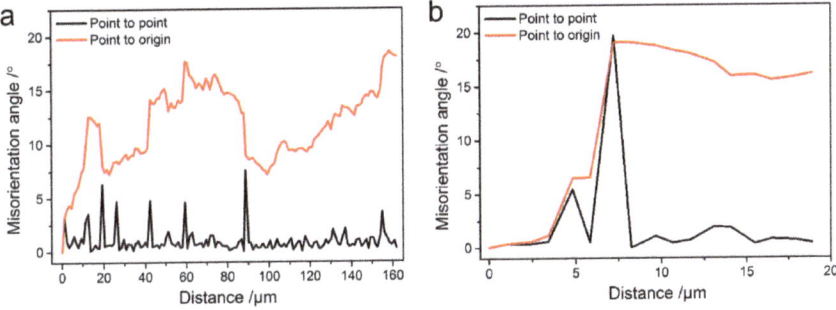

Figure 7. Misorientation profiles obtained along the lines marked in Figure 6a. (**a**) L1; (**b**) L2.

At 1023 K (Figure 6b), as expected, the extensive DRX can be observed. The degree of DRX reaches 42%. Obviously, the DRX process is facilitated when the temperature increases from 873 to 1023 K at the strain rate of 1.7×10^{-4} s^{-1}, and thus DRX becomes a main softening mechanism. However, no necklace structure like the one in Figure 6a is observed, and numerous equiaxed grains can be found. Meanwhile, sub-structures are formed in the deformed grains, especially near the grain boundaries. The misorientation change across a typical deformed grain from left to right is shown in Figure 8a. Similarly, this grain is divided into several well-defined sub-grains with different misorientations. The cumulative misorientation exceeds 10° over the distance of 17–80 μm. These sub-grain boundaries will easily develop into HAGBs with the increase of strain through the progressive sub-grain rotation [36]. Many sub-grains have been partly surrounded by the HAGBs (black arrows in Figure 6b), indicating the CDRX nucleation mechanism. Hence, the softening behavior at 1023 K is mainly controlled by DRV and CDRX. In addition, it is observed that the size of DRX grains increases in comparison to that at 873 K. This is because the increased temperature provides a greater driving force for migration of sub-grain/grain boundaries. Because of this, some new grains can still be nucleated by the bend of the grain boundary. Figure 8b shows the misorientation change across a special local bulging area from left to right, marked L4 in Figure 6b. The misorientation cumulates up to 10° at the bottom of bulging area, indicating that a well-developed sub-grain boundary has formed.

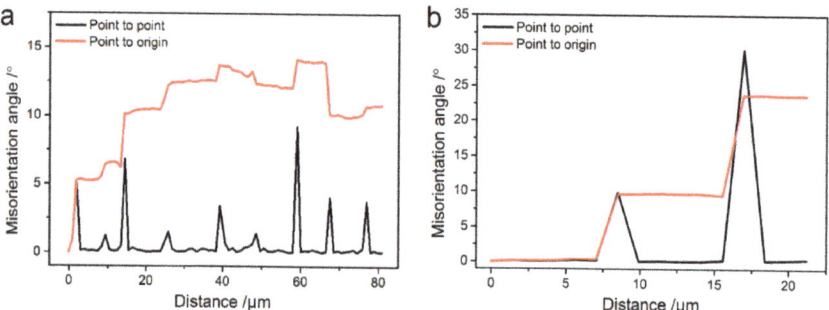

Figure 8. Misorientation profiles obtained along the lines marked in Figure 6b. (a) L3; (b) L4.

At 1123 K (Figure 6c), it is seen that the area of DRX region is ~46%, which has a slight increase when compared to that obtained at 1023 K. Coarse grains can be observed, and the average size of DRX grains is increased to ~32 µm. However, the amount of sub-structures decreases significantly, and therefore, the density of dislocations is reduced. Most of sub-boundaries are straight and regular, indicating that DRV becomes sufficient because of high driving force for dislocations climb and cross-slip [25]. The reduction of sub-structures accompanied with low dislocation density lowers the amount of stored energy, and thus reducing the driving force for lattice rotation. It is inferred that high temperature can contribute to the coalescence of sub-grains and the migration of sub-grain/grain boundaries. Figure 9 shows the variation in misorientation across several sub-grains from left to right, which reveals the development of sub-grain coalescence. From the lines L5 and L6, the misorientation angles between grains A and B or D and E are less than 2°, and the cumulative misorientations do not exceed 5°. This indicates that the adjacent grains gradually coalesce into a single grain through rotating to reduce the misorientation. It is noted that some new equiaxed grains have been formed on the boundaries of coarse grains (black arrows in Figure 6c). These new grains may be nucleated through grain boundary bulging. However, they are very small in size, and thus results in a less uniform microstructure. In summary, at high temperature of 1123 K, the effect of progressive sub-grain rotation induced HAGBs is weakened, and CDRX caused by sub-grain coalescence becomes the main mechanism of DRX.

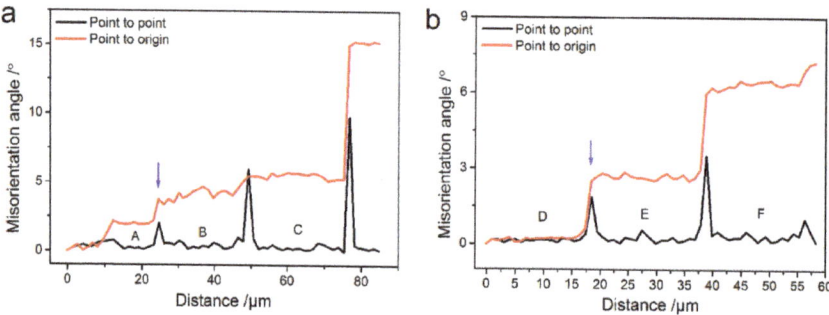

Figure 9. Misorientation profiles obtained along the lines marked in Figure 6c. (a) L5; (b) L6.

In order to analyze the strain rate effect on the microstructures during hot tensile deformation, IPF-coloring orientation map of the studied steel deformed at a high strain rate of 1.7×10^{-2} s^{-1} with 1123 K are presented in Figure 6d. The elongated original grains contained sub-structures as evident from the color variations can be observed, and a considerable HAGBs fragments are formed within these deformed grains. In addition, the new recrystallized grains can be seen inside the grains. It is inferred that these new grains nucleate through CDRX with the mechanism of sub-grain rotation. The HAGBs fragments will be able to contribute the CDRX nucleation. However, the fraction of DRX decreases to 38% as compared with that under the strain rate of 1.7×10^{-4} s^{-1}. The DRX grain size also decreases. The high strain rate can increase the dislocation density, and generate more stored deformation energy [34,37]. Hence, the sub-grain rotation can be accelerated, leading to a quick nucleation of DRX. However, the growth of new grains is restricted due to the insufficient time for the migration of grain boundaries at the high strain rate. As a result, a finer microstructure is obtained.

The full grain boundary distribution corresponding to the Figure 6 is shown in Figure 10. Black and green lines represent HAGBs and LAGBs respectively. It is clear that a large number of LAGBs composed of dislocation-rich layers exist in the deformed grains at 873 K (Figure 10a,b). Interestingly, in large deformed grains, the closer to the original HAGBs, the higher the density of sub-grain boundaries. It can therefore be inferred that the area near original HAGBs has a high dislocation density. With increasing the deformation temperature (Figure 10c–f), as expected, the fraction of LAGBs decreases, while the fraction of HAGBs increases. This indicates that the dynamic softening accompanied with CDRX is reinforced with the increase of temperature. However, at high temperature of 1123 K, the fraction of sub-grain boundaries with misorientation angles less than 2° increases to 34%. This is related to the decrease in the boundary misorientation caused by a large number of sub-grain coalescence. In addition, according to Refs [38,39], the high fraction of 10–15° misorientation angles in sub-structures means that it may produce more HAGBs from these LAGBs by sub-grain rotation. For the strain rate of 1.7×10^{-4} s^{-1}, the fractions of 10–15° misorientation angles at temperatures of 873, 1023, and 1123 K are 6%, 7% and 3.5%, respectively, which illustrates that the transition and migration from LAGBs to HAGBs is reduced at 1123 K. Therefore, the sub-grain coalescence becomes the main mode of CDRX gradually. With increasing the strain rate up to 1.7×10^{-2} s^{-1} at 1123 K (Figure 10g,h), the distribution of sub-grain structure trends to uniform, which results in the fragment of the original grains. A high fraction of LAGBs (45%) can be obtained, which is much higher than that at 1.7×10^{-4} s^{-1}. The presence of more sub-grains with LAGBs and segments of HAGBs in the interiors of grains will be able to contribute the CDRX nucleation by sub-grain rotation.

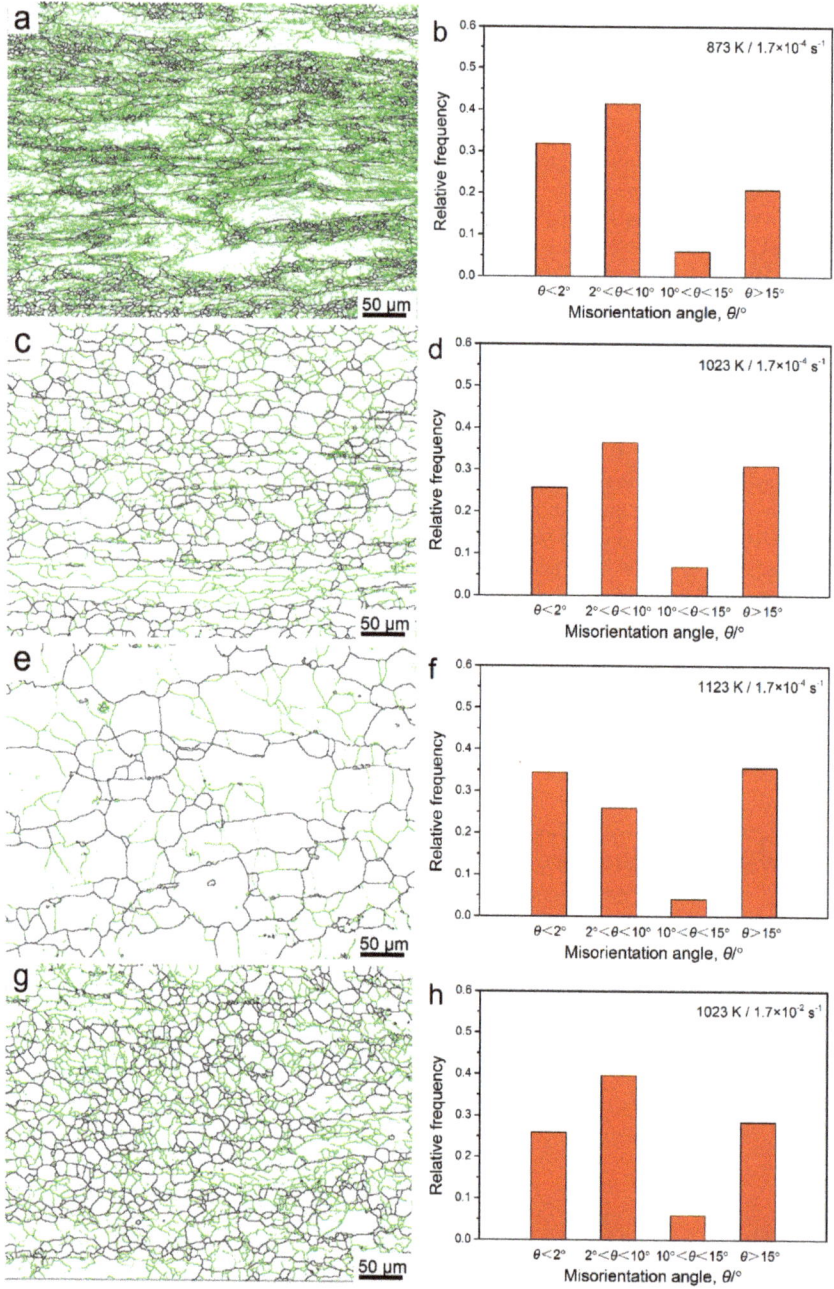

Figure 10. The distribution of grain boundary for the 0Cr18AlSi steel tested under different deformation conditions: (**a**) and (**b**) 1.7×10^{-4} s^{-1} and 873 K; (**c**) and (**d**) 1.7×10^{-4} s^{-1} and 1023 K; (**e**) and (**f**) 1.7×10^{-4} s^{-1} and 1123 K; (**g**) and (**h**) 1.7×10^{-2} s^{-1} and 1123 K. Green lines correspond to boundaries with low misorientation $2° < \theta < 15°$, and black lines $\theta > 15°$ represent high-angle boundaries.

Kernel average misorientation (KAM) maps corresponding to the Figure 6 are shown in Figure 11. The KAM value can be recognized as an indicator of dislocation density. The higher the KAM value, the higher the dislocation density. At 1.7×10^{-4} s^{-1} and 873 K (Figure 11a), it is observed that the average KAM value is 1.27°. At this stage, DRV, rather than DRX, is the dominant softening mechanism. Furthermore, the high KAM values can be found in the areas surrounding the new DRX grains, indicating that a high density of dislocation has formed in these areas. The limited deformation between the adjacent new grains due to the high strain nearby the grain boundaries cannot offset the stress concentration caused by dislocations pilling up, and as a result, the cavities can be induced. This explains why the cavities preferentially nucleate nearby the original grain boundaries where DRX apparently takes place (Figure 6a). The high density of spherical cavities can be responsible for the low ductility and the significant necking. The average KAM value decreases with increasing the deformation temperature (Figure 11b,c), 0.40° for 1023 K and 0.37° for 1123 K. The dislocation density decreases significantly due to DRX, resulting in an obvious decrease in strength during the hot tensile process. However, a slight decrease in KAM value is observed at 1123 K, which is attributed to limited increase of DRX and sufficient DRV, thus leading to the slow decrease of flow stress in the range of 1023–1123 K. Moreover, at 1123 K, the area on the side of the bulging for new grains nucleated at the original grain boundaries has a relatively high KAM value. The deformation incompatibility in this area due to large difference in grain size generally can provide the sites for the nucleation of cavities (Figure 5c). The coarse grains caused by fast migration of boundaries and uneven grain distribution lead to the decrease of ductility. Figure 11d shows the effect of strain rate on the KAM value. It is found that the density of dislocations is increased at 1.7×10^{-2} s^{-1}, and the average KAM value is 0.53°, which is slightly higher than that at 1.7×10^{-4} s^{-1}. The distribution of KAM value is relatively uniform. Therefore, the strength of the studied steel is increased.

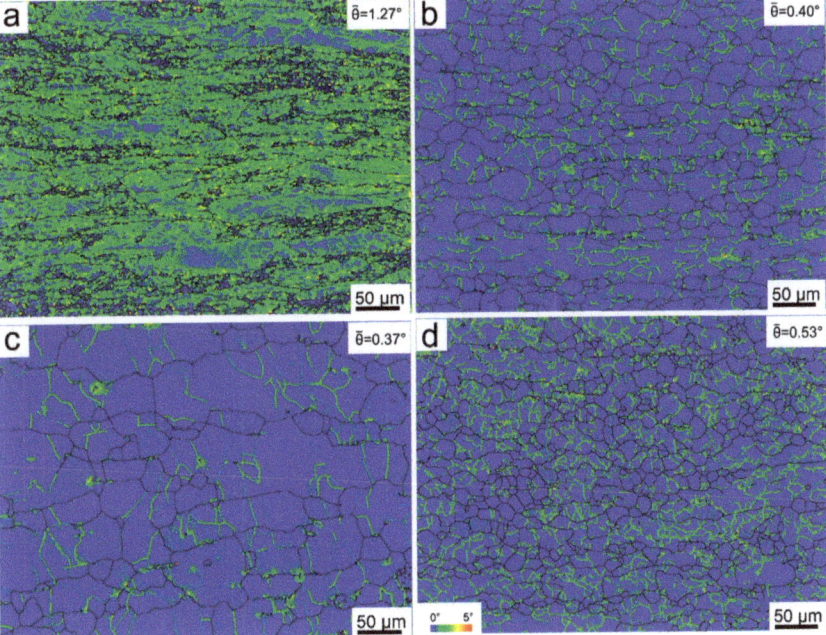

Figure 11. Kernel average misorientation (KAM) maps for the 0Cr18AlSi steel tested under different deformation conditions: (**a**) 1.7×10^{-4} s^{-1} and 873 K; (**b**) 1.7×10^{-4} s^{-1} and 1023 K; (**c**) 1.7×10^{-4} s^{-1} and 1123 K; (**d**) 1.7×10^{-2} s^{-1} and 1123 K.

3.4. Kinetic Analysis

The hot tensile flow behavior of the 0Cr18AlSi steel can be presented through the correlation among true stress, deformation temperature and strain rate, which is generally expressed by the Arrhenius type equation [40,41]:

$$\dot{\varepsilon} = A\sigma^n \exp\left(-\frac{Q}{RT}\right), \quad (1)$$

where $\dot{\varepsilon}$ the strain rate (s^{-1}), A a constant (s^{-1}), σ the true stress (MPa), n the stress exponent, Q the deformation activation energy (kJ·mol^{-1}), R the universal gas constant (8.314 J mol^{-1} K^{-1}), and T the temperature in K.

In Equation (1), the stress exponent and the deformation activation energy are considered to be the significant parameters with physical meaning. The stress exponent correlates the high temperature deformation mechanism, while the deformation activation energy reflects how the temperature affects the kinetics or the rate of the process, i.e., diffusion. In order to estimate the n value of the 0Cr18AlSi steel at the current state, Equation (1) can be rewritten in the following expression:

$$\ln \sigma = \frac{\ln \dot{\varepsilon}}{n} + \frac{Q}{nRT} - \frac{\ln A}{n}. \quad (2)$$

Here, the stable true stresses are taken from the flow curves. Remarkably, the n value can be obtained from the reciprocal of the slope of the linear regression line in the lnσ–ln$\dot{\varepsilon}$ plot at a particular temperature. In fact, such slope of the regressive line, the reciprocal of the stress exponent, represents strain rate sensitivity (m, $m = 1/n$). Ideally, a larger m value indicated a larger ductility and higher necking resistance [40]. Figure 12a shows the correlation between the flow stress and strain rate in a natural log. It is clear that at the low temperature of 873 K, m has the lowest value (~0.12), indicating a low ductility, which is consistent with the results in Figure 3c. While increasing the temperature, the value of m rises first and then declines. The maximum m value (~0.19) can be obtained at temperatures between 1023 and 1073 K. After averaging the m values, the stress exponent is estimated to be 5.9. It is accepted that, when $n = 4$–6, the motion of dislocations controlled by climb would be the dominant deformation mechanism [42].

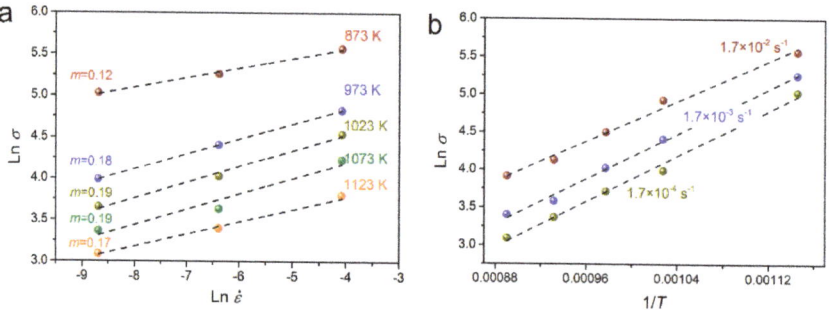

Figure 12. Plots of lnσ–ln$\dot{\varepsilon}$ (a) and lnσ–1/T (b) for the 0Cr18AlSi steel.

Similarly, for a particular strain rate, the partial derivatives from both sides of Equation (2) are taken to $1/T$, Q can be expressed as:

$$Q = nR\left[\frac{\partial \ln \sigma}{\partial (1/T)}\right]_{\dot{\varepsilon}} = nRS, \quad (3)$$

where S the slope of lnσ–1/T plot at various strain rates, as shown in Figure 12b. The Q value of the present steel is calculated to be 355 kJ·mol^{-1}. This value is found to be much higher than the activation energy of self-diffusion in α–Fe (239 kJ/mol) [43], indicating that the softening mechanism of the

0Cr18AlSi steel at high temperatures is controlled by DRV and DRX rather than diffusion. However, it is lower than the previously determined values of 385 kJ/mol for high purity 17Cr ferritic stainless steel [44], 405 kJ/mol for 21Cr ferritic stainless steel [45], or 424 kJ/mol for Ti and Nb containing 17Cr ferritic stainless steel [46]. Such relatively low deformation activation energy for the 0Cr18AlSi steel means that the movement of dislocations, such as climbing of edge dislocations and cross-slipping of screw dislocations, only requires a small driving force during hot deformation. Therefore, DRV and DRX occur more easily in comparison with the above conventional ferritic stainless steels.

4. Conclusions

The hot tensile tests were conducted at temperatures ranging from 873 to 1123 K and strain rates of 1.7×10^{-4}–1.7×10^{-2} s^{-1} for the 0Cr18AlSi ferritic stainless steel. The tensile properties and mechanism of microstructural evolution were investigated. The main results are as follows:

(1) The maximum elongation to fracture can be obtained at 1023 K for the strain rates of 1.7×10^{-3} and 1.7×10^{-2} s^{-1}. The temperature for the maximum elongation to fracture is delayed to 1073 K when decreasing the strain rate to 1.7×10^{-4} s^{-1}. The decrease in necking deformation resistance caused by fast migration of boundaries and uneven grain distribution results in the low ductility at high temperature of 1123 K.

(2) At low temperature, DRV dominates the softening process though a small number of DDRX occurs. As increasing the temperature, the flow softening by CDRX is reinforced due to the acceleration of rotation and coalescence of sub-grains. The main reason of ductility decrease at high temperature and low strain rate is the grain coarsening and the deformation discordance between small new DRX grains and coarse grains.

(3) The stress exponent and the deformation activation energy are 5.9 and 355 kJ·mol^{-1} respectively. The dominant deformation mechanism is the dislocations motion controlled by climb.

Author Contributions: Y.H. conceived and designed the experiments; J.S. (Jiaqi Sun) performed the experiments; Y.H. and Y.S. analyzed the data; J.S. (Jiapeng Sun) contributed analysis tools; Y.H. and X.R. wrote the paper. All authors have read and agreed to the published version of the manuscript.

Funding: This work is supported by the National Natural Science Foundation of China (No. 51604034 and 51974032) and the Science and Technology Development Program of Jilin Province (No. 20190302003GX).

Conflicts of Interest: The authors declare no conflict of interest. The founding sponsors had no role in the design of the study; in the collection, analyses, or interpretation of data; in the writing of the manuscript, and in the decision to publish the results.

References

1. Fu, J.; Li, F.; Sun, J.; Wu, Y. Texture, orientation, and mechanical properties of Ti-stabilized Fe-17Cr ferritic stainless steel. *Mater. Sci. Eng. A* **2018**, *738*, 335–343. [CrossRef]
2. Han, J.; Li, H.; Zhu, Z.; Jiang, L.; Xu, H.; Ma, L. Effects of processing optimization on microstructure, texture, grain boundary and mechanical properties of Fe-17Cr ferritic stainless steel thick plates. *Mater. Sci. Eng. A* **2014**, *616*, 20–28. [CrossRef]
3. Wang, W.; Han, P.; Peng, P.; Zhang, T.; Liu, Q.; Yuan, S.-N.; Huang, L.-Y.; Yu, H.-L.; Qiao, K.; Wang, K.-S. Friction Stir Processing of Magnesium Alloys: A Review. *Acta Met. Sin. Engl. Lett.* **2019**, *32*, 1–15. [CrossRef]
4. Wang, W.; Han, P.; Yuan, J.; Peng, P.; Liu, Q.; Qiang, F.; Qiao, K.; Wang, K.-S. Enhanced Mechanical Properties of Pure Zirconium via Friction Stir Processing. *Acta Met. Sin. Engl. Lett.* **2019**, *32*, 1–7. [CrossRef]
5. Hu, W.Q.; Dong, Z.; Yu, L.M.; Ma, Z.Q.; Liu, Y.C. Synthesis of W-Y2O3 alloys by freeze-drying and subsequent low temperature sintering: Microstructure refinement and second phase particles regulation. *J. Mater. Res. Technol.* **2020**, *36*, 84–90. [CrossRef]
6. Chiu, Y.-T.; Lin, C.-K.; Wu, J.-C. High-temperature tensile and creep properties of a ferritic stainless steel for interconnect in solid oxide fuel cell. *J. Power Sources* **2011**, *196*, 2005–2012. [CrossRef]

7. Gurram, M.; Adepu, K.; Pinninti, R.R.; Gankidi, M.R. Effect of copper and aluminium addition on mechanical properties and corrosion behaviour of AISI 430 ferritic stainless steel gas tungsten arc welds. *J. Mater. Res. Technol.* **2013**, *2*, 238–249. [CrossRef]
8. Zhang, C.; Liu, Z.; Xu, Y.; Wang, G. The sticking behavior of an ultra purified ferritic stainless steel during hot strip rolling. *J. Mater. Process. Technol.* **2012**, *212*, 2183–2192. [CrossRef]
9. Fujita, N.; Ohmura, K.; Kikuchi, M.; Suzuki, T.; Funaki, S.; Hiroshige, I. Effect of niobium on high-temperature properties for ferritic stainless steel. *Scr. Mater.* **1996**, *35*, 705–710. [CrossRef]
10. Sim, G.M.; Ahn, J.C.; Hong, S.C.; Lee, K.J.; Lee, K.S. Effect of Nb precipitate coarsening on the high temperature strength in Nb containing ferritic stainless steels. *Mater. Sci. Eng. A* **2005**, *396*, 159–165. [CrossRef]
11. Han, J.; Li, H.; Xu, H. Microalloying effects on microstructure and mechanical properties of 18Cr–2Mo ferritic stainless steel heavy plates. *Mater. Des.* **2014**, *58*, 518–526. [CrossRef]
12. Chiu, Y.-T.; Lin, C.-K. Effects of Nb and W additions on high-temperature creep properties of ferritic stainless steels for solid oxide fuel cell interconnect. *J. Power Sources* **2012**, *198*, 149–157. [CrossRef]
13. Wei, L.; Zheng, J.; Chen, L.; Misra, R.D.K. High temperature oxidation behavior of ferritic stainless steel containing W and Ce. *Corros. Sci.* **2018**, *142*, 79–92. [CrossRef]
14. DeMeyer, M.; Vanderschueren, D.; DeCoon, B.C. The influence of the substitutionof Si by Al on the properties of cold rolled C–Mn–Si TRIP steels. *ISIJ Int.* **1999**, *39*, 813–822.
15. Hojo, T.; Sugimoto, K.-I.; Mukai, Y.; Ikeda, S. Effects of Aluminum on Delayed Fracture Properties of Ultra High Strength Low Alloy TRIP-aided Steels. *ISIJ Int.* **2008**, *48*, 824–829. [CrossRef]
16. Qian, L.; Zhou, Q.; Zhang, F.; Meng, J.; Zhang, M.; Tian, Y. Microstructure and mechanical properties of a low carbon carbide-free bainitic steel co-alloyed with Al and Si. *Mater. Des.* **2012**, *39*, 264–268. [CrossRef]
17. Chen, S.; Rana, R.; Haldar, A.; Ray, R.K. Current state of Fe-Mn-Al-C low density steels. *Prog. Mater. Sci.* **2017**, *89*, 345–391. [CrossRef]
18. Wu, Z.; Tang, Y.; Chen, W.; Lu, L.; Li, E.; Li, Z.; Ding, H. Exploring the influence of Al content on the hot deformation behavior of Fe-Mn-Al-C steels through 3D processing map. *Vacuum* **2019**, *159*, 447–455. [CrossRef]
19. Xu, S.; Zhou, Z.; Long, F.; Jia, H.; Guo, N.; Yao, Z.; Daymond, M.R. Combination of back stress strengthening and Orowan strengthening in bimodal structured Fe–9Cr–Al ODS steel with high Al addition. *Mater. Sci. Eng. A* **2019**, *739*, 45–52. [CrossRef]
20. Ota, H.; Nakamura, T.; Maruyama, K. Effect of solute atoms on thermal fatigue properties in ferritic stainless steels. *Mater. Sci. Eng. A* **2013**, *586*, 133–141. [CrossRef]
21. Zou, D.; Zhou, Y.; Zhang, X.; Zhang, W.; Han, Y. High temperature oxidation behavior of a high Al-containing ferritic heat-resistant stainless steel. *Mater. Charact.* **2018**, *136*, 435–443. [CrossRef]
22. Sastry, S.D.; Rohatgi, P.K.; Abraham, K.P.; Prasad, Y.V.R.K. Influence of heat treatment on the strength and fracture behavior of Fe-12Cr-6Al ferritic stainless steel. *J. Mater. Sci.* **1982**, *17*, 3009–3016. [CrossRef]
23. Zhang, X.; Fan, L.; Xu, Y.; Li, J.; Xiao, X.; Jiang, L. Texture, microstructure and mechanical properties of aluminum modified ultra-pure 429 ferritic stainless steels. *Mater. Des.* **2016**, *89*, 626–635. [CrossRef]
24. Gao, F.; Liu, Z.Y.; Misra, R.D.K.; Liu, H.T.; Yu, F.X. Constitutive modeling and dynamic softening mechanism during hotdeformation of an ultra-pure 17%Cr ferritic stainless steel stabilized with Nb. *Met. Mater. Int.* **2014**, *20*, 939–951. [CrossRef]
25. Gao, F.; Song, B.; Xu, Y.; Xia, K. Substructural changes during hot deformation of an Fe-26Cr ferritic stainless steel. *Met. Mater. Trans. A* **2000**, *31*, 21–27. [CrossRef]
26. Kim, S.-L.; Yoo, Y.-C. Continuous dynamic recrystallization of AISI 430 ferritic stainless steel. *Met. Mater. Int.* **2002**, *8*, 7–13. [CrossRef]
27. Chauhan, A.; Litvinov, D.; Aktaa, J. High temperature tensile properties and fracture characteristics of bimodal 12Cr-ODS steel. *J. Nucl. Mater.* **2016**, *468*, 1–8. [CrossRef]
28. Pan, L.; Zheng, L.; Han, W.; Zhou, L.; Hu, Z.; Zhang, H. High-temperature tensile properties of a NiTi–Al-based alloy prepared by directional solidification and homogenizing treatment. *Mater. Des.* **2012**, *39*, 192–199. [CrossRef]
29. Liu, Y.; Ning, Y.; Nan, Y.; Liang, H.; Li, Y.; Zhao, Z.; Yao, Z.; Guo, H. Characterization of hot deformation behavior and processing map of FGH4096–GH4133B dual alloys. *J. Alloy. Compd.* **2015**, *633*, 505–515. [CrossRef]

30. Jiang, Z.H.; Han, J.P.; Li, Y.; He, P. High temperature ductility and corrosion resistance property of novel tin-bearing economic 17Cr-xSn ferritic stainless steel. *Ironmak Steelmak* **2015**, *42*, 504–511. [CrossRef]
31. Belyakov, A.; Sakai, T.; Kaibyshev, R. New grain formation during warm deformation of ferritic stainless steel. *Met. Mater. Trans. A* **1998**, *29*, 161–167. [CrossRef]
32. Mehtonen, S.; Palmiere, E.; Misra, R.; Karjalainen, L.; Porter, D. Dynamic restoration mechanisms in a Ti–Nb stabilized ferritic stainless steel during hot deformation. *Mater. Sci. Eng. A* **2014**, *601*, 7–19. [CrossRef]
33. Ebied, S.; Hamada, A.; Borek, W.; Gepreel, M.; Chiba, A. High-temperature deformation behavior and microstructural characterization of high-Mn bearing titanium-based alloy. *Mater. Charact.* **2018**, *139*, 176–185. [CrossRef]
34. Xie, B.; Zhang, B.; Ning, Y.; Fu, M. Mechanisms of DRX nucleation with grain boundary bulging and subgrain rotation during the hot working of nickel-based superalloys with columnar grains. *J. Alloy. Compd.* **2019**, *786*, 636–647. [CrossRef]
35. Han, Y.; Liu, G.W.; Zou, D.N.; Liu, R.; Qiao, G.J. Deformation behavior and microstructural evolution of as-cast 904L austenitic stainless steel under hot compression. *Mater. Sci. Eng. A* **2013**, *565*, 342–350. [CrossRef]
36. Xiao, W.; Wang, B.; Wu, Y.; Yang, X. Constitutive modeling of flow behavior and microstructure evolution of AA7075 in hot tensile deformation. *Mater. Sci. Eng. A* **2018**, *712*, 704–713. [CrossRef]
37. Lin, Y.; Jiang, X.-Y.; Shuai, C.-J.; Zhao, C.-Y.; He, D.-G.; Chen, M.-S.; Chen, C. Effects of initial microstructures on hot tensile deformation behaviors and fracture characteristics of Ti-6Al-4V alloy. *Mater. Sci. Eng. A* **2018**, *711*, 293–302. [CrossRef]
38. Liu, Z.; Li, P.; Xiong, L.; Liu, T.; He, L. High-temperature tensile deformation behavior and microstructure evolution of Ti55 titanium alloy. *Mater. Sci. Eng. A* **2017**, *680*, 259–269. [CrossRef]
39. Mandal, S.; Bhaduri, A.K.; Sarma, V.S. A study on microstructural evolution and dynamic recrystallization during isothermal deformation of a Ti-modified austenitic stainless steel. *Metall. Mater. Trans. A* **2010**, *42*, 1062–1072. [CrossRef]
40. Alsagabi, S.; Shrestha, T.; Charit, I. High temperature tensile deformation behavior of Grade 92 steel. *J. Nucl. Mater.* **2014**, *453*, 151–157. [CrossRef]
41. Lin, Y.C.; Huang, J.; Li, H.-B.; Chen, D.-D. Phase transformation and constitutive models of a hot compressed TC18 titanium alloy in the α+β regime. *Vacuum* **2018**, *157*, 83–91. [CrossRef]
42. Shrestha, T.; Basirat, M.; Charit, I.; Potirniche, G.P.; Rink, K.K.; Sahaym, U. Creep deformation mechanisms in modified 9Cr–1Mo steel. *J. Nucl. Mater.* **2012**, *423*, 110–119. [CrossRef]
43. Shewmon, P. *Diffusion in Solids*, 2nd ed.; TMS: Warrendale, PA, USA, 1989.
44. Gao, F.; Liu, Z.-Y.; Wang, G.-D. Hot deformation behavior of high-purified 17%Cr ferritic stainless steel. *J. Northeast Univ.* **2011**, *32*, 1406–1409.
45. Mehtonen, S.; Karjalainen, L.; Porter, D. Hot deformation behavior and microstructure evolution of a stabilized high-Cr ferritic stainless steel. *Mater. Sci. Eng. A* **2013**, *571*, 1–12. [CrossRef]
46. Gao, F.; Yu, F.-X.; Liu, H.-T.; Liu, Z.-Y.; Liu, F.-T. Hot Deformation Behavior and Flow Stress Prediction of Ultra Purified 17% Cr Ferritic Stainless Steel Stabilized with Nb and Ti. *J. Iron Steel Res. Int.* **2015**, *22*, 827–836. [CrossRef]

© 2020 by the authors. Licensee MDPI, Basel, Switzerland. This article is an open access article distributed under the terms and conditions of the Creative Commons Attribution (CC BY) license (http://creativecommons.org/licenses/by/4.0/).

Article

Effect of High Ti Contents on Austenite Microstructural Evolution During Hot Deformation in Low Carbon Nb Microalloyed Steels

Leire García-Sesma [1,2], Beatriz López [1,2] and Beatriz Pereda [1,2],*

[1] Materials and Manufacturing Division, Ceit, Manuel Lardizabal 15, 20018 Donostia/San Sebastián, Spain; leiregartzi81@gmail.com (L.G.-S.); blopez@ceit.es (B.L.)
[2] Mechanical and Materials Engineering Department, University of Navarra, Tecnun, Manuel Lardizabal 13, 20018 Donostia/San Sebastián, Spain
* Correspondence: bpereda@ceit.es; Tel.: +34-943-212-800

Received: 25 November 2019; Accepted: 18 January 2020; Published: 22 January 2020

Abstract: This work has focused on the study of hot working behavior of Ti-Nb microalloyed steels with high Ti contents (> 0.05%). The role of Nb during the hot deformation of low carbon steels is well known: it mainly retards austenite recrystallization, leading to pancaked austenite microstructures before phase transformation and to refined room temperature microstructures. However, to design rolling schedules that result in properly conditioned austenite microstructures, it is necessary to develop models that take into account the effect of high Ti concentrations on the microstructural evolution of austenite. To that end, in this work torsion tests were performed to investigate the microstructural evolution during hot deformation of steels microalloyed with 0.03% Nb and different high Ti concentrations (0.05%, 0.1%, 0.15%). It was observed that the 0.1% and 0.15% Ti additions resulted in retarded softening kinetics at all the temperatures. This retardation can be mainly attributed to the solute drag effect exerted by Ti in solid solution. The precipitation state of the steels after reheating and after deformation was characterized and the applicability of existing microstructural evolution models was also evaluated. Determined recrystallization kinetics and recrystallized grain sizes reasonably agree with those predicted by equations previously developed for Nb-Ti microalloyed steels with lower Ti concentrations (<0.05%).

Keywords: high Ti steels; Nb microalloying; recrystallization kinetics; strain-induced precipitation

---

1. Introduction

The steel industry is under continuous pressure to improve the mechanical properties of steel while simultaneously decreasing its price. Due to this, there is a permanent need for the development of new steel compositions, and when possible, adapted processing parameters. In low carbon steels, tailored mechanical properties can be obtained by adding microalloying elements, such as Nb, V or Ti. Nb leads to retarded softening kinetics during hot deformation [1–3]. As a result, if adequately designed rolling strategies are applied, deformed austenite microstructures before phase transformation and refined room temperature microstructures with improved mechanical strength can be obtained [4–6]. Ti, in conventional concentrations (<0.025%), results in the formation of TiN precipitates at very high temperatures, even during solidification. Models have been developed to predict the amount of these precipitates in both binary and multicomponent alloys [7–9]. If adequate in size and volume fraction, these precipitates can prevent austenite grain growth during reheating and after recrystallization and also provide room temperature microstructural refinement [10]. V is added due to the nanometer-sized precipitates that are formed during and/or after phase transformation, which results in precipitation hardening [11]. To some extent, Ti and Nb can also lead to a similar effect [12–14]. In the case of Ti, the large hardening potential of high Ti additions (>0.05%) that occur due to precipitation during or after phase transformation has been long known [15], although this has not been conventionally applied at industrial scale.

However, over the last years, the use of high Ti additions has attracted increasing interest. Many works have investigated the mechanical properties of these types of steels and shown that it is possible to produce high precipitation hardening contributions and yield strength values over 700 MPa [16]. High Ti additions are typically used in combination with other microalloying elements, such as Mo, V or Nb [17–19]. In a previous work, very high precipitation hardening contributions, over 300 MPa, were estimated for a 0.1% Ti–0.03 Nb steel at some of the coiling simulation conditions investigated [20]. This suggests that high Ti-Nb microalloying can be an interesting alternative to achieve steels with very high mechanical strength. In good agreement with these results, yield strengths of 700 MPa have been reported for industrially produced high Ti-Nb microalloyed steels [21].

However, in addition to the effect of precipitation, another factor to consider for high Ti-Nb steels is the supplementary grain size hardening effect that can be attained if pancaked austenite microstructures are obtained before phase transformation. Microstructural evolution models can be a good tool for evaluating and designing rolling schedules that result in such austenite conditioning. However, to use the models, it is necessary to have experimental data and develop equations that consider the effect that Ti and Nb have on the austenite microstructure evolution. Many works have studied the effect of these elements on the austenite recrystallization, grain growth or strain-induced precipitation kinetics [1–3,22,23]. However, most of them have focused on conventional Ti additions, while the data on high Ti steels is very scarce and mostly concentrated on the study of TiC precipitation during hot deformation. For instance, Akben et al. [24] investigated the effect of Mn on the dynamic precipitation kinetics of 0.1%Ti steels, while Liu et al. [25,26] and Wang et al. [27,28] analyzed the strain-induced precipitation behavior of different high Ti steels. However, none of these works considered high Ti-Nb microalloying combinations. It is necessary to take into account that, even for conventional additions, complex interactions have been observed in the precipitate dissolution behavior or precipitation kinetics when both Nb and Ti are present [29]. In the presence of Ti and Nb, it has been observed that they are mutually soluble and the precipitation of coarse particles starts at very high temperatures as (Ti,Nb)N. Okaguchi et al. [30] provided some data on the composition of undissolved precipitates in high Ti-Nb steels and studied the strain-induced precipitation kinetics of a 0.05% Ti–0.02 Nb steel. However, information about the effect of high Ti on other austenite microstructural evolution mechanisms, such as recrystallization or austenite grain growth, is scarcer. Taking this into account, the aim of this work was to characterize the austenite microstructure evolution processes that occur during hot deformation in 0.03% Nb microalloyed steels with high Ti concentrations (0.05%, 0.1%, 0.15%). Different types of torsion tests were performed to study the recrystallization kinetics in these steels. In addition, quenched samples were characterized to analyze the recrystallized microstructures, the grain growth kinetics and the morphology of the precipitates present after reheating and after deformation. The applicability of available microstructural evolution models for these steels was also considered.

2. Materials and Methods

The composition of the steels investigated is detailed in Table 1. All of them were laboratory-casted in a vacuum induction melting furnace and contain similar C, Mn and residual element levels. Aluminium and Sulphur were added in an industrially relevant level. The compositions include a reference plain C-Mn steel (Ti0Nb0), a 0.03% Nb microalloyed steel (Ti0Nb3), and three steels with the same 0.03%Nb level, and different high Ti concentrations of 0.05%, 0.1% and 0.15% (Ti5Nb3, Ti10Nb3 and Ti15Nb3 steels, respectively).

Table 1. Chemical composition of the steels investigated in this work (wt.%).

Steel	C	Mn	Nb	Ti	P	S	Al	Si	Cr	Ni	N
Ti0Nb0	0.058	1.82	0.001	0.001	0.020	0.003	0.019	0.011	0.019	0.006	0.0064
Ti0Nb3	0.059	1.79	0.033	0.003	0.020	0.003	0.036	0.026	0.013	0.005	0.0036
Ti5Nb3	0.058	1.81	0.033	0.044	0.020	0.004	0.024	0.012	0.019	0.006	0.0064
Ti10Nb3	0.059	1.80	0.034	0.091	0.021	0.003	0.036	0.016	0.017	0.006	0.0054
Ti15Nb3	0.059	1.82	0.033	0.139	0.021	0.003	0.040	0.014	0.016	0.006	0.0054

To study their hot deformation and microstructural evolution behavior, different types of torsion tests were performed. The specimens were a reduced central gauge section of 16.5 mm in length and 7.5 mm in diameter. A schematic representation of the thermomechanical schedule used in the tests is shown in Figure 1. In all cases, a soaking treatment was applied at 1250 °C for 10 min. To determine the softening kinetics, double-hit torsion tests were carried out. In some of the tests (type I in Figure 1), after reheating, the samples were cooled directly to the deformation temperature, where two deformation passes (ε_1 and $\varepsilon_2 = 0.1$) with different interpass times (t_{ip}) between them were applied. In the rest of the tests (type II in Figure 1), a roughing deformation pass with $\varepsilon = 0.35$, $\dot{\varepsilon} = 1 s^{-1}$ was first applied at 1100 °C, followed by a holding time that was determined to result in a fully recrystallized microstructure. The aim was to obtain refined initial austenite microstructures, more representative of those present during hot rolling. In the double-hit tests, both deformation passes were applied at constant temperature and strain rate conditions. The range of deformation parameters used was: ε_1 from 0.1 to 0.35, strain rate from 1 s^{-1} to 5 s^{-1} and deformation temperatures from 1150 °C to 850 °C. The fractional softening was determined using the 2% offset method, which is reported as the method that most accurately excludes the effect of recovery in the absence of strain-induced precipitation [31,32].

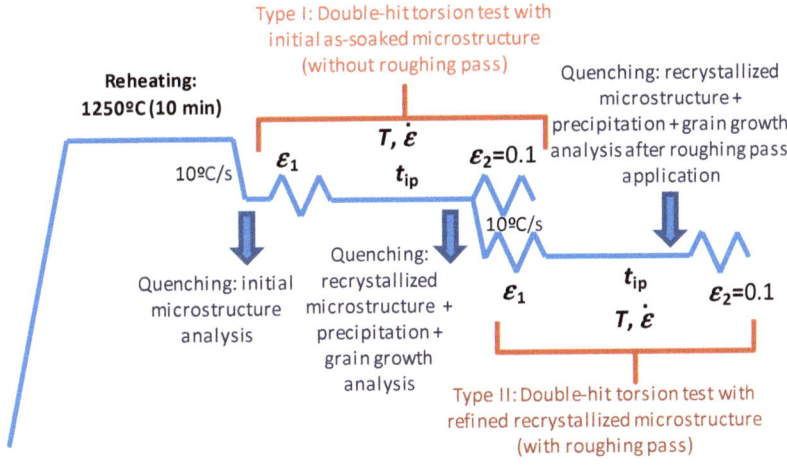

Figure 1. Thermomechanical cycles employed in the double-hit torsion tests and quenching treatments performed for microstructural analysis.

To analyze the initial microstructure present before the tests, samples were quenched after reheating and cooling to 1100 °C. Samples were also quenched before applying the second deformation pass of the double-hit torsion tests at different conditions: after holding times corresponding to $t_{0.95}$ (time for 95% fractional softening) to analyze the recrystallized microstructures, at longer times to analyze the austenite grain growth behavior ($t_{0.95}$ + 50 s and + 250 s), or at shorter or longer times to investigate the strain-induced precipitation evolution.

Metallographic measurements were performed on a section corresponding to 0.9 of the outer radius of the torsion specimen, also known as the sub-surface section. Bechet-Beaujard etching [33] was used to reveal the previous austenite grain boundaries in the quenched samples. The etched samples were examined via optical microscopy, and austenite grain size measurements were performed in terms of the mean equivalent diameter (MED). To analyze the coarsest precipitates, some specimens were also examined via optical microscopy or by field-emission gun scanning electron microscopy (FEG-SEM, JEOL JMS 7100F, JEOL Ltd., Tokyo, Japan) after conventional mechanical polishing. To examine the smallest precipitates, carbon extraction replicas were prepared and examined using a transmission electron microscope operated at 200 kV with a LaB6 filament (TEM Jeol 2100, JEOL Ltd., Tokyo, Japan). To prepare the carbon replica, the quenched samples were polished following standard metallographic techniques and etched with 2% Nital (2% nitric acid in ethanol). Next, a carbon film was deposited onto the samples and cut into ≅ 2 mm squares. The replicas were extracted by etching the samples again in Nital solutions with variable concentration and collected using Ni or Cu grids.

3. Results

3.1. Initial Precipitation and Microstructural State

The presence of coarse undissolved precipitates was first analyzed using optical microscopy in the high Ti specimens quenched for initial microstructure analysis. An example of the micrographs obtained for the Ti5Nb3 steel is shown in Figure 2. A significant number of coarse particles, in the size range of μm, can be detected. While some correspond to inclusions, precipitates containing Ti can be distinguished in the micrograph due to their characteristic orange contrast (see particles marked with arrows). The optical micrographs obtained for the rest of the high Ti steels showed similar features, and no significant differences were observed between the Ti5Nb3, Ti10Nb3 and Ti15Nb3 samples regarding precipitate amount or size.

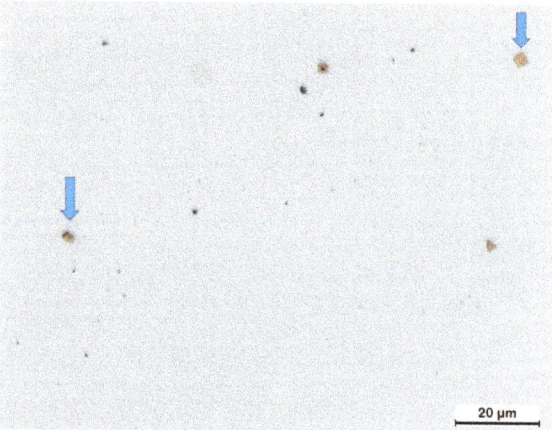

Figure 2. Examples of the large precipitates observed by optical microscopy in the Ti10Nb3 steel quenched after reheating for initial microstructural analysis.

To characterize the nature of these coarse precipitates in more detail, a FEG-SEM analysis was performed. Although in some cases the particles were identified only as TiN, in most of the cases different constituents were detected. Figure 3 shows a representative example of the precipitates found in the Ti5Nb3 and Ti10Nb3 steel samples. In the particles, a rounded area with dark contrast can be observed (Energy-dispersive spectroscopy analysis (EDS) number 1 in Figure 3). The EDS analysis shows that this is most likely Al_2O_3. The cube-shaped lighter contrast areas were identified as TiN (EDS number 3) and round lighter contrast areas were MnS (EDS number 2) co-precipitates. In the EDS of the TiN precipitates (number 3), a low Nb signal can also be detected.

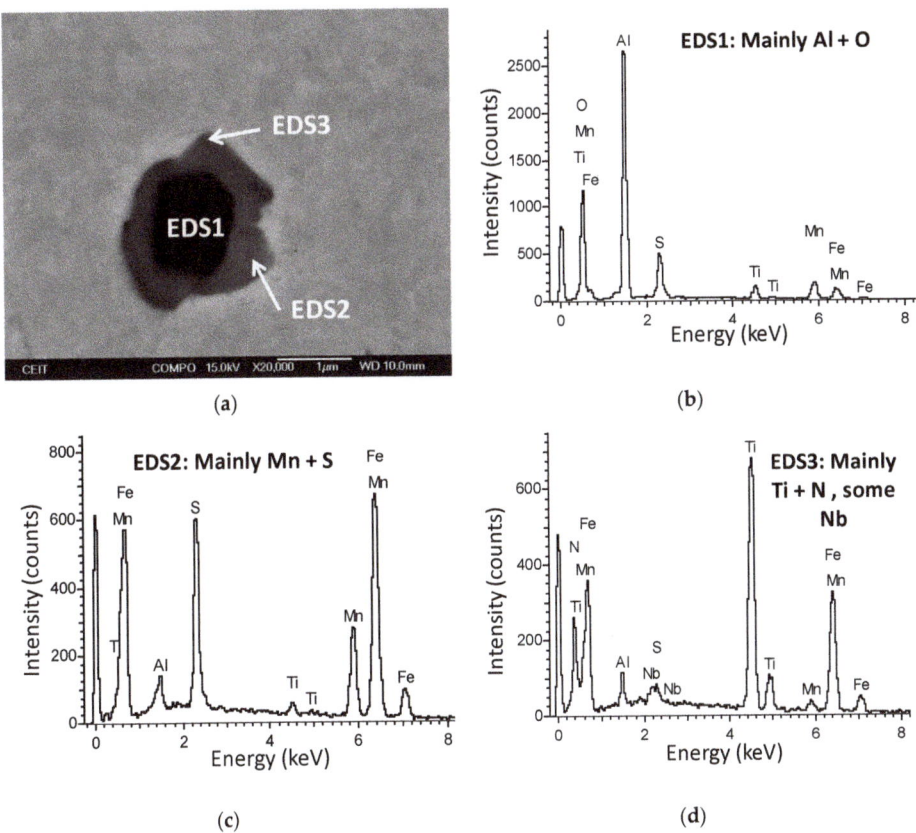

Figure 3. (a) Example of a Ti-complex precipitate observed by field-emission gun scanning electron microscopy (FEG-SEM) in the Ti5Nb3 sample quenched for initial microstructure analysis and (b–d), corresponding EDS analysis.

In the case of the Ti15Nb3 steel, precipitates with similar morphology were also found. However, in most cases, instead of MnS, the rounded areas were found to be rich in Ti and S. In addition, large particles containing only Ti and S were also found dispersed in the matrix for this steel.

Examples of the TEM results for the carbon extraction replicas extracted from the specimens quenched after soaking are shown in Figure 4. Nanometer-sized precipitates could be found dispersed in the replicas of all the microalloyed steels analyzed, even in the case of the Ti0Nb3 steel. From the EDS, it can be observed that in all cases the precipitates contain Ti and some Nb (the Cu and Ni signals are due to the grid used to support the replicas and the C signal can be due to both the replica and the precipitates and their contributions cannot be separated). In the Ti0Nb3 steel, since the Ti content is residual and the N level lower, the presence of these Nb-Ti precipitates was less expected. In good agreement with this, the precipitate amount was much lower than in the rest of the high Ti replicas analyzed. In addition, average precipitate size for this steel ($D = 71 \pm 10$ nm) was smaller than for the rest of the high Ti steels (D = 131 ± 7 nm, 135 ± 15 nm and 138 ± 8 nm for the Ti5Nb3, Ti10Nb3, Ti15Nb3 steels, respectively).

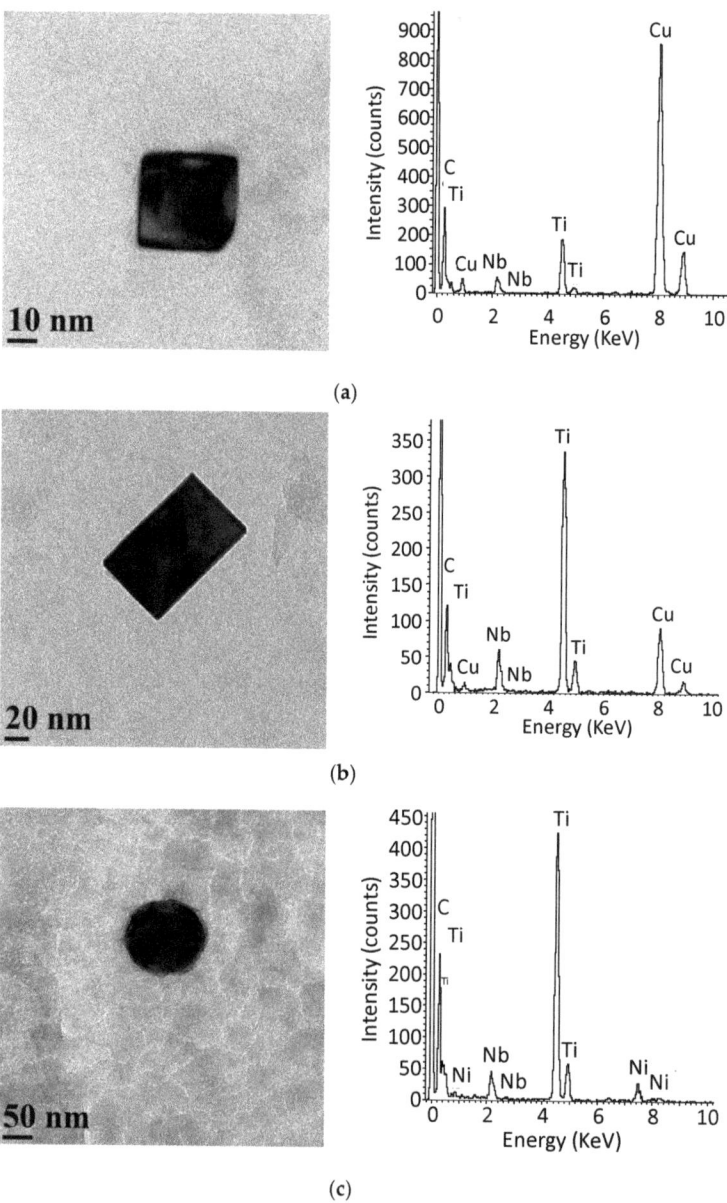

Figure 4. Examples of Nb-Ti precipitates found in the replicas extracted from the initial microstructure and corresponding EDS analysis. (**a**) Ti0Nb3 steel, (**b**) Ti5Nb3 steel, (**c**) Ti15Nb3 steel.

For the high Ti steels, although very similar precipitate sizes were measured, greater precipitate size heterogeneity was observed for the Ti10Nb3 and Ti15Nb3 steels. In addition, as shown in Figure 4, the precipitates tend to have a more rounded shape in the Ti10Nb3 and Ti15Nb3 steels, while in the

Ti5Nb3 steel the shape is more cubic/rectangular. Although the precipitate density in the replicas was not measured, it must be mentioned that the amount of precipitates found in the Ti10Nb3 replicas was lower than in the Ti5Nb3 and Ti15Nb3 ones. Finally, in the case of the Ti10Nb3 and Ti15Nb3 steels, particles containing S and Ti could also be detected. Although in some cases these could be identified as S and Ti co-precipitates on Nb-Ti undissolved precipitates, these were mainly found as individual particles (see Figure 5) and in some cases, some Nb was detected in their composition. The size distribution of both the Nb-Ti- and S-containing precipitates overlap, and it can be difficult to distinguish them without EDS analysis.

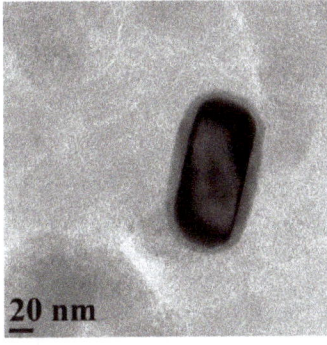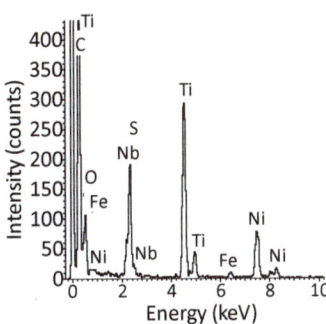

Figure 5. S-, Nb- and Ti- containing precipitate found in the carbon replicas extracted from the Ti10Nb3 specimen quenched for initial microstructure analysis.

The grain size was measured from specimens quenched after reheating and after application of the roughing pass using Bechet-Beaujard etching [33]. The results are summarized in Table 2. After reheating, similar austenite grain sizes of 230–240 µm were measured for the Ti0Nb0 and Ti0Nb3 steels. This indicates that the amount of undissolved Nb-Ti precipitates detected for Ti0Nb3 was too low to prevent grain growth during reheating, although some improvement in grain size homogeneity was observed in the latter. For the Ti5Nb3 and Ti15Nb3 steels, some grain refinement was observed, which could be attributed to the pinning effect exerted by the smallest undissolved precipitates [10]. On the other hand, this was not observed for the Ti10Nb3 steel, which is in good agreement with the lower density of the nanometer-sized undissolved precipitates observed in the replicas for this steel. After the roughing pass, the average grain sizes decreased in all cases, and differences between the steels can also be observed. It is worth noting the significant grain size refinement obtained for the Ti0Nb3 steel, while the coarsest grain size corresponds again to the reference Ti0Nb0.

Table 2. Average austenite grain sizes measured for samples quenched for initial microstructure analysis before and after roughing pass application.

Condition	Ti0Nb0	Ti0Nb3	Ti5Nb3	Ti10Nb3	Ti15Nb3
Grain size after Reheating (µm)	230 ± 15	241 ± 12	163 ± 8	258 ± 4	193 ± 10
Grain size with Roughing (µm)	126 ± 4	88 ± 4	70 ± 2	114 ± 4	96 ± 3

3.2. Softening Kinetics

Figure 6 shows examples of the fractional softening data obtained for the all the steels at constant strain and strain-rate conditions ($\varepsilon = 0.35$, $\dot{\varepsilon} = 1\ \text{s}^{-1}$) in tests carried out after the application of a roughing pass. For the Ti0Nb0 steel at all temperatures, for the Ti5Nb3 steel at temperatures from

1100 to 950 °C, and for the rest of the steels from 1100 to 1050 °C, the fractional softening follows a sigmoidal shape and can be fitted to an Avrami curve of the type:

$$FS = 1 - exp\left(-0.693\left(\frac{t}{t_{0.5}}\right)^n\right) \quad (1)$$

where FS is the fractional softening corresponding to a time t, $t_{0.5}$ is the time to reach a 50% fractional softening and n is the Avrami exponent. At lower temperatures, the behavior varies depending on the steel considered. For the Ti0Nb3 steel at 950 °C and 900 °C (Figure 6b) and for the Ti15Nb3 steel at 900 °C (Figure 6e), softening stagnations, or plateaux, which are usually related to the onset of strain-induced precipitation [2,23], are detected. In the rest of the cases, although a larger softening retardation is observed relative to higher temperatures, the softening increases continuously as the interpass time increases within the time range investigated.

Examples of the effect of strain and of roughing pass application are illustrated in Figure 7. As is usually reported, increasing strain results in faster softening kinetics. The same is observed with the application of a roughing pass, due to initial microstructure refinement, although the effect is lower than it is for increasing strain. Similar trends were observed for the rest of the steels investigated.

Examples of the effect of steel composition on the softening kinetics at similar deformation and in high temperature conditions are illustrated in Figure 8. It can be observed that in both cases, the Ti0Nb0 steel showed the fastest softening kinetics, followed by Ti5Nb3 and then Ti0Nb3. The slowest softening kinetics correspond to the Ti10Nb3 and Ti15Nb3 steels. For the Ti0Nb0 steel, the initial austenite grain sizes were determined to be coarser than or in the same range of those measured for Ti0Nb3 (Table 2). Since coarser initial microstructures are known to delay the recrystallization kinetics [3], the softening retardation observed for Ti0Nb3 compared to Ti0Nb0 can only be explained due to the solute drag exerted by Nb [1–3,23]. On the other hand, for the Ti5Nb3 steel, the initial microstructure is slightly finer than for Ti0Nb3. Although the difference is very small after roughing, refinement of the initial microstructure can contribute to the acceleration of the softening kinetics observed for the Ti5Nb3 steel compared to Ti0Nb3. Additionally, a decrease in the Nb content in solid solution due to its presence in undissolved precipitates in Ti5Nb3 (Figure 4) could also contribute to this acceleration. On the other hand, for the Ti10Nb3 and Ti15Nb3 steels a higher amount of Ti in solid solution can explain the softening retardation observed for both.

The softening curves determined for all steels at lower temperatures of 950 °C and 900 °C are compared in Figure 9. It can be observed that at 950 °C, for the Ti0Nb0 and Ti5Nb3 steels, the softening curve also follows a sigmoidal shape. For the Ti0Nb3 steel a plateau, which is usually considered to be caused by strain-induced precipitation start, is observed at $\cong$ 180 s. For the Ti10Nb3 and Ti15Nb3 steels, in contrast, there are no clear plateaux. However, the very retarded softening kinetics are similar to what is observed for Ti0Nb3. At 900 °C, for the Ti5Nb3 steel, a very slow but continuous increase in softening is observed within the interpass time range investigated. For the Ti0Nb3, Ti5Nb3 and Ti15Nb3 steels, the softening is more retarded and tends to saturate at 20–30% levels up to interpass times as long as 1000 s.

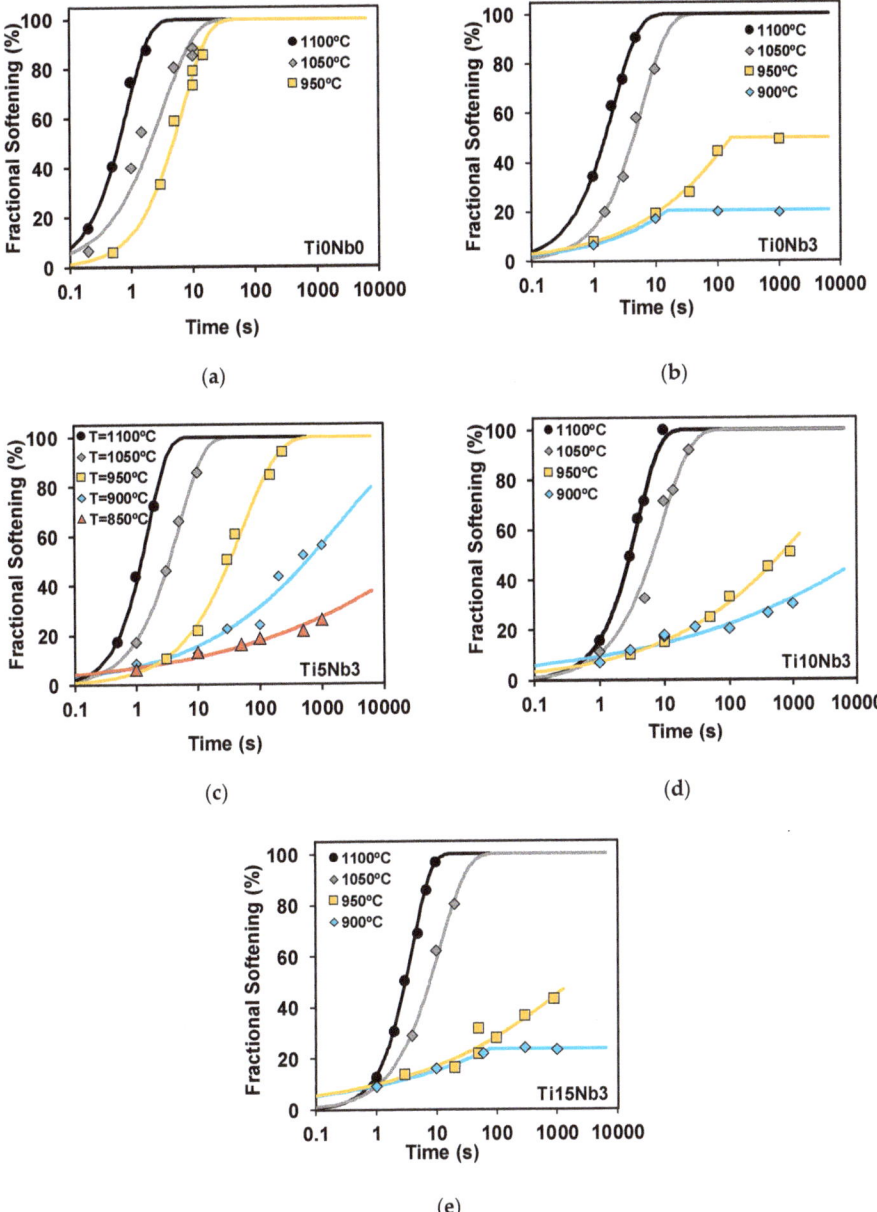

Figure 6. Softening curves obtained for the steels investigated at constant strain and strain-rate conditions ($\varepsilon = 0.35$, $\dot{\varepsilon} = 1\ \text{s}^{-1}$) and different temperatures after roughing pass application. (**a**) Ti0Nb0 steel, (**b**) Ti0Nb3 steel, (**c**) Ti5Nb3 steel, (**d**) Ti10Nb3 steel and (**e**) Ti15Nb3 steel.

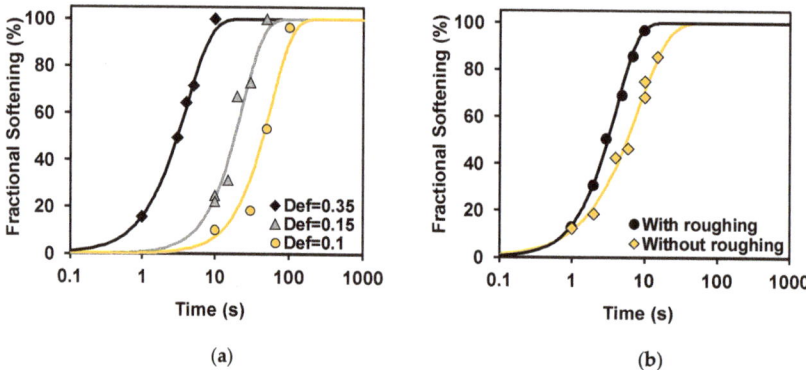

Figure 7. Examples of the effect of strain and of initial microstructural refinement on the softening curves. (**a**) Ti10Nb3 steel, $T = 1100\ °C$, roughing pass applied, $\dot{\varepsilon} = 1\ s^{-1}$ and (**b**) Ti10Nb3 steel, $T = 1100\ °C$, $\varepsilon = 0.35$, $\dot{\varepsilon} = 1\ s^{-1}$.

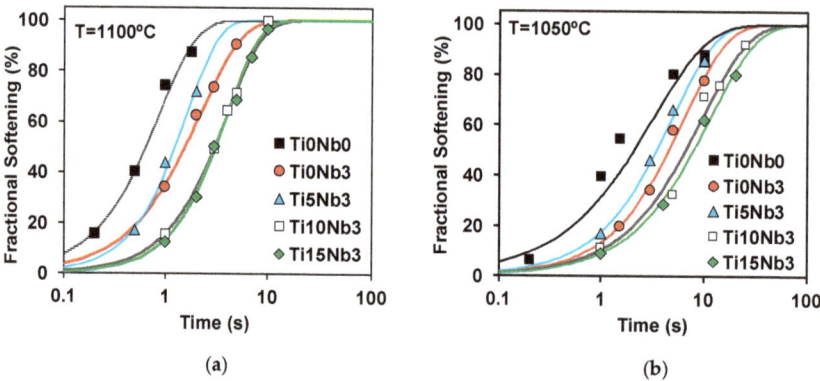

Figure 8. Effect of steel composition on the softening curves at high temperatures, roughing pass applied, $\varepsilon = 0.35$, $\dot{\varepsilon} = 1\ s^{-1}$. (**a**) $T = 1100\ °C$, (**b**) $1050\ °C$.

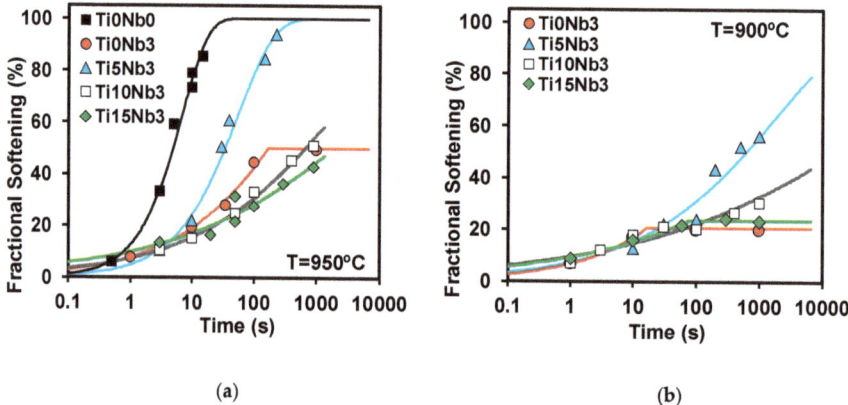

Figure 9. Softening kinetics for the steels investigated at low temperatures, roughing pass applied, $\varepsilon = 0.35$, $\dot{\varepsilon} = 1\ s^{-1}$. (**a**) $T = 950\ °C$, (**b**) $900\ °C$.

3.3. Strain-Induced Precipitation Behavior

From the softening data, it is difficult to accurately determine the start time for strain-induced precipitation, especially for the high Ti steels. Therefore, to study the precipitation behavior of these steels, quenching treatments were performed for all the microalloyed steels after deformation at 900 °C (roughing pass applied, $\varepsilon = 0.35$, $\dot{\varepsilon} = 1\,\text{s}^{-1}$) and interpass times of 100 s and 1000 s (fractional softening results corresponding to these conditions are shown in Figure 9b). For Ti0Nb3, an additional quenching treatment was performed at similar deformation conditions for 950 °C, t = 180 s (time for plateau start, Figure 9a). An analysis of the carbon extraction replicas prepared from this specimen showed the presence of some Nb strain-induced precipitates with an average size of $D = 15.5 \pm 1.8$ nm. Although these were scarce and found to be heterogeneously distributed, the presence of these particles suggests that the plateau observed at 950 °C could be related to strain-induced precipitation start.

Regarding the 900 °C condition, for the Ti5Nb3 steel, strain-induced precipitates were not detected at neither at 100 or 1000 s holding times. However, as Figure 10 shows, the co-precipitation of Nb-rich caps on the undissolved precipitates was observed in the replicas extracted from this steel.

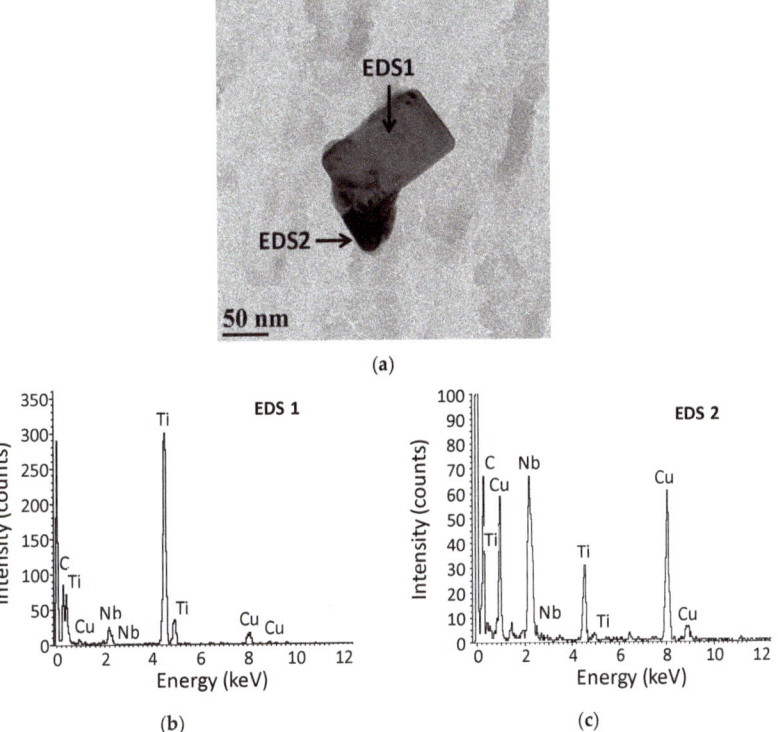

Figure 10. (a) Example of Ti-Nb undissolved precipitate with Nb-Ti co-precipitate observed in the Ti5Nb3 900 °C-1000 s sample and, (b,c) corresponding EDS analysis.

For the Ti10Nb3 and Ti15Nb3 steels, some co-precipitation was also observed on the undissolved precipitates, while for the Ti0Nb3 steel this was more difficult to evaluate since a low number of undissolved precipitates was detected. However, more homogeneously distributed strain-induced precipitates could be observed in all the Ti0Nb3, Ti10Nb3 and Ti15Nb3 samples analyzed.

Figure 11 shows examples of the precipitation state and an EDS analysis observed for the 100 s condition for the three steels. Additionally, while very fine Nb-containing precipitates ($D = 7.2$ nm) were detected for the Ti0Nb3, for both the Ti10Nb3 and Ti15Nb3 steels, the precipitates were coarser ($D \cong 15$–16 nm) and both Nb and Ti could be clearly detected in their composition. It must be mentioned that in carbon extraction replicas the smallest precipitates may not be extracted [12], although in the case of Figure 11a precipitates as small as 3.3 nm have been measured.

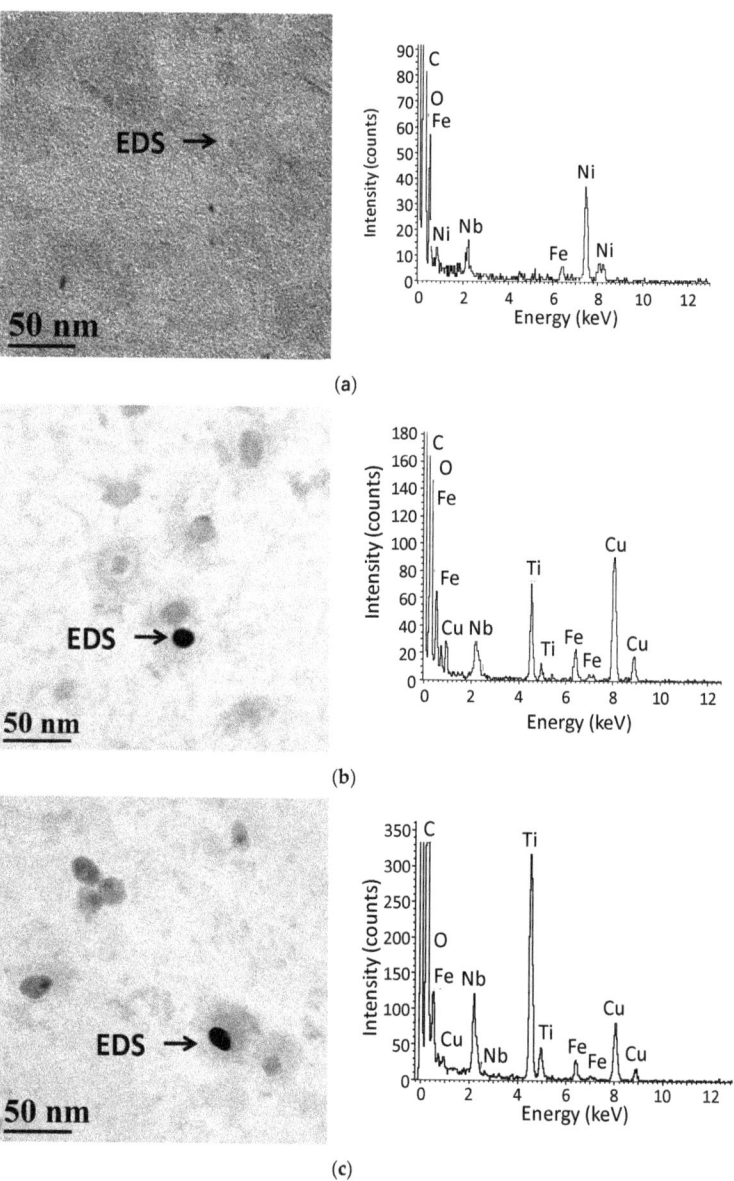

Figure 11. Strain-induced precipitates detected in the carbon replicas after deformation at 900 °C and corresponding EDS analysis (tests performed with roughing, $\varepsilon = 0.35$, $\dot{\varepsilon} = 1\ \mathrm{s}^{-1}$ and 100 s holding time). (**a**) Ti0Nb3 steel, $D = 7.2 \pm 0.2$ nm, (**b**) Ti10Nb3 steel, $D = 15.0 \pm 0.4$ nm, (**c**) Ti15Nb3 steel, $D = 16.0 \pm 0.5$ nm.

Figure 12 summarizes the precipitate size distributions measured after deformation at the same temperature, but with a longer holding time of 1000 s. The EDS analysis showed no significant changes in the composition of the precipitates compared to the ones extracted after 100 s. However, it can be observed that for all the steels, the precipitate size increases during the interpass time. Nevertheless, while precipitate size growth is quite limited for the Ti0Nb3 steel (D increases from 7 to 13 nm), this is much larger for the Ti15Nb3 (D from 16 to 33 nm) and especially the Ti10Nb3 steel (D from 16 to 52 nm).

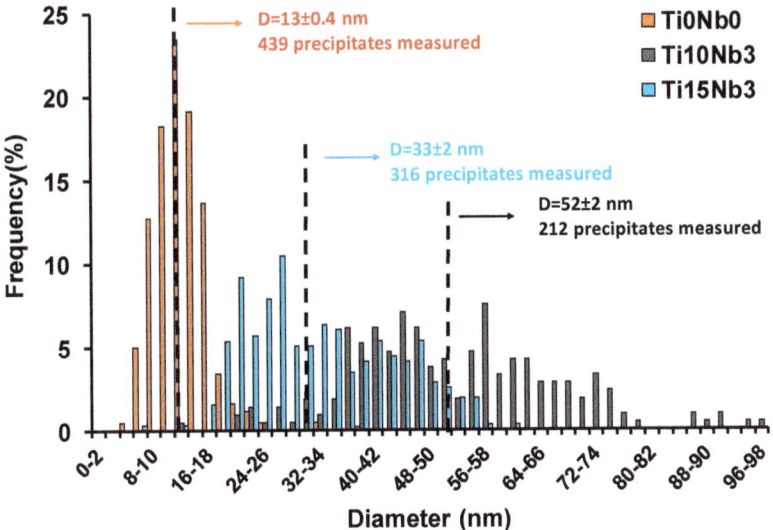

Figure 12. Precipitate size distributions measured for the Ti0Nb3, Ti10Nb3 and Ti15Nb3 steels in samples quenched after deformation at 900 °C (roughing pass applied, $\varepsilon = 0.35$, $\dot{\varepsilon} = 1\,\text{s}^{-1}$) and 1000 s holding time.

3.4. Recrystallized Grain Size

Figure 13 shows examples of the recrystallized microstructures obtained from the Ti0Nb3 and Ti5Nb3 steels at $T = 1100\,°\text{C}$ and $1150\,°\text{C}$, $\dot{\varepsilon} = 1\,\text{s}^{-1}$ in specimens quenched at $t_{0.95}$. The micrographs were obtained after reheating and applying different deformation conditions: a single deformation pass of $\varepsilon = 0.35$ at 1100 °C in the case of Figure 13a,b, a single deformation pass of $\varepsilon = 0.15$ at 1100 °C–1150 °C in Figure 13c,d, and a roughing pass plus a deformation pass of $\varepsilon = 0.35$ at 1100 °C in Figure 13e,f. As expected, equiaxed austenite microstructures are obtained in all cases. When considering the effect of deformation conditions, similar behavior is observed for both steels: the application of a single deformation pass of $\varepsilon = 0.35$ leads to significant microstructural refinement of the initial as-soaked microstructure (from 241 to 88 µm for Ti0Nb3 and from 163 to 64 µm for Ti5Nb3). On the other hand, when a lower strain level of $\varepsilon = 0.15$ is applied, for both steels the grain size remains approximately constant or even slightly increases for the Ti0Nb3 steel. Finally, the application of another $\varepsilon = 0.35$ deformation pass also results in some grain size refinement (from 88 to 65 µm for the Ti0Nb3 and from 70 to 56 µm for the Ti5Nb3 steel), although the effect is less marked than when the coarser as-soaked microstructure was deformed.

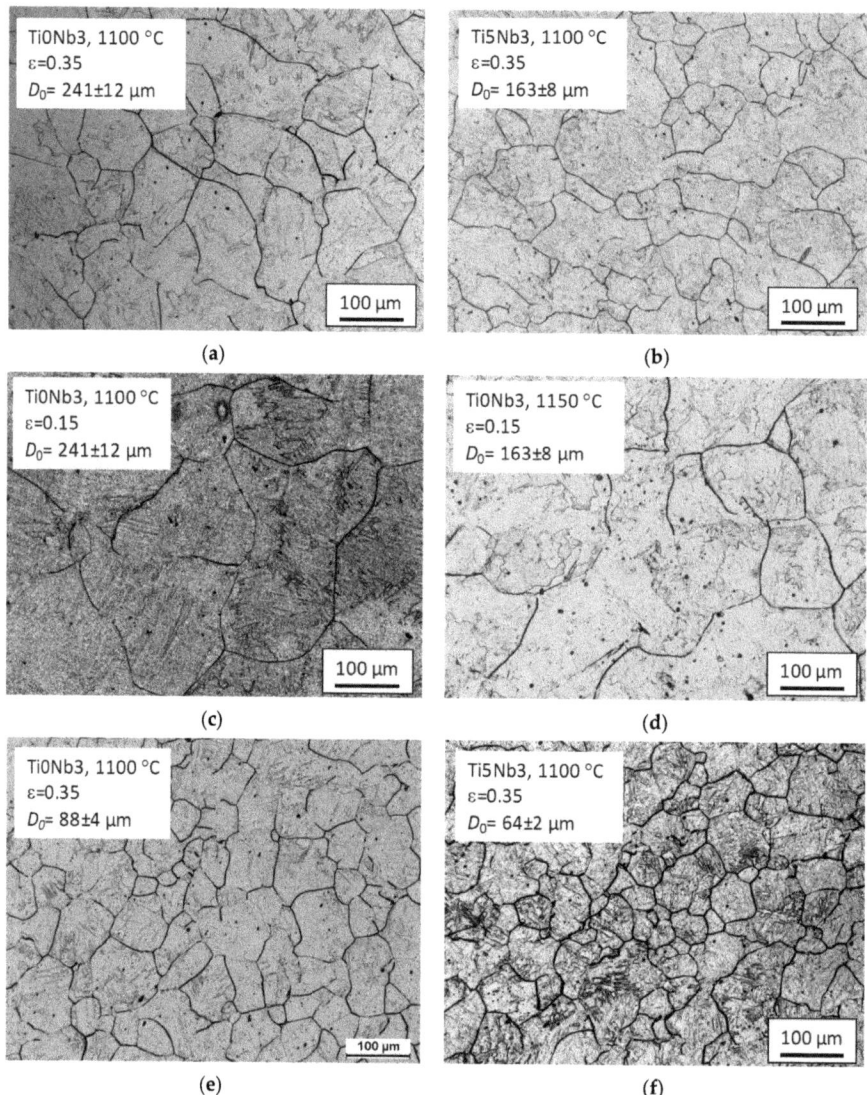

Figure 13. Examples of recrystallized austenite microstructures obtained at different deformation conditions. (a) $d_{SRX} = 88 \pm 4$ μm, (b) $d_{SRX} = 70 \pm 2$ μm, (c) $d_{SRX} = 256 \pm 4$ μm, (d) $d_{SRX} = 162 \pm 9$ μm, (e) $d_{SRX} = 65 \pm 4$ μm, (f) $d_{SRX} = 56 \pm 1$ μm.

Table 3 summarizes the recrystallized grain size (d_{SRX}) measurements performed at the different conditions under study, together with the initial austenite grain size (D_0) and deformation conditions employed in the tests. The trends are similar for all the steels investigated: applying deformation levels of $\varepsilon = 0.35$ led to microstructural refinement, while lower levels of $\varepsilon = 0.15$ led to similar or slightly larger austenite grain sizes. The results obtained for the Ti0Nb3 steel after a roughing pass show that, as is usually reported, modifying the deformation temperature from 1100 to 1050 °C does not significantly affect the recrystallized grain size. Similarly, data corresponding to the Ti5Nb3 at $\varepsilon = 0.35$ and without a roughing pass shows a small effect on variation in strain-rate from 1 to 5 s^{-1}.

Table 3. Experimental d_{SRX} determined at different deformation conditions.

Steel	ε	D_0 (μm)	T_{def} (°C)	$\dot{\varepsilon}$ (s^{-1})	d_{SRX} (μm)	Steel	ε	D_0 (μm)	T_{def} (°C)	$\dot{\varepsilon}$ (s^{-1})	d_{SRX} (μm)
TiONb0	0.35	230 ± 15	1100	1	126 ± 4	Ti5Nb3	0.35	163 ± 8	1100	1	70 ± 2
	0.35/0.35	126 ± 4	1100	1	62 ± 1		0.35	163 ± 8	1100	5	64 ± 2
	0.35/0.35	126 ± 4	1050	1	70 ± 2		0.15	163 ± 8	1150	1	162 ± 9
							0.35/0.35	70 ± 2	1100	1	56 ± 1
TiONb3	0.35	241 ± 12	1100	1	88 ± 4						
	0.15	241 ± 12	1100	1	256 ± 14	Ti10Nb3	0.35	258 ± 4	1100	1	114 ± 4
	0.35/0.35	88 ± 4	1100	1	65 ± 2		0.35/0.35	114 ± 4	1100	1	79 ± 2
	0.35/0.35	88 ± 4	1050	1	69 ± 2		0.35	193 ± 10	1100	1	96 ± 3
						Ti15Nb3	0.35/0.35	96 ± 3	1100	1	81 ± 2
							0.35/0.15	96 ± 3	1100	1	131 ± 7

3.5. Grain Growth Behavior

Figure 14 shows examples of microstructures quenched at $t_{0.95}$ or at longer holding times ($t_{0.95}$ + 50 s or 250 s) to study the grain growth behavior. The microstructures correspond to the Ti0Nb0, Ti0Nb3 and Ti15Nb3 steels and a constant holding temperature of 1100 °C. For Ti0Nb0, as time increases, a significant grain size increase is observed, from 62 to 120 μm after 250 s holding time. For the Ti0Nb3, grain growth is more limited, from 65 to 105 μm, and for the Ti15Nb3, the grain size remains approximately constant at ≅ 85 μm.

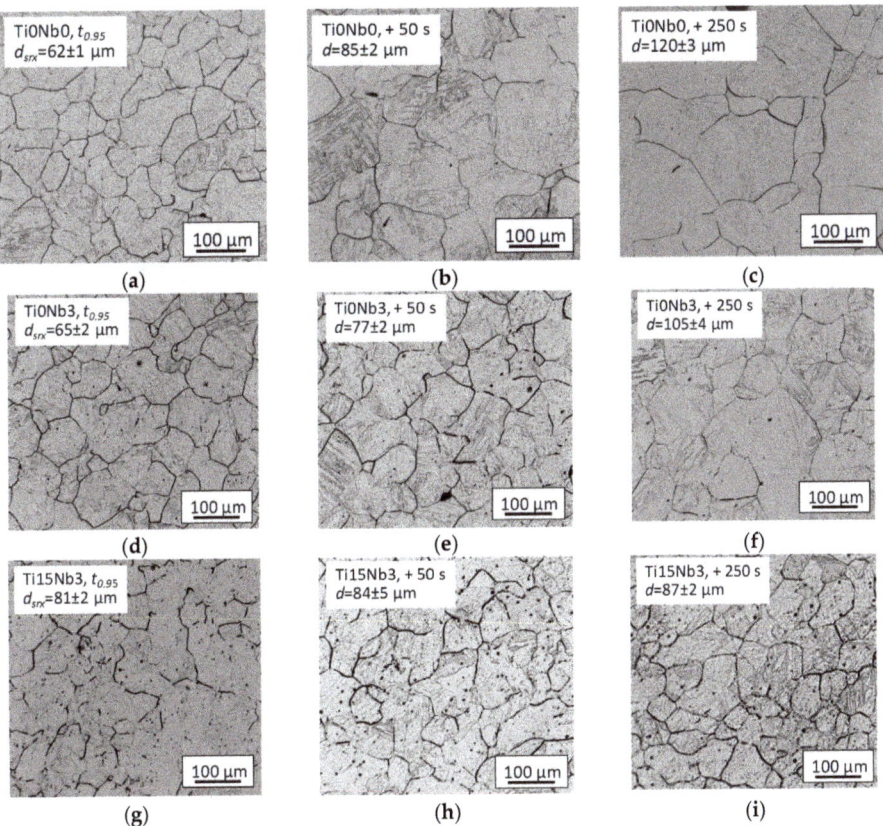

Figure 14. Examples of recrystallized microstructures quenched at $t_{0.95}$ (T = 1100 °C, ε = 0.35, $\dot{\varepsilon}$ = 1 s^{-1}) or after $t_{0.95}$ and longer holding times of 50 s and 250 s for the Ti0Nb0, Ti0Nb3 and Ti15Nb3 steels.

4. Discussion

4.1. Initial Microstructural and Precipitation State

To further understand the precipitation sequence at the conditions investigated, calculations were performed using the Thermo-Calc software (version 2017a, Thermo-Calc software, Solna Sweden) [34] (Thermo-Calc Software TCFE8 Steels/Fe-alloys database [35]). Table 4 summarizes the amount of Nb and Ti in solid solution predicted by the software at 1250 °C in the austenite phase, and in Figure 15 the Ti concentration present in each of the equilibrium phases (in wt.%) is plotted as a function of temperature for the high Ti steels. The table shows that for all the steels, Nb is expected to be in solid solution after soaking. With regard to Ti, the situation is more complex. As observed in Figure 15, in addition to the austenite, it is predicted that at 1250 °C some Ti will be present in three other phases: $Ti_4C_2S_2$, N-rich fcc precipitates and C-rich fcc precipitates. As a result, at 1250 °C, only slightly less than half of the nominal Ti concentration is calculated to be in solid solution in the austenite at equilibrium conditions.

Table 4. Amount of Nb, Ti in solid solution in austenite (wt.%) at 1250 °C for the microalloyed steels calculated using Thermo-Calc software [34,35].

Microalloying Element	Ti0Nb3	Ti5Nb3	Ti10Nb3	Ti15Nb3
Nb(%)	0.033	0.033	0.034	0.032
Ti (%)	0.001	0.012	0.044	0.073

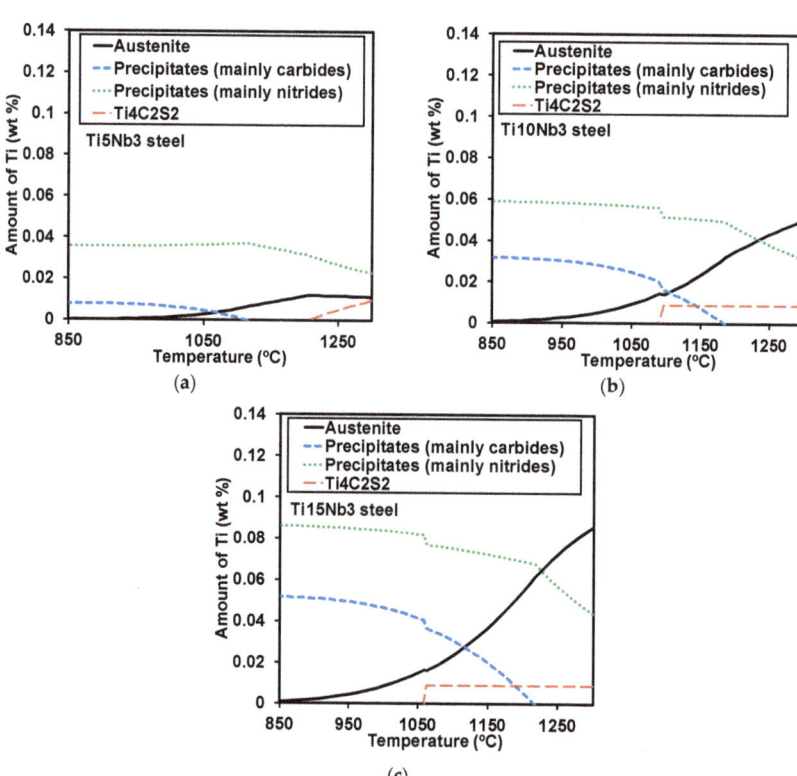

Figure 15. Amount of Ti (wt.%) in each of the phases at equilibrium conditions calculated with Thermo-Calc software [34,35] for the high Ti steels. (a) Ti5Nb3, (b) Ti10Nb3, (c) Ti15Nb3.

For Ti0Nb3 steel, in contrast to the Thermo-Calc predictions, some Nb-Ti precipitates were found. However, these were scarce, only found via TEM, small in size (D = 71 nm) and were probably related to the presence of some residual Ti content in the steel. This means that the amount of Nb in these scarce precipitates is expected to be very low or even negligible.

On the other hand, for the high Ti steels a large number of precipitates, with a wide range of sizes from microns to several nanometers, were detected. This reduces the amount of Ti in solid solution in austenite that is available for further precipitation, which is in good agreement with the predictions of Thermo-Calc software. The coarsest Ti containing precipitates were in most cases found nucleated over oxides (Al_2O_3), which suggests that the amount of these particles can be related to the number of oxides and therefore influenced by steel cleanliness. This can be critical, since these coarse Ti nitrides (in the micron range) are very difficult to dissolve during reheating. Nanometer-sized particles with similar average size (D ≅ 135 nm) were also found for all the high Ti specimens quenched for initial microstructure analysis, although the density tends to be lower for the Ti10Nb3 steel. The reasons for the reduced precipitate amount are not clear, but they may be related to casting or solidification conditions. Regarding precipitate composition, for the three high Ti steels some N could be detected in the coarsest particles observed via FEG-SEM, which suggests that these are mostly nitrides. This is in good agreement with the predictions of Thermo-Calc, which shows that at 1250°C a large amount of Ti is expected to be pinned in the form of TiN. In the case of the smallest precipitates detected by TEM, some Nb was found in their composition (Figure 4). This suggests that some loss of Nb in solid solution not predicted by Thermo-Calc could be taking place. For the Ti10Nb3 and Ti15Nb3 steels, S- and Ti-containing precipitates, mainly small for the former steel and of all sizes for the latter, were also detected. In other types of steels, it has been reported that Ti and S can combine to form TiS or $Ti_4C_2S_2$ [36]. It is difficult to detect C by EDS in the FEG-SEM, and it is not possible in the carbon replicas examined in the TEM. However, the predictions of Thermo-Calc suggest that these could be of the $Ti_4C_2S_2$ type. Finally, although these were the least frequent type of particles, in some of the precipitates found via TEM, Nb and S was also detected. However, this was not investigated in detail.

4.2. Static Softening and Strain-Iduced Precipitation Behavior

From the softening curves, the $t_{0.5}$ and n Avrami exponent were calculated for all the conditions investigated. The results obtained for all those cases where interaction with strain-induced precipitation was not observed are summarized in Table 5.

Table 5. Experimental $t_{0.5}$ and n values determined from the softening curves.

Steel	D_0 (μm)	ε	T_{def} (°C)	$\dot{\varepsilon}(s^{-1})$	$t_{0.5}$ (s)	n	Steel	D_0 (μm)	ε	T_{def} (°C)	$\dot{\varepsilon}(s^{-1})$	$t_{0.5}$ (s)	n
TiONb0	230 ± 15	0.35	1100	1	1.2	1.0	Ti10Nb3	258 ± 4	0.35	1100	1	4.9	1.1
		0.35	1100		0.7	1.2			0.35	1050		6.7	1.0
	126 ± 4	0.15	1100		4.6	2.0		114 ± 4	0.35	1100		3.0	1.3
		0.35	1050		2.1	0.8			0.15	1100		18.6	1.5
		0.35	950		5.0	1.0			0.1	1100		44.3	1.5
TiONb3	241 ± 12	0.35	1100	1	3.4	1.1	Ti15Nb3	193 ± 10	0.35	1100	1	5.7	1.0
		0.15	1100		23.1	0.9			0.35	1050		7.9	1.0
		0.35	1050		4.5	1.0		96 ± 3	0.35	1100		3.1	1.4
	88 ± 4	0.35	1100		1.6	1.1			0.15	1100		22.5	1.5
		0.15	1100		11	1.4			0.1	1100		57.3	2.1
		0.1	1100		14	2.5		-	-	-	-	-	-
Ti5Nb3	163 ± 8	0.35	1100	1	2.4	1.2	-	-	-	-	-	-	-
		0.35	1100	5	2.2	0.8	-	-	-	-	-	-	-
		0.35	1150		1.6	1.0	-	-	-	-	-	-	-
		0.15	1150	1	17.1	1.1	-	-	-	-	-	-	-
	70 ± 2	0.35	950		34.1	0.7	-	-	-	-	-	-	-
		0.35	1050		3.5	1.0	-	-	-	-	-	-	-
		0.35	1100		1.2	1.4	-	-	-	-	-	-	-
		0.15	1100		10.3	1.9	-	-	-	-	-	-	-

In a previous work [3], the following equation was developed to predict $t_{0.5}$ as a function of initial grain size (D_0), deformation conditions (ε, $\dot{\varepsilon}$ and T) and the amount of microalloying elements in solid solution:

$$t_{0.5} = 9.92 \times 10^{-11} D_0 \varepsilon^{-5.6} D_0^{-0.15} \dot{\varepsilon}^{-0.53} \exp\left(\frac{180000}{RT}\right) \exp\left[\left(\frac{275000}{T} - 185\right)([Nb] + 0.374[Ti])\right] \quad (2)$$

where the amount of microalloying elements is given in wt%, D_0 in µm, $\dot{\varepsilon}$ in s^{-1} and T in K. The equation was developed considering steels microalloyed with Nb, Ti and Nb-Ti with 0.035%Nb and 0.067%Ti as maximum levels [3]. Although not applicable in the present work, the equation has been extended to steels with high Mo [37], Al [38] and B [39] additions. In Figure 16 the $t_{0.5}$ times calculated from this equation are plotted against the experimentally determined values. The amount of microalloying element in solid solution predicted by Thermo-Calc (Table 4) was used in the calculations and only those cases in which interaction with strain-induced precipitation was not detected were considered. From the figure, on average, a good fit is observed for all the steels investigated. Although the above results indicate that Thermo-Calc may slightly overestimate values of Nb in solid solution for the high Ti steels, this does not seem to lead to significant error in $t_{0.5}$ prediction. In fact, the fit is slightly more accurate for the microalloyed steels than for TiONb0. In some cases, some overestimation of the calculated $t_{0.5}$ can be observed. However, it must be mentioned that this was observed for the longest $t_{0.5}$ values, which correspond mainly to the lowest applied strain values ($\varepsilon = 0.1$ and $\varepsilon = 0.15$). The lower accuracy of these type of models when low strain levels are applied has also been observed in previous works [38].

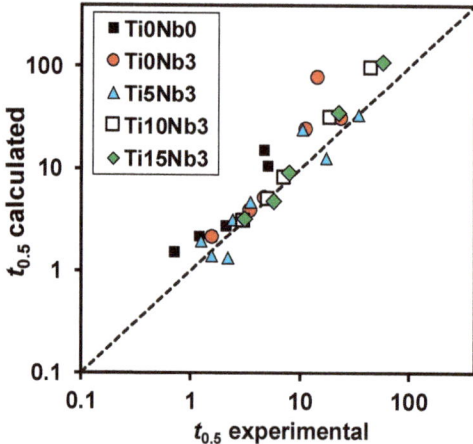

Figure 16. Comparison between the experimental $t_{0.5}$ and the values predicted using Equation (2).

Regarding the experimental n Avrami exponent, in some works this parameter's dependence on deformation temperature has been reported [2,23,40]. However, within the range of conditions investigated in this work (most of the available data corresponding to $T = 1100$ °C or 1050 °C) and taking into account that this parameter is rather noisy [40], it is difficult to assess such an effect with the available data. However, the average value determined from the available results, $n = 1.26$, is within the range of values reported for other microalloyed steels [3], and also close to the predictions made by the equations developed by Medina et al. [2] for low alloy steels ($n = 0.94$ to 0.98) and for microalloyed steels ($n = 1.07$ to 1.21) in the 1050–1100 °C temperature range.

The strain-induced precipitation behavior shows a more complex dependence on steel composition. The carbon replica and softening analysis show that, within the microalloyed steels, the Ti5Nb3 one presents the most retarded precipitation kinetics. For this steel, some loss of Nb in solid solution into the undissolved precipitates can take place. However, the presence of coarse undissolved precipitates in the microstructure of the high Ti steels must also be considered. The activation energy for nucleation of precipitates depends both on the driving force and the energy barrier. The driving force is a function of the amount of microalloying content in solid solution relative to the equilibrium values, while the energy barrier includes the interfacial energy and the strain mismatch between precipitate and matrix [41]. For the Ti5Nb3 steel, lower amount of Nb in solid solution compared to Ti0Nb3 can reduce the driving force and make less favorable precipitate nucleation. However, another factor to take into account is that heterogeneous nucleation of Nb-Ti precipitates on undissolved precipitates takes place following a coherent cube-cube orientation relationship, leading to a decrease of the interfacial energy [29]. As a result, for this steel, co-precipitation on undissolved particles could become the most favorable nucleation mechanism for the Ti5Nb3. Similar results were observed by Hong et al. [29] for 0.04%C-0.05%Nb steels with 0.016%Ti addition. For the Ti10Nb3 and Ti15Nb3 steels (Figure 11), some co-precipitation over undissolved precipitates was also observed. However, for these steels, more homogeneously distributed precipitates were observed in the replicas (Figure 11b,c). A higher amount of Ti in solid solution in these steels leads to higher driving force for precipitation. As result, both precipitation mechanisms, co-precipitation and precipitation in the austenite lattice defects, such as dislocation or sub-grain boundaries [32,41], is observed to take place, leading to larger softening retardation levels than for the Ti5Nb3 steel.

Some differences can also be observed regarding the composition and size of the strain-induced precipitates. While EDS analysis of Ti0Nb3 steel shows that they contain mostly Nb, for the Ti10Nb3 and Ti15Nb3 steels, both Nb and Ti could be clearly detected in the composition. In addition, a much higher strain-induced precipitate growth/coarsening rate was determined in the high Ti steels compared to Ti0Nb3. This could be related to the higher amount of microalloying element in solid solution present in these steels [42]. After a 1000s holding time, the precipitate size increase was especially enhanced for the Ti10Nb3. The lower amount of nanometer-sized precipitates found in the initial microstructure analysis suggests that for this steel more Ti could be put into solution after reheating. However, although the effect is small, more retarded softening kinetics were observed for Ti15Nb3 than for Ti10Nb3, which can only be explained due to larger amount of Ti in solid solution (Figure 8). Given the small amount of material that it is analyzed by TEM means, local segregation effects could explain this result.

4.3. Recrystallized Grain Size

As shown in Table 3 and commonly reported in the literature, the effect of deformation temperature or strain rate on the recrystallized grain size is low. As a result, to calculate the recrystallized grain size, the following type of dependence is usually applied [43–45]:

$$d_{SRX} = AD_0^p \varepsilon^{-q} \tag{3}$$

where A, p and q are material dependent constants. Table 6 summarizes some of the values proposed in the literature for these constants for C-Mn and Nb microalloyed steels [43–45], while in Figure 17 the experimental recrystallized grain sizes measured in this work have been compared to the predictions of these models.

Table 6. Parameters proposed for the calculation of the recrystallized grain size for different steels using equation (3).

Steel	A	p	q	Reference
CMn - Nb (1)	1.4	0.56	1	Abad et al. [43]
CMn - Nb (2)	1.1	0.67	0.67	Sellars [44]
CMn	0.743	0.67	1	Beynon et al. [45]

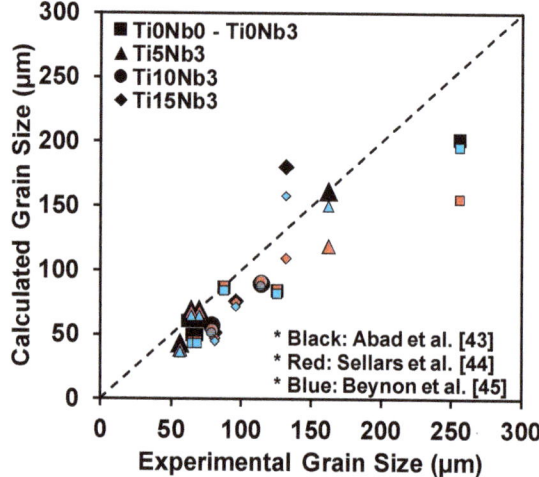

Figure 17. Comparison between the experimental recrystallized grain size measured for the steels investigated and predictions from several models proposed in the literature [43–45].

It can be observed from Figure 17 that for the finest recrystallized grain sizes, the fit is similar for the three equations considered. In addition, no significant differences can be appreciated in the fit as a function of high Ti content. Equation from reference [44] tends in most cases to underestimate the recrystallized grain size, while for the coarsest recrystallized grain sizes, equations from references [43] and [45] provide more accurate results.

4.4. Grain Growth Behavior

The following equation is usually applied to low/medium carbon steels to predict growth as a function of holding time (in s) and temperature (in K):

$$d^m = d_{srx}^m + kt \cdot exp(-Q_{gg}/RT) \quad (4)$$

where Q_{gg} is the apparent activation for grain growth, m is the growth exponent and k is a constant. The parameters of some of the models developed for different steels are summarized in Table 7:

Table 7. Values of the parameters proposed for some grain growth models in the literature.

Steel	m (s⁻¹)	k	Q_{gg} (J/mol)	Reference
Nb, Hogson	4.5	4.1×10^{23}	435000	[22]
Ti, Hodgson	10	2.6×10^{28}	437000	[22]
CMn, Sellars	10	3.9×10^{32}	400000	[46]
CMn, Hodgson	7	1.45×10^{27}	400000	[22]

Figure 18 compares the predictions of these models and the experimental measurements performed in this work. Regarding the experimental data, it can be noted that the Ti0Nb0 steel shows the fastest growth kinetics, while this is slower for Ti0Nb3 and much more limited for the high Ti steels. On the other hand, when considering the predicted values, these trends were not always true; higher grain growth rates were predicted for the Nb microalloyed steels relative to the CMn ones by the equations developed. At 1050 °C and at 1100 °C, when the coarsest initial recrystallized grain size is considered (Figure 18b,c), good agreement is observed between the experimental data for the Ti0Nb3 steel and the values predicted by the equation developed by Hodgson for Nb microalloyed steels. On the other hand, at 1100 °C, when finer initial recrystallized grain sizes are considered (Figure 18a), the experimental values tend to be lower than the ones predicted. In the same conditions the grain growth kinetics for the Ti0Nb0 steel was best predicted by the Hodgson CMn equation, while at 1050 °C, this was underestimated by all the reported models. This makes difficult to determine which is the most adequate equation for the Ti0Nb0 and Ti0Nb3 steels. However, when considering the high Ti steels, especially Ti15Nb3, applying the equation developed by Hodgson for Ti microalloyed steels, which predicts very limited grain growth, would provide the most reliable results.

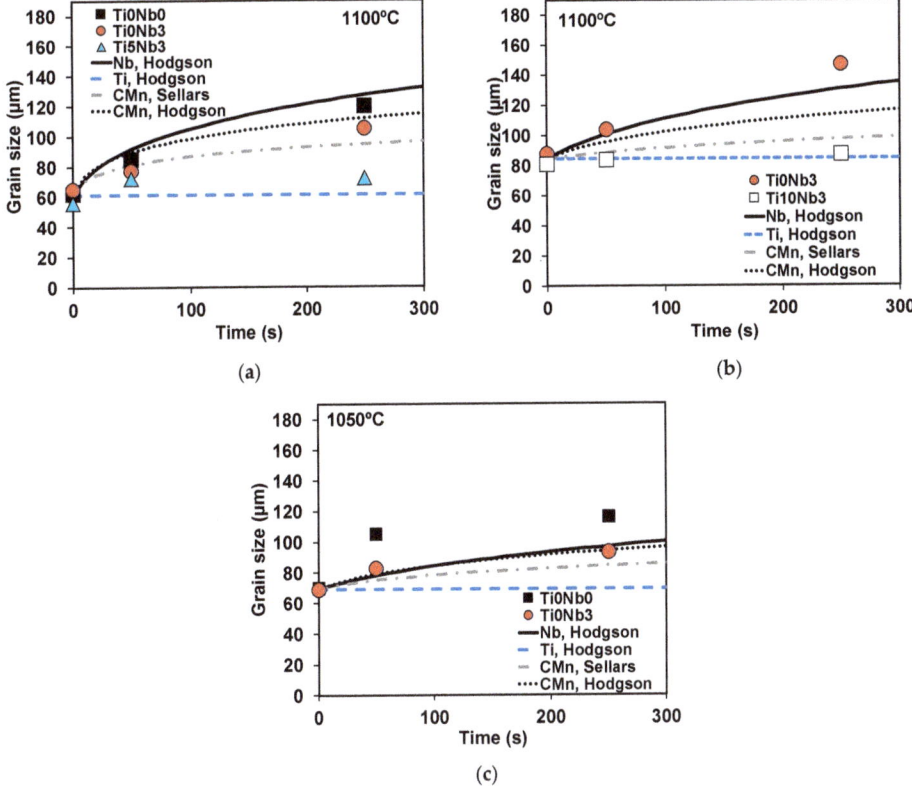

Figure 18. Comparison between the experimental grain growth data and the results of several models reported in the literature (symbols represent experimental data and lines model predictions) at (**a**), (**b**) 1100 °C and (**c**) 1050 °C.

5. Conclusions

This work analyzed the austenite microstructural evolution of 0.03%Nb microalloyed steels with different high Ti additions (0.05%, 0.1%, 0.15%) during hot working. The main conclusions that can be extracted are as follows:

For the reference 0.03%Nb microalloyed steel, after reheating at 1250 °C, all the Nb is in solid solution in the austenite. In contrast, for the high Ti steels, according to thermodynamic calculations, only slightly half of the Ti concentration available can be dissolved. The coarsest Ti containing precipitates (in the range of microns) often appear co-precipitated over Al_2O_3 particles. For the Ti5Nb3 and Ti10Nb3 steels, these co-precipitates are mainly TiN, although they can also contain some Nb, while for the Ti15Nb3 steel coarse Ti- and S-containing particles were also detected. Additionally, nanometer-sized precipitates were present for all the steels. Some Nb could also be detected in these particles, which can result in lower Nb in solid solution than the values predicted by the Thermo-Calc software.

The fastest recrystallization kinetics was determined for the Ti0Nb0 steel, followed by the Ti5Nb3 and then Ti0Nb3 steels. The most retarded softening kinetics was observed for the Ti10Nb3 and Ti15Nb3 steels. Grain size and the amount of microalloying in solid solution can explain the observed trends. Experimental $t_{0.5}$ and n Avrami exponent values determined for all high Ti steels are in good agreement with those predicted by available models developed for Nb, Nb-low Ti and CMn microalloyed steels.

Within the microalloyed steels investigated, the most retarded strain-induced precipitation kinetics was determined for the Ti5Nb3 steel. In this steel mainly co-precipitation over undissolved precipitates occurs. This effect could be explained by a lower amount of microalloying element in solution compared to the Ti0Nb3 or to the reduction of interfacial energy for nucleation. On the other hand, for the Ti10Nb3 and Ti15Nb3 steels, both co-precipitation and more dispersed precipitates were detected, which can be attributed to larger amount of Ti in solid solution, resulting in higher precipitation driving force.

The strain-induced precipitates found in the Ti10Nb3 and Ti15Nb3 steels presented both Nb and Ti in their composition. These precipitates resulted in a softening retardation effect comparable to or even larger than that observed for the Ti0Nb3 steel. However, the size of these precipitates was also coarser than for Ti0Nb3 and they presented a higher growth rate.

For the high Ti steels, the recrystallized grain sizes could be reasonably well predicted by the models developed for other microalloyed steels. Regarding grain growth, for the Ti0Nb0 and Ti0Nb3 steels it is not clear which of the existing models is the most accurate one. However, for the high Ti steels, grain growth was much more limited.

Author Contributions: L.G.-S., B.L. and B.P. conceived and designed the experiments; L.G.-S. performed the experiments; L.G.-S., B.L. and B.P. analyzed the data; B.L. and B.P wrote the paper. All authors have read and agreed to the published version of the manuscript.

Funding: The authors acknowledge financial support to carry out this work from the European Commision's Research Fund for Coal and Steel (RFSR-CT-2015-00013).

Conflicts of Interest: The authors declare no conflict of interest.

References

1. Andrade, H.L.; Akben, M.G.; Jonas, J.J. Effect of Molybdenum, Niobium, and Vanadium on Static Recovery and Recrystallization and on Solute Strengthening in Microalloyed Steels. *Metall. Trans. A* **1983**, *14*, 1967–1977. [CrossRef]
2. Medina, S.F.; Quispe, A. Improved Model for Static Recrysallization Kinetics of Hot Deformed Austenite in Low Alloy and Nb/V Microalloyed Steels. *ISIJ Int.* **2001**, *41*, 774–781. [CrossRef]
3. Fernandez, A.I.; Uranga, P.; López, B.; Rodriguez-Ibabe, J.M. Static Recrystallization Behaviour of a Wide Range of Austenite Grain Sizes in Microalloyed Steels. *ISIJ Int.* **2000**, *40*, 893–901. [CrossRef]
4. Aretxabaleta, Z.; Pereda, B.; Lopez, B. Multipass Hot Deformation Behaviour of High Al and Al-Nb Steels. *Mater. Sci. Eng. A* **2014**, *600*, 37–46. [CrossRef]
5. Bengochea, R.; Lopez, B.; Gutierrez, I. Influence of the Prior Austenite Microstructure on the Transformation Products obtained for C-Mn-Nb Steels after Continuous Cooling. *ISIJ Int.* **1999**, *39*, 583–591. [CrossRef]
6. DeArdo, A.J. Niobium in Modern Steels. *Int. Mater. Rev.* **2013**, *48*, 371–402. [CrossRef]
7. Wolczynski, W. Back-Diffusion in Crystal Growth. Eutectics. *Arch. Metall. Mater.* **2015**, *60*, 2403–2407. [CrossRef]
8. Wolczynski, W. Back-Diffusion in Crystal Growth. Peritectics. *Arch. Metall. Mater.* **2015**, *60*, 2409–2414. [CrossRef]
9. Himemiya, T.; Wolczynski, W. Solidification Path and Solute Redistribution of an Iron-Based MultiComponent Allow with Solute Diffusion in the Solid. *Mater. Trans.* **2002**, *43*, 2409–2414. [CrossRef]

10. Medina, S.F.; Chapa, P.; Valles, A.; Quispe, M.I.; Vega, M.I. Influence of Ti and N Contents on Austenite Grain Control and Precipitate Size in Structural Steels. *ISIJ Int.* **1999**, *39*, 930–936. [CrossRef]
11. Lagneborg, R.; Siwecki, T.; Zajac, S. Role of Vanadium in Microalloyed Steels. *Scand. J. Metall.* **1999**, *28*, 186–241.
12. Sanz, L.; Pereda, B.; Lopez, B. Effect of Thermomechanical Treatment and Coiling Temperature on the Strengthening Mechanisms of Low Carbon Steels Microalloyed with Nb. *Mater. Sci. Eng. A* **2017**, *685*, 377–390. [CrossRef]
13. Iza-Mendia, A.; Altuna, M.A.; Pereda, B.; Gutierrez, I. Precipitation of Nb in Ferrite after Austenite Conditioning. Part I: Microstructural Characterization. *Metall. Trans. A* **2012**, *43*, 4553–4570. [CrossRef]
14. Altuna, M.A.; Iza-Mendia, A.; Gutierrez, I. Precipitation of Nb in Ferrite after Austenite Conditioning. Part II: Strengthening Contribution in High-Strength Low-Alloy (HSLA) Steels. *Metall. Mater. Trans. A* **2012**, *43*, 4571–4586. [CrossRef]
15. Freeman, S.; Honeycombe, R.W.K. Strengthening of Titanium Steels by Carbide Precipitation. *Met. Sci.* **1977**, *11*, 59–64. [CrossRef]
16. Funakawa, Y.; Shiozaki, T.; Tomita, K.; Yamamoto, T.; Maeda, E. Development of High Strength Hot-Rolled Sheet Steel consisting of Ferrite and Nanometer-Sized Carbides. *ISIJ Int.* **2004**, *44*, 1945–1951. [CrossRef]
17. Bu, F.Z.; Wang, X.M.; Yang, S.W.; Shang, C.J.; Misra, R.D.K. Contribution of Interphase Precipitation on Yield Strength in Thermomechanically Simulated Ti-Nb and Ti-Nb-Mo Microalloyed Steels. *Mater. Sci. Eng. A* **2015**, *620*, 22–29. [CrossRef]
18. Chen, C.Y.; Yen, H.W.; Kao, F.H.; Li, W.C.; Huang, C.Y.; Yang, J.R.; Wang, S.H. Precipitation Hardening of High-Strength Low-Alloy Steels by Nanometer-Sized Carbides. *Mater. Sci. Eng. A* **2009**, *449*, 162–166. [CrossRef]
19. Chen, J.; Wang, J.D. Precipitation Characteristics in a Low-Carbon Vanadium-Titanium-Bearing Steel. *Steel Res. Int.* **2015**, *86*, 821–824. [CrossRef]
20. Garcia-Sesma, L.; Lopez, B.; Lopez, B.; Pereda, B. Effect of Coiling Conditions on the Strengthening Mechanisms of Nb Microalloyed Steels with High Ti Addition Levels. *Mater. Sci. Eng. A* **2019**, *748*, 386–395. [CrossRef]
21. Patra, P.K.; Sam, S.; Singhai, M.; Hazra, S.S.; Ram, G.D.J.; Bakshi, S.R. Effect of Coiling Temperature on the Microstructure and Mechanical Properties of Hot-Rolled Ti-Nb Microalloyed Ultra High Strength Steel. *Trans. Indian Inst. Met.* **2017**, *70*, 1773–1781. [CrossRef]
22. Hodgson, P.D.; Gibbs, R.K. A Mathematical Model to Predict the Mechanical Properties of Hot Rolled C-Mn and Microalloyed Steels. *ISIJ Int.* **1992**, *32*, 1329–1338. [CrossRef]
23. Pereda, B.; Rodriguez-Ibabe, J.M.; Lopez, B. I Improved Model of Kinetics of Strain Induced Precipitation and Microstructure Evolution of Nb Microalloyed Steels during Multipass Rolling. *ISIJ Int.* **2008**, *48*, 1457–1466. [CrossRef]
24. Akben, M.G.; Chandra, T.; Plassiard, P.; Jonas, J.J. Dynamic Precipitation and Solute Hardening in a Titanium Microalloyed Steel Containing Three Levels of Manganese. *Acta Metall.* **1984**, *32*, 591–601. [CrossRef]
25. Liu, W.J.; Jonas, J.J. A Stress Relaxation Method for Following Carbonitride Precipitation in Austenite at Hot Working Temperatures. *Metall. Trans. A* **1988**, *19*, 1403–1413. [CrossRef]
26. Liu, W.J.; Jonas, J.J. Ti(CN) Precipitation in Microalloyed Austenite during Stress Relaxation. *Metall. Trans. A* **1988**, *19*, 1415–1424. [CrossRef]
27. Wang, Z.; Mao, X.; Yang, Z.; Sun, X.; Yong, Q.; Li, Z.; Wen, Y. Strain-induced Precipitation in a Ti Micro-Alloyed HSLA Steel. *Mater. Sci. Eng. A* **2011**, *529*, 459–467. [CrossRef]
28. Wang, Z.; Sun, X.; Yang, Z.; Yong, Q.; Zhang, C.; Li, Z.; Weng, Y. Effect of Mn Concentration on the Kinetics of Strain Induced Precipitation in Ti Microalloyed Steels. *Mater. Sci. Eng. A* **2013**, *561*, 212–219. [CrossRef]
29. Hong, S.G.; Kang, K.B.; Park, C.G. Strain-induced Precipitation of NbC in Nb and Nb-Ti Microalloyed HSLA Steels. *Scripta Mater.* **2002**, *46*, 163–168. [CrossRef]
30. Okaguchi, S.; Hashimoto, T. Computer Model for Prediction of Carbonitride Precipitation during Hot Working in Nb-Ti Bearing HSLA Steels. *ISIJ Int.* **1992**, *32*, 283–290. [CrossRef]
31. Fernandez, A.I.; Lopez, B.; Rodriguez-Ibabe, J.M. Relationship Between the Austenite Recrystallized Fraction and the Softening Measured from the Interrupted Torsion Test Technique. *Scripta Mater.* **1999**, *40*, 543–549. [CrossRef]

32. Llanos, L.; Pereda, B.; Lopez, B. Interaction Between Recovery, Recrystallization and NbC Strain-Induced Precipitation in High-Mn Steels. *Metall. Mater. Trans. A* **2015**, *46*, 5248–5265. [CrossRef]
33. Bechet, S.; Beaujard, L. New Reagent for the micrographical demonstration of the austenite grain of hardened or hardened-tempered steels. *Rev. Met.* **1955**, *52*, 830–836. [CrossRef]
34. Andersson, J.O.; Helander, T.; Höglund, L.; Shi, P.F.; Sundman, B. Thermo-Calc and DICTRA, Computational Tools for Materials Science. *Calphad* **2002**, *26*, 273–312. [CrossRef]
35. Thermo-Calc Software. *TCFE9 Steels/Fe-Alloys Database*; Thermo-Calc Software: Solna, Sweden.
36. Hua, M.; Garcia, C.I.; DeArdo, A.J. Multi-phase Precipitates in Interstitial-Free Steels. *Scripta Metall. Mater.* **1993**, *28*, 973–978. [CrossRef]
37. Pereda, B.; Lopez, B.; Rodriguez-Ibabe, J.M. Increasing the Non-Recrystallization Temperature of Nb Microalloyed Steels by Mo Addition. In Proceedings of the International Conference on Microalloyed Steels: Processing, Microstructure, Properties and Performance Proceedings, Pittsburgh, PA, USA, 16–19 July 2007; pp. 151–159.
38. Pereda, B.; Aretxabaleta, Z.; Lopez, B. Softening Kinetics in High Al and High Al-Nb-Microalloyed Steels. *J. Mater. Eng. Perform.* **2015**, *24*, 1279–1293. [CrossRef]
39. Larrañaga-Otegui, A.; Pereda, B.; Jorge-Badiola, D.; Gutierrez, I. Austenite Static Recrystallization Kinetics in Microalloyed, B. Steels. *Metall. Mater. Trans. A* **2016**, *47*, 3150–3164. [CrossRef]
40. Llanos, L.; Pereda, B.; Lopez, B.; Rodriguez-Ibabe, J.M. Modelling of Static Recrystallization Behavior of High Manganese Austenitic Steels with Different Alloying Contents. *ISIJ Int.* **2016**, *56*, 1038–1047. [CrossRef]
41. Dutta, B.; Palmiere, E.J.; Sellars, C.M. Modelling the Kinetics of Strain Induced Precipitation in Nb Microalloyed Steels. *Acta Mater.* **2001**, *49*, 785–794. [CrossRef]
42. Nagata, M.T.; Speer, J.G.; Matlock, D.K. Titanium Nitride Precipitation Behaviour in Thin-Slab Cast High-Strength Low-Allow Steels. *Metall. Mater. Trans. A* **2002**, *33*, 3099–3110. [CrossRef]
43. Abad, R.; Fernandez, A.I.; Lopez, B. Interaction Between Recrystallization and Precipitation during Multipass Rolling in a Low Carbon Niobium Microalloyed Steel. *ISIJ Int.* **2001**, *41*, 1373–1382. [CrossRef]
44. Sellars, C.M. Hot Working and Forming Processes. In Proceedings of the International Conference, Sheffield, UK, 17–20 July 1979; Sellars, C.M., Davies, G.J., Eds.; Metals Society: London, UK, 1980; pp. 3–15.
45. Beynon, J.H.; Sellars, C.M. Modelling Microstructure and Its Effects during Multipass Hot Rolling. *ISIJ Int.* **1992**, *32*, 359–367. [CrossRef]
46. Sellars, C.M.; Whiteman, J.A. Recrystallization and Grain Growth in Hot Working. *Met. Sci.* **1979**, *13*, 187–194. [CrossRef]

© 2020 by the authors. Licensee MDPI, Basel, Switzerland. This article is an open access article distributed under the terms and conditions of the Creative Commons Attribution (CC BY) license (http://creativecommons.org/licenses/by/4.0/).

Article

Defect Reduction and Quality Optimization by Modeling Plastic Deformation and Metallurgical Evolution in Ferritic Stainless Steels

Silvia Mancini [1], Luigi Langellotto [1], Paolo Emilio Di Nunzio [1], Chiara Zitelli [1] and Andrea Di Schino [2],*

[1] RINA Consulting Centro Sviluppo Materiali, Via di Castel Romano 100, 00128 Roma, Italy; silvia.mancini@rina.org (S.M.); luigi.langellotto@rina.org (L.L.); paolo.dinunzio@rina.org (P.E.D.N.); chiara.zitelli@rina.org (C.Z.)
[2] Dipartimento di Ingegneria, Università degli Studi di Perugia, Via G. Duranti 93, 06125 Perugia, Italy
* Correspondence: andrea.dischino@unipg.it

Received: 9 December 2019; Accepted: 25 January 2020; Published: 27 January 2020

Abstract: Manufacturing of ferritic stainless steels flat bars is an important industrial topic and the steel 1.4512 is one of the most commonly used grades for producing this component. In this paper, the origin of some edge defects occurring during hot rolling of flat bars of this grade is analyzed and thermomechanical and microstructural calculations have been carried out to enhance the quality of the finished products by reducing the jagged borders defect on hot rolled bars. An accurate investigation has been carried out by analyzing the defects on the final product from both the macroscopic and microstructural point of view through the implementation of thermomechanical and metallurgical models in a finite element (FE) MSC Marc commercial code. Coupled metallurgical and damage models have been implemented to investigate the microstructural evolution of ferritic grain size and material damaging. Three levels of prior ferritic grain size (PFGS) and three furnace discharge temperatures have been considered in the thermo-mechanical simulations of the roughing passes. Rheological laws for modeling the evolution of ferritic grain have been modified to describe the specific cases simulated. Results have shown that the defect is caused by processing conditions that trigger an anomalous heating which, in turn, induces an uncontrolled grain growth on the edges. The work-hardened and elongated grains do not recrystallize during hot deformation. Consequently, they tend to squeeze out the surrounding softer and recrystallized matrix towards the edges of the bar where the fractures that characterizes the surface defect occur.

Keywords: rheological law modeling; rolling; plastic deformation; microstructural and mechanical coupling; defect reduction

---

1. Introduction

Owing to their lower cost with respect to austenitic stainless steels, ferritic stainless steels are more and more requested. They are nowadays used in many applications facing with strength/ductility requirements coupled with high targets of corrosion resistance [1]. In particular, they are employed in automotive [2], construction and building [3,4], energy [5,6], aeronautical [7], food [8], and 3D printing [9] applications.

Manufacturing ferritic stainless steels flat bars is an important topic in steel industry. The steel grade EN 1.4512 is widely used in the production of ferritic bars. Flat bar products made of this material can exhibit some defects such as jagged edges when subjected to hot rolling. In order to study and identify the origin of this type of defect, a study on the rolling conditions of ferritic flat bars made of steel grade 1.4512 steel has been carried out.

The evolution of the steel microstructure is the key element to understand the nature of the abovementioned defects. For this reason, a model to simulate the effect of recrystallization and grain growth during hot deformation has been developed. Due to the high temperature and the intense strain field characterizing the roughing process, the solidification structure is fully replaced by a new recrystallized microstructure.

The roughing step has been identified as a crucial part of the hot rolling process. As a matter of fact, local hot working conditions of the material change strongly depending on the applied strain, strain rate, temperature, and also on the interpass time during the roughing process [10]. It is well known that recrystallization is influenced by temperature, strain amount and soaking time [11]. In case of hot rolling, two processes can be distinguished: the static recrystallization which takes place after deformation during the interpass time, and the dynamic or meta-dynamic recrystallization that takes place during or just after the hot deformation and is promoted by small strain rate and high temperature. A semi-empirical model considering either static and meta dynamic recrystallization has been developed and calibrated on the steel grade 1.4512 to describe all the possible elementary processes. It is coupled with the thermomechanical FE model of roughing in order to describe accurately the local conditions through the thickness and along the bar width on 2D sections.

Almost all the semi-empirical models for describing the microstructural evolution during hot deformation that can be found in the literature refer to austenitic stainless steels [12–14]. The same holds also for the equations for predicting the austenite grain size as a function of the hot working conditions [15]. Instead, it is rather difficult to find consistent literature data about the evolution of ferritic microstructure during the hot deformation processes. Some data can be found for interstitial free (IF) steels [16] and with low-carbon steels content during plastic deformation [17,18]. Therefore, in this work, a comprehensive model for recrystallization and grain growth of the ferritic stainless-steel grade EN 1.4512 is proposed.

2. Materials and Methods

The focus of this paper is on the jagged defect on the edges of hot rolled EN 1.4512 steel grade (see for example Figure 1). In order to investigate such defect, a detailed analysis on the border of selected specimen of bars has been performed by a scanning electron microscope (SEM) (Figure 2).

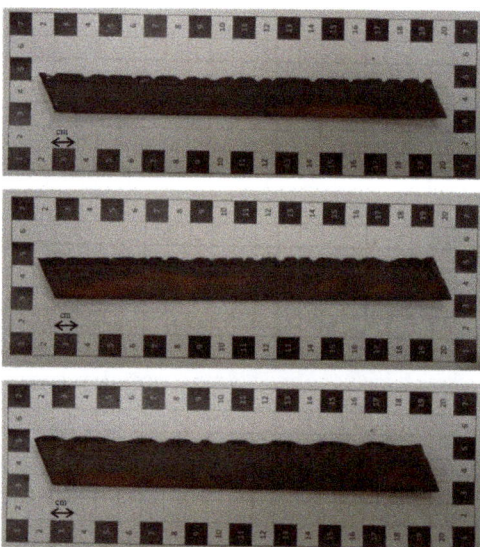

Figure 1. Specimens of flat bar in steel grade 1.4512 affected by jagged borders.

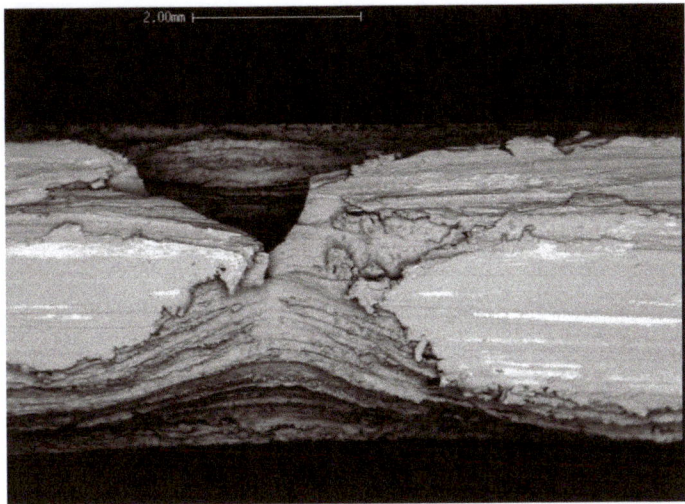

Figure 2. SEM images of the surface of a jagged bar.

The image in Figure 3 has been the starting point for the understanding of the mechanism of the defect formation. It shows the presence of elongated and not recrystallized abnormal grains "squeezed out" of the edge of the bar which represent a possible cause of the macroscopically observed jagged border.

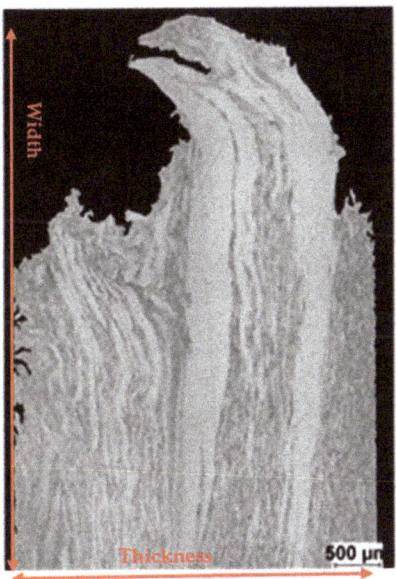

Figure 3. Optical micrograph showing a detail of the bar edge with the macroscopic defect. The elongated and work hardened grains at mid thickness appear to be extruded from the microstructure during hot working (2% Nital Etching).

The original microstructure of the bar is characterized by a central zone, at mid thickness, with an average grain size of approximatively 5 mm as measured according to ASTM E112 specification.

Going towards the surfaces the grain size decreases continuously down to 0.15 mm. This information, converted into the analytical function describing the prior ferritic grain size (PFGS) shown in Figure 4, has been used as input in the FE model of the roughing mill. The initial microstructure has been considered representative of the as-cast microstructure illustrated in Figure 5 and composed by a fine cortical grain size and a coarser columnar structure in the core region.

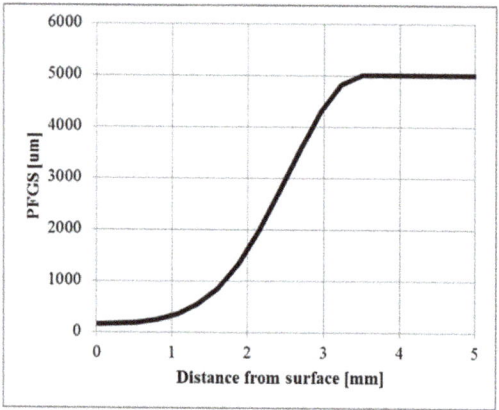

Figure 4. Mathematical function used to represent the through thickness variation of the ferrite grain size prior to hot rolling based on the as-cast microstructure.

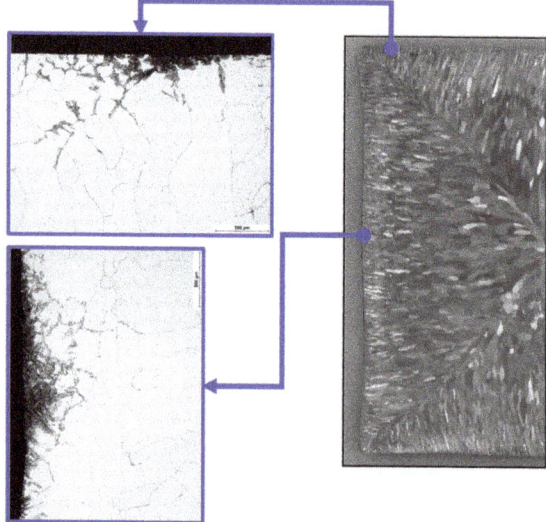

Figure 5. Typical macrostructure of an as-cast billet of ferritic stainless steel.

In order to understand the effect of the initial grain size distribution through the thickness on the evolution of the microstructure during the hot rolling process, simulations with the following grain distributions have been considered:

- from 5000 µm (center) to 0.15 µm (surface)-reference condition
- uniform grain size of 5000 µm
- from 5000 µm (center) to 1000 µm (surface)

In order to calibrate the material and damage model tensile tests have been performed on EN 1.4512 steel grade in the temperature range 750–1200 °C, at strain rate of 0.1 s^{-1} and 5.0 s^{-1}. For the tensile tests a standard round specimen for tensile test following ISO 6892-1:2009 with a diameter of 9 mm and a gauge length of 45 mm.

3. Rheological and Microstructural Models

In order to obtain the parameters for predicting the material microstructural evolution, an ad-hoc 3D thermo-mechanical and metallurgical FE models has been developed and tuned on specific experimental measurements and on literature data when available.

3.1. Hot Deformation Model

The simulation of the material flow when the bar undergoes plastic deformation process has been carried out by implementing a rheological model of the steel. The optimized equation used in this work is Equation (1), was originally proposed by Hansel–Spittel [19]. In such approach the flow stress is expressed by the product of strain, strain rate, and a power of temperature

$$\sigma_F = A * (\varepsilon + \varepsilon_0)^{m_2} * (\dot{\varepsilon} + \dot{\varepsilon}_0)^{m_8 T} * e^{m_1 T} \tag{1}$$

where σ_F, ε, $\dot{\varepsilon}$ are respectively flow stress, strain and strain rate, T is the temperature in Celsius degree, A is a scaling coefficient for the flow stress curve, and the coefficients m_1, m_2, m_8 and are unknown and optimizable parameters: coefficient m_1 is related to the material's sensitivity to temperature, m_2 models material's sensitivity to strain, m_8 combines the effect of temperature and strain rate.

The Hansel–Spittel model in Equation (1) can be used for strains below peak deformation. Above that threshold value the damage mechanism has to be taken into account in order to represent the lowering of the flow stress curve once the peak stress is reached. At this stage, recrystallization effects were not considered.

The peak deformation ε_p is expressed as in Equation (2), where α and β are coefficients which needs to be calibrated.

$$\varepsilon_p = \alpha(Z)^\beta \tag{2}$$

where the Zener–Hollomon parameter Z is evaluated according to [19] as

$$Z = (\dot{\varepsilon} + \dot{\varepsilon}_0) \exp\left(\frac{Q}{RT}\right) \tag{3}$$

Coefficients in Equation (3) are the Zener–Hollomon parameter Z, the strain rate and the initial strain rate $\dot{\varepsilon}$, $\dot{\varepsilon}_0$ and the coefficients at the exponent are respectively the activation energy Q, the gas constant R, and temperature T.

α and β parameters in Equation (2) have been set to 0.0184 and 0.8, respectively after a further analysis on experimental curves and the activation energy Q has been set to 277 kJ/mol.

The damage model, based on a modified Lemaitre equation [20], has been implemented in the FE model by means of external subroutines, as well as the implementation of the yield stress. The law governing the damage $\dot{D}$ is

$$\dot{D} = \left(-\frac{h_c y}{S_0}\right)^{s_1 - s_2 D_{in}} \overline{\dot{\varepsilon}_p} \tag{4}$$

where $\overline{\dot{\varepsilon}_p}$ is the plastic deformation strain rate, the material-dependent constants S_0, s_1, s_2 take into account the strength of the damage and the coefficient h_c is set to 1 if the material is in state of tension and 0.2 if it is in state of compression. The quantities y and D_{in} represent respectively the energy released during deformation and the initial damage. All the parameters have been largely discussed in [20]. For the last two parameters, the following expressions have been adopted

$$-y = \frac{\bar{\sigma}}{2E(1-D)^2}\left[\frac{2}{3}(1+\vartheta) + 3(1-2\vartheta)\left(\frac{\sigma_H}{\bar{\sigma}}\right)^2\right] \quad (5)$$

$$D_{in} = ae^{-b\frac{r}{R}} \quad (6)$$

where $\bar{\sigma}$ is the equivalent stress and $\sigma_H/\bar{\sigma}$ the stress triaxiality factor, i.e., the ratio between the hydrostatic tension and the equivalent stress. The damage variable D depends on the initial value of the damage on the axis of the bar, a, and r is a radial coordinate ranging from 0 to the bar radius R. The coefficient b is calculated from the condition of no initial damage located at a certain radius r in the core area, which is determined through an analysis of the void's distribution.

Tensile tests performed on EN 1.4512 steel grade in the temperature range 750–1200 °C, at strain rate of 0.1 s^{-1} and 5.0 s^{-1} have been reproduced by means of non-linear finite element modeling. The tensile tests were modeled with axisymmetric elements. MSC. Marc was used for all the simulations. Isotropic work hardening law, Equation (1), and von Mises yield criterion have been adopted by using external YIELD subroutine. Damage, formulations from Equation (4) to Equation (6), have been introduced in the FE model by means of UDAMAG user routine.

As far as material stress–strain curve modeling concerns, the after necking extended stress–strain curve was determined by using a modification of the inverse calibration approach proposed by [21,22]. The basic data are the load–displacement curves obtained from conventional tensile tests, fitted by the same results obtained from finite element simulations of the tensile test. The initial stress–strain curve, estimated as a simple power law from experimental data up to necking, was iteratively modified in the post-necking regime until the error between the experimental and numerical load–displacement curve has been reduced below a defined tolerance. Iterations were performed by using the automatic procedure described in [23]. Automatic iterative optimization procedure considers modify all coefficients of Equation (1) considering the total error between all experimental and numerical load–displacement curve in the temperature range 750–1200 °C, at strain rate of 0.1 s^{-1} and 5.0 s^{-1}. For this multi-objective optimization problem, no single solution exists that simultaneously optimizes each objective, so a nondominated optimal solution that allow to reach the tolerance defined in the optimization procedure has been chosen.

The optimized coefficients of Equation (1) can be found in Table 1 and for Equation (4) are reported in Table 2. Examples of final, calibrated stress–strain curves are reproduced in Figure 6 and compared with experimental data of tensile tests. Comparisons have been reported for tensile tests carried out at: (a) T = 1200 °C $\dot{\varepsilon}$ = 5s^{-1}; (b) T = 1050 °C $\dot{\varepsilon}$ = 5 s^{-1}; (c) T = 900°C $\dot{\varepsilon}$ = 1 s^{-1}.

Table 1. Coefficients of first attempt and final value of the same coefficients for the rheological model of the EN 1.4512 steel in Equation (1).

Coefficient	First Attempt Value [19] Chapter 109	Final Value
A	4422.71	4650
m_1	−0.0029	−0.0032
m_2	0.48151	0.3
m_8	0.000202	0.00017
ε_0	0.0	0.001
$\dot{\varepsilon}_0$	0.0	0.01

Table 2. Coefficients of the damage model in Equation (4).

Coefficient	Value
s_0	1.1
s_1	0.1
s_2	0.5

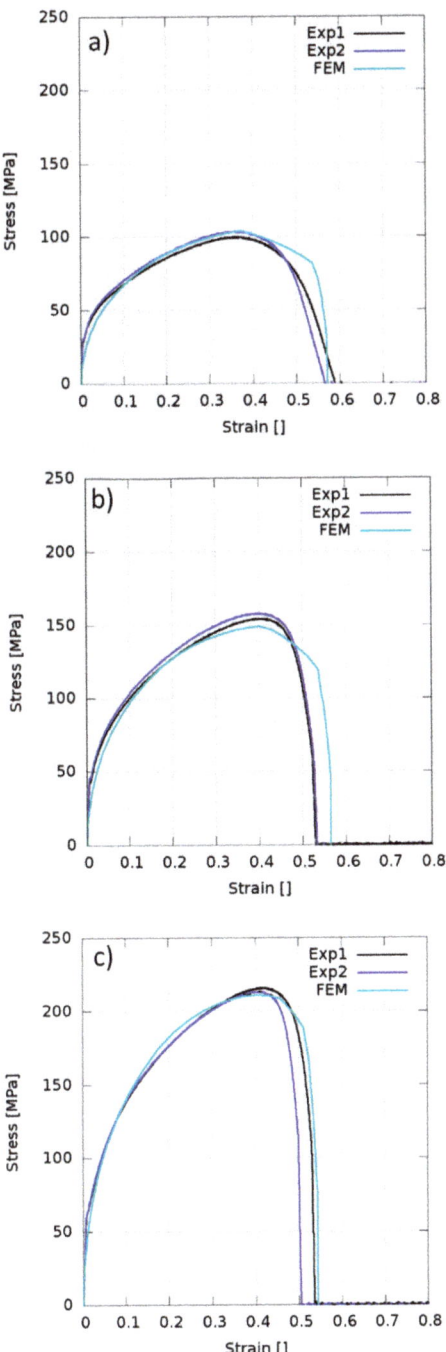

Figure 6. Comparison between numerical and experimental data of tensile tests simulation after the optimization procedure of rheological and damage coefficients. Comparisons have been reported for tensile tests carried out at: (a) T = 1200 °C $\dot{\varepsilon}$ = 5 s^{-1}; (b) T = 1050 °C $\dot{\varepsilon}$ = 5 s^{-1}; (c) T = 900 °C $\dot{\varepsilon}$ = 1 s^{-1}.

Once the material model has been tuned in terms of Equation (1) and Equation (4) parameters, a thermo-mechanical hot rolling 3D model has been implemented. A screenshot of the thermo-mechanical 3D model is reported in Figure 7. The FE model of industrial rolling has been calibrated in terms of heat transfer coefficient between material and surrounding and material and rolls thanks the thermal image. Inter-pass time has been evaluated through the analysis of the rolling force signals acquired during hot rolling. The calculated temperatures are in good agreement with the experimental data obtained by a thermal imaging camera as revealed by Figure 7.

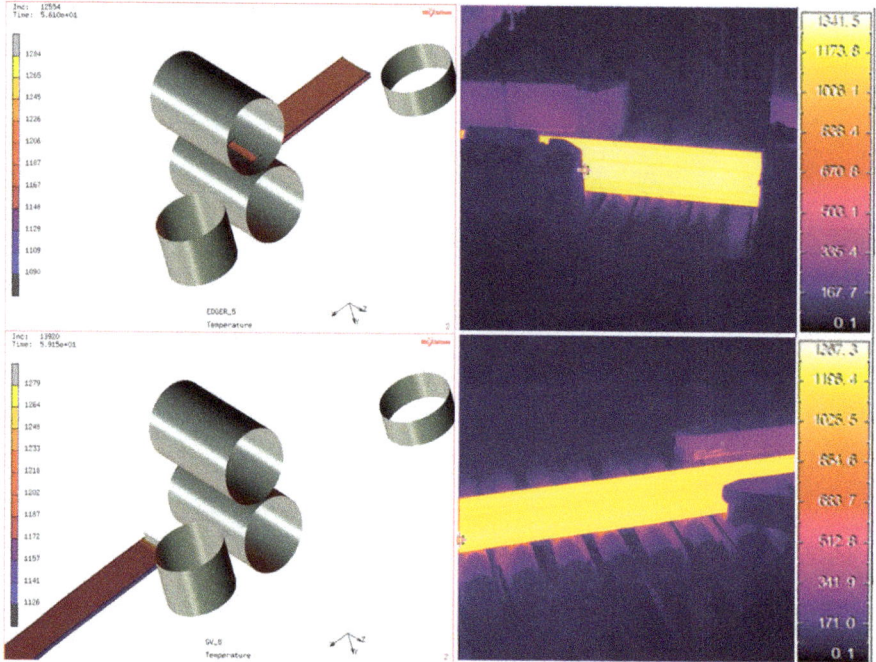

Figure 7. FE Thermomechanical 3D model of the hot rolling of flat bar compared to temperature experimental measures.

3.2. Static Recrystalization and Grain Growth Models

In this work the static recrystallization (which takes place during the interpass time between roughing passes) and three metadynamic recrystallization (more typical of ferritic steels) processes and the grain growth after recrystallization are considered. As previously mentioned, in the literature there is plenty of data on hot deformation of austenite but very few on ferrite. Therefore, the mathematical form of the equations describing the recrystallization kinetics and grain growth has been taken from the literature starting from and in particular from some works on hot deformation of IF and low-carbon steels [17,24] and the calibrated thermo-mechanical FE simulations have been used as a basis for refining the phenomenological coefficients of the metallurgical model.

The static recrystallization (SRX) has been modeled by an Avrami-like equation expressing recrystallized fraction X_{rex} as a function of time in isothermal conditions [25]

$$X_{rex} = 1 - exp\left[-0.693\left(\frac{t}{t_{0.5}}\right)^n\right] \quad (7)$$

where the parameter $t_{0.5}$ is the time to reach 50% SRX and n a kinetic exponent. The $t_{0.5}$ depends in turn on the initial grain size, temperature, strain, and strain rate as

$$t_{0.5} = C\varepsilon^p \dot{\varepsilon}^q d^s \exp\left(\frac{Q_{app}}{RT}\right) \quad (8)$$

where C is a constant related to the chemical composition, the exponents p, q and s express the effect of strain ε, strain rate $\dot{\varepsilon}$ and initial grain size d, R and T are the gas constant and the temperature respectively, and Q_{app} is the activation energy needed for the process. Such parameters were determined starting from literature data reported in [26] and are listed in Table 3.

Table 3. Coefficients for SRX process from [26].

Coefficient	Value
N	1.5
C (s)	1.69 × 10^{-10}
P	−1.7
d_0	157
Q	0
S	1.5
Q (kJ/mol)	143

If the steel undergoes low strain and low strain rate deformation, SRX mechanism is activated [27]. In this case, the recrystallized grain size d_{srx} can be expressed as

$$d_{SRX} = c_1 + c_2 d_\alpha{}^{c_3} \varepsilon^{c_4} \left(\dot{\varepsilon} \exp\left(\frac{Q}{RT}\right)\right)^{c_5} \quad (9)$$

c_1, c_2, c_3, c_4 and c_5 parameters and activation energy Q were determined starting from literature data reported in [27] and are listed in Table 4.

Table 4. Coefficient used in Equation (9) for predicting the ferrite grain size after SRX [27].

Coefficient	Value
c_1 (s)	28.26
c_2 (s)	18.24
c_3	0
c_4	−0.6
c_5	−0.05
Q (kJ/mol)	267

Concerning the grain growth after SRX, the following equation used:

$$d_{gg} = \left(d_\alpha{}^{c_1} + c_2 \exp\left(\frac{Q}{RT}\right) t_{gg}\right)^{c_3} \quad (10)$$

c_1, c_2, c_3 parameters, the activation energy Q and ferritic grain size d_α were determined starting from literature data reported in [27] and are listed in Table 5.

Table 5. Coefficient used in Equation (10) for grain growth during static recrystallization [27].

Coefficient	Value
c_1	7
c_2	3 × 10^{23}
c_3	7
Q (kJ/mol)	−356
d_α	3.6

3.3. Metadynamic Recrystalization and Grain Growth Models

According to the experimental data in literature on high chromium stainless steels, at high strain levels and high temperature the dominant mechanism is the metadynamic recrystallization (MDRX) [27]. This process is activated when the imposed strain exceeds a critical level ε_C which in turn, is proportional to the peak strain according to the relationship in Equation (11).

$$\varepsilon_C = A\, \varepsilon_p \tag{11}$$

In the present case, the estimation of the peak strain has been carried out by exploiting a collection of literature data [27]. The peak deformation is a function of the Zener–Hollomon parameter as shown in Equation (2). For the MDRX the α_{MDRX} and β_{MDRX} are 0.025 and 0.59, respectively. The coefficient A in Equation (12), ranging between 0.7 and 0.8, has been set equal to 0.75.

The recrystallized fraction after MDRX fraction is calculated using the same Equation (7) as in the case of SRX. Instead, the ferrite grain size after MDRX is

$$d_{MDRX} = c_1 \dot\varepsilon \exp\!\left(\frac{Q}{RT}\right)^{c_2} \tag{12}$$

c1, c2 coefficients, properly calibrated on experimental data from the literature [27], are reported in Table 6.

Table 6. Coefficient for grain evolution during MDRX [27].

Coefficient	Value
c_1	18277
c_2	−0.246
Q (kJ/mol)	267
d_0	157

The local strain, strain rate, and temperature conditions predicted by the thermo-mechanical FE calculation are used to identify whether SRX or a MDRX process is activated.

4. Results

The output of the FE thermo-mechanical simulations and the microstructural model have been coupled in order to simulate the evolution of the microstructure during roughing of the bar. In this section, a sensitivity analysis of the main factors affecting the microstructural evolution is reported and discussed.

Results are given in the form of maps of total equivalent plastic strain, equivalent Von Mises stress and temperature fields, taken on the transversal section of the hot rolled bar, as shown in Figure 8. Additionally, the chosen section must not suffer of border defect (associated for example to the finite nature of the computational domain), and stress–strain and temperature fields must be stationary.

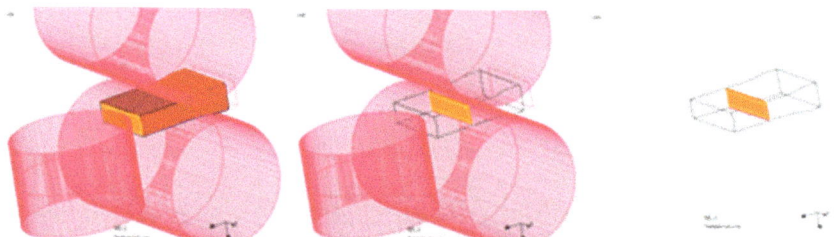

Figure 8. Selection of transversal section of the bar.

The simulations have been carried out to evaluate the influence of reheating temperature, the effect of different initial ferrite grain sizes (PFGS) and the effect of damage on the microstructure evolution.

The effect of the reheating temperature on the ferrite grain size are summarized in Figure 9. The data are represented as maps of the ferrite grain size on the transversal section of the bar after the second roughing pass. They show the presence of bigger grains in the center of the bar compared to the sub-surface regions but, in all cases, there is no evidence of abnormal grains. It is possible to conclude that at this stage the microstructure evolution does not favor the formation of the jagged border defect.

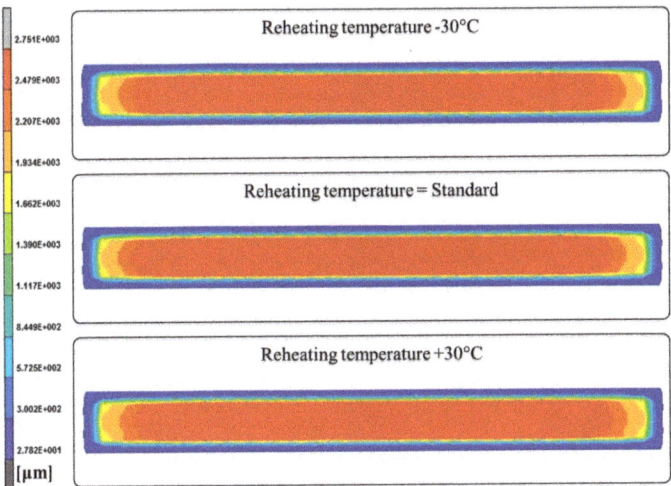

Figure 9. Effect of the reheating temperature on the ferrite grain size after the second roughing pass.

The same data after the fifth pass are shown in Figure 10. Now the effect of the reheating temperature on the grain size is apparent. High reheating temperatures produces smaller grains in the center of the bar and a more homogeneous microstructure. Nevertheless, the relationship with the jagged border defect is still not clear because grains on the edge are smaller than those in the core with no formation of abnormal grains.

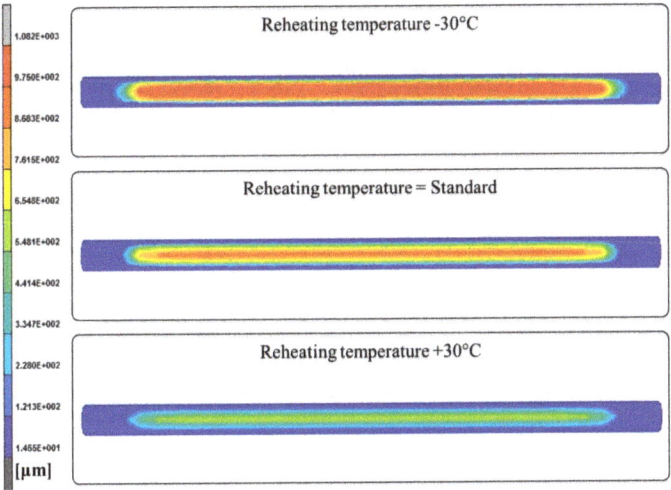

Figure 10. Effect of the reheating temperature on the ferrite grain size after the fifth roughing pass.

In order to take into account the effect of the grain size at prior to hot rolling, calculations with different distributions of the PFGS have been carried out The grain size maps in Figure 11 summarize the results after the second pass. The map calculated from the uniform PFGS equal to 5000 µm is substantially different from the two maps calculated using a gradient of grain sizes form center to surface. The effect is already apparent just after the beginning of the hot rolling process. In the first map of Figure 11, it can be observed that smaller grains are formed on the edges of the bar. On the contrary, in the other maps smaller grains are uniformly distributed along the whole surface.

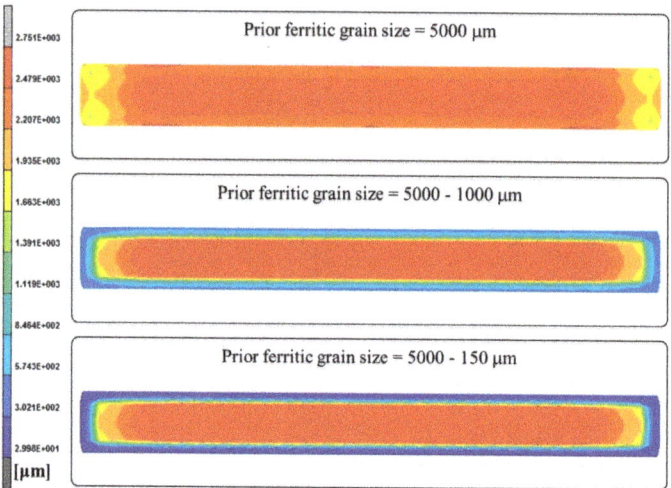

Figure 11. Effect of different PFGS on the ferrite grain size after the second roughing pass.

Even greater differences can be appreciated after the fifth pass, as shown in Figure 12. Also, in this case, the starting in condition with a uniform PFGS equal to 5000 µm differs from the others. It is possible to observe that of smaller grains are formed on the edges of the bar but there are bigger grains just on the corners.

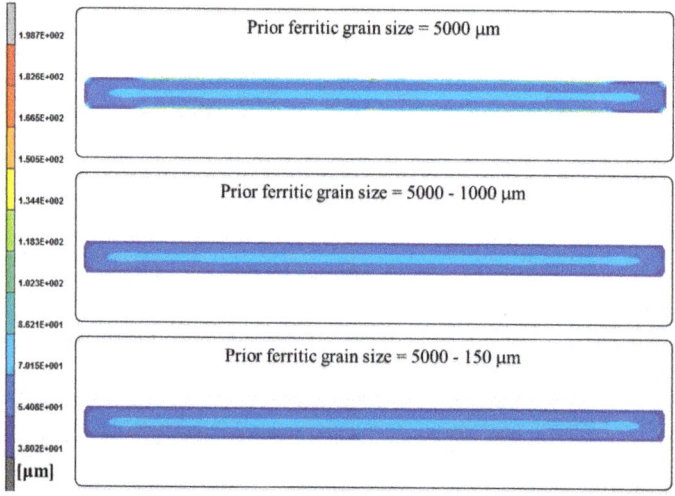

Figure 12. Effect of different PFGS on the ferrite grain size after the fifth roughing pass.

The maps showing the influence of the PFGS on damage after the fifth pass are reported in Figure 13. Damage increases going towards the edges in all cases. there is a minor presence of the damage on the center of the slabs, while its value increases on the edge. It can be noticed that the distribution of the damage parameter on the bar section is similar to the profile of the jagged border defect.

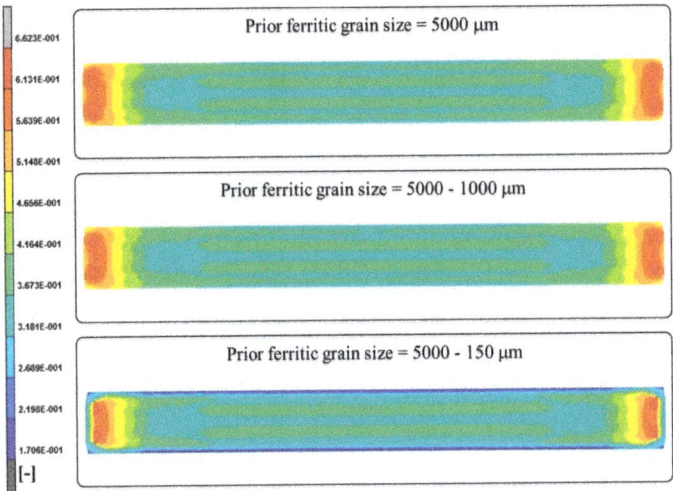

Figure 13. Effect of PFGS on damage after the fifth roughing pass.

After the fifth roughing pass, even if the recrystallization interferes with the increasing of the damage, an increased density of damage is observed close to the edges. As shown in Figure 14, this effect is particularly intense for the configuration with a uniform PFGS equal to 5000 µm.

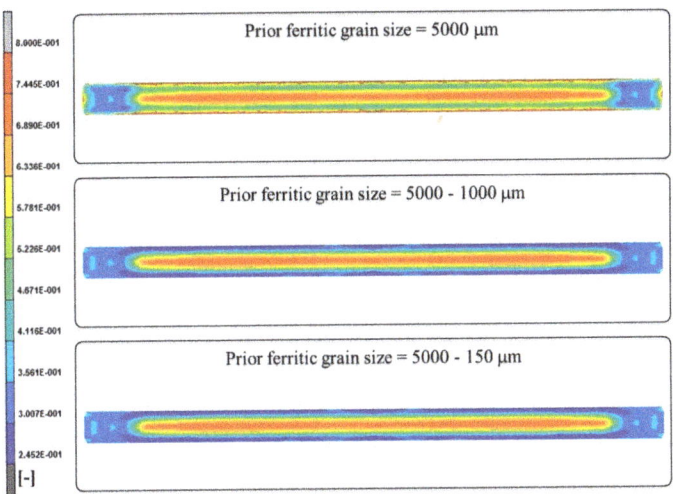

Figure 14. Effect of PFGS on damage after the fifth roughing pass.

A synthesis of the previous observations after the fifth pass for the case of a uniform PFGS equal to 5000 µm is schematized in Figure 15 in comparison with a macrograph of a hot rolled bar. It is apparent that the positions of the edge defects observed in the specimen correspond to the regions that, in the calculations, exhibit the larger grain size and the higher level of damage. Figure 16 shows another view of the bar, where is possible to observe the border affected by a coarser grain.

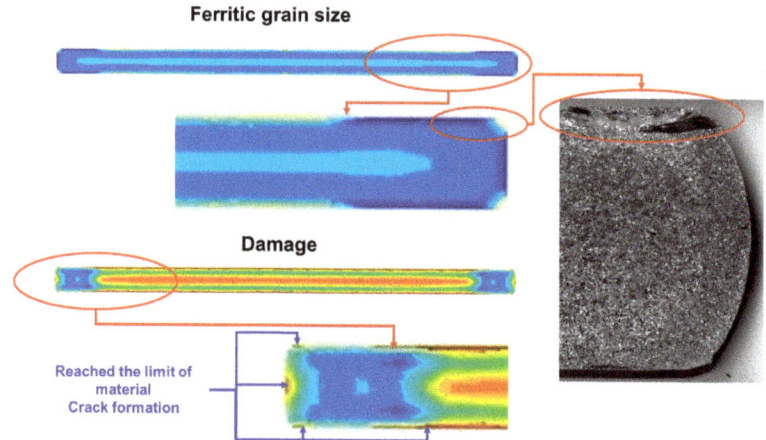

Figure 15. Comparison between macrography of the bar and the grain growth resulting from the maps of ferrite grain size and damage.

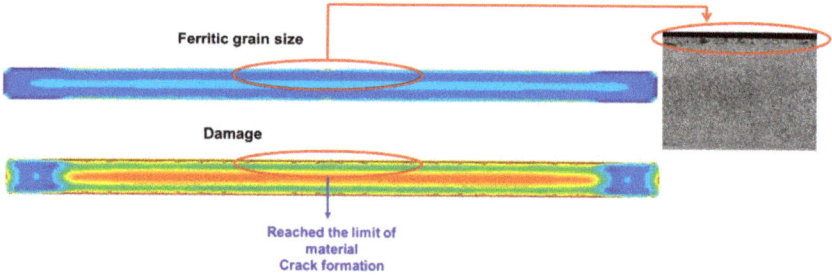

Figure 16. Comparison between macrography of the bar and the grain growth resulting from the maps of ferrite grain size and damage.

5. Discussion

The results of FE simulations have shown the evolution of ferritic grain size during the roughing stage of hot rolling due to recrystallization as a function of different initial conditions regarding temperature and through-thickness distribution of the prior ferrite grain size. The calculated maps have highlighted that the recrystallization reduces the damage in all the examined cases and also on the bar edges. However, when a uniform initial ferrite grain size equal to 5000 µm is considered, the damage on the edges is more intense compared to the other analyzed conditions. During the hot rolling process, the ferrite grain is continuously refined and, in all cases, a homogeneous grain size is attained at the center of the bar. Instead, some differences are observed around the edges. In the surface region of the bar a localized presence of coarser grains (up to five time than the average size) is observed especially when the initial grain size is coarse. Instead, a finer grain size is predicted at the surface of the bar when the PFGS includes grains in the range 5000 to 150 µm.

The configuration with uniform PFGS = 5000 µm is representative of an anomalous grain growth, both superficial and sub-superficial. Such PFGS values in ferritic steels are usually related to high reheating temperature [28] which in industrial processing should happen following uncontrolled reheating in the furnace. It is well known [29] that the presence of such large grains opposes to the recrystallization behavior according to Equation (7) and Equation (8). This will result in the presence of a deformed fiber characterized by work-hardened grains surrounded by recrystallized grains: such topological arrangement will result in grain boundaries sliding rather than further deformation [30].

As a consequence, when the work-hardened grains are pushed in proximity of the edges, they will expel the recrystallized grains out of the bar. The material spreading due to the hot rolling condition is accentuated and the edges exhibit a jagged aspect. Moreover, the slower the recrystallization kinetics the higher the damaging of the material, thus inducing a higher probability of cracks and micro cracks formation on the bar surfaces.

6. Conclusions

An application of standard and advanced modeling techniques to the solution of industrial problems during hot rolling has been described. A special focus has been set to the use of combined experimental and numerical analysis on materials and products, in particular on hot rolling of flat bars made in 1.4512 stainless steel. The developed modeling tools have demonstrated the required accuracy to successfully predict the state of the material during the entire processing route and to help the manufacturer to improve the final product, identifying the cause that may lead to the formation of the jagged border defect on the edges of the rolled bar.

In order to investigate the formation of the defect, the material has been characterized by a mechanical point of view. Moreover, metallurgical models and material damaging theory have been implemented and coupled with FE calculation results in order to predict the microstructural evolution of the ferrite grain structure during hot deformation.

Important parameters such as strain, stress, and temperature have been considered in order to analyze the problem and to investigate their influence on the development of the steel microstructure.

Results have shown that the jagged border defect could be caused by uncontrolled or improper reheating or cooling stage of the flat bars during the process. This can induce the formation of large and elongated grains that produce a 'squeezing' of the grains on the edge of the flat bar, thus originating the jagged border defect. The influence of PGFS is also relevant. FE analysis has shown that when a smaller initial grain size is considered the defect is not produced and the bar is characterized by smaller grains near the surface coarser grains at the core.

On the contrary, a uniform and coarse prior ferrite grain size of 5000 µm is more prone to develop the defect in conditions of uncontrolled reheating. The abnormally coarse grains result from work-hardening and recovery rather than recrystallization, their behavior being substantially different from that of the surrounding recrystallized matrix. In these conditions, when the strain hardened grains are in proximity of the skin, they can squeeze out the recrystallized grains, thus generating the defect on the bar edges.

Author Contributions: Conceptualization, S.M. and L.L.; methodology, S.M. and L.L.; software, S.M. and L.L.; validation, S.M. and L.L.; formal analysis, S.M. and L.L.; investigation, S.M. and L.L.; resources, S.M. and L.L.; data curation, S.M. and L.L.; writing—original draft preparation, S.M.; writing—review and editing, S.M., A.D.S., L.L., P.E.D.N. and C.Z.; visualization, A.D.S., L.L., P.E.D.N. and C.Z.; supervision, A.D.S. All authors have read and agreed to the published version of the manuscript.

Funding: This research received no external funding.

Conflicts of Interest: The authors declare no conflict of interest.

References

1. Marshall, P. *Austenitic Stainless Steels: Microstructure and Mechanical Properties*; Springer: Berlin, Germany, 1984.
2. Rufini, R.; Di Pietro, O.; Di Schino, A. Predictive Simulation of Plastic Processing of Welded Stainless Steel Pipes. *Metals* **2018**, *8*, 519. [CrossRef]
3. Corradi, M.; Di Schino, A.; Borri, A.; Rufini, R. A review of the use of stainless steel for masonry repair and reinforcement. *Constr. Build. Mater.* **2018**, *181*, 335–346. [CrossRef]

4. Di Schino, A.; Kenny, J.M.; Abbruzzese, G. Analysis pf the recrystallization and grain growth processes in AISI 316 stainless steel. *J. Mat. Sci.* **2002**, *37*, 5291–5298. [CrossRef]
5. Di Schino, A.; Porcu, G.; Longobardo, M.; Turconi, G.L.; Scoppio, L. Metallurgical design and development of C125 grade for mild sour service application. In Proceedings of the NACE—International Corrosion Conference Series, San Diego, CA, USA, 12–16 March 2006; pp. 061251–0612514.
6. Di Schino, A. Analysis of heat treatment effect on microstructural features evolution in a micro-alloyed martensitic steel. *Acta Metall. Slovaca* **2016**, *22*, 266–270. [CrossRef]
7. Di Schino, A.; Valentini, L.; Kenny, J.M.; Gerbig, Y.; Ahmed, I.; Hefke, H. Wear resistance of high-nitrogen austenitic stainless steel coated with nitrogenated amorphous carbon films. *Surf. Coat. Technol.* **2002**, *161*, 224–231. [CrossRef]
8. Bregliozzi, G.; Ahmed, S.I.-U.; Di Schino, A.; Kenny, J.M.; Haefke, H. Friction and Wear Behavior of Austenitic Stainless Steel: Influence of Atmospheric Humidity, Load Range, and Grain Size. *Tribol. Lett.* **2004**, *17*, 697–704. [CrossRef]
9. Zitelli, C.; Folgarait, P.; Di Schino, A. Laser powder bed fusion of stainless-steel grades: A review. *Metals* **2019**, *9*, 731. [CrossRef]
10. Jhonas, J. Effect of Quench and Interpass Time on Dynamic and Static Softening during Hot Rolling. *Steel Res. Int.* **2005**, *76*, 392–398. [CrossRef]
11. Hapmhreys, F.J.; Hatherly, M. *Recrystallization and Related Annealing Phenomena*; Elsevier: Amsterdam, The Netherlands, 2004.
12. Sung, K.; Yeon-Chul, Y. Dynamic recrystallization behavior of AISI 304 stainless steel. *Adv. Mater. Sci. Eng.* **2001**, *311*, 108–113.
13. Dehghan-Manshadi, A.; Barnett, M.R.; Hodgson, P.D. Hot deformation and recrystallization of austenitic stainless steel. Part I: Dynamic recrystallization. *Metall. Mater. Trans. A* **2008**, *39*, 1359–1370. [CrossRef]
14. Stanley, J.K.; Perrotta, J. Grain Growth in Austenitic Stainless Steels. *Metallography* **1969**, *11*, 349–362. [CrossRef]
15. Marchattiwar, A.; Sarkar, A.; Chakravartty, J.K.; Kashyap, B.P. Dynamic Recrystallization during Hot Deformation of 304 Austenitic Stainless Steel. *J. Mater. Eng. Perform.* **2013**, *22*, 2168–2175. [CrossRef]
16. Duggan, B.J.; Tse, Y.Y.; Lam, G.; Quadir, M.Z. Deformation and Recrystallization of Interstitial Free (IF) Steel. *Mater. Manuf. Processes* **2011**, *26*, 51–57. [CrossRef]
17. Barnett, M.R.; Jonas, J.J. Influence of ferrite rolling temperature on microstructure and texture in deformed low C and IF steels. *ISIJ Int.* **1997**, *37*, 697–705. [CrossRef]
18. Shin, D.H.; Byung, C.K.; Yong-Seog, K.; Kyung-Tae, P. Microstructural evolution in a commercial low carbon steel by equal channel angular pressing. *Acta Mater.* **2000**, *48*, 2247–2255. [CrossRef]
19. Spittel, M.; Spittel, T. Flow stress of steel. Metal forming data of ferrous alloys-deformation behavior-advanced materials and technologies (numerical data and functional relationships in science and technology). Group VIII Advanced Materials and Technologies. In *Landolt-börnstein*; Springer: Berlin, Germany, 2009; Volume 2C1.
20. Ghiotti, A.; Fanini, S.; Bruschi, S.; Bariani, P. Modeling of the Mannesman effect. *CIRP Ann. Manuf. Technol.* **2009**, *58*, 255–258. [CrossRef]
21. Ling, Y. Uniaxial true stress-strain after necking. *AMP J. Technol.* **1996**, *5*, 37–48.
22. Choung, J.M.; Cho, S.R. Study on true stress correction from tensile tests. *J. Mech. Sci. Technol.* **2008**, *22*, 1039–1051. [CrossRef]
23. Cortese, L.; Coppola, T.; Caserta, L. Calibration of material damage models using a multi-test inverse approach. In Proceedings of the XII International Symposium on Plasticity and Its Current Applications, Halifax, NS, Canada, 17–22 July 2006; Neat Press: Fulton, ML, USA, 2006; pp. 607–609.
24. Barnett, M.R.; Jonas, J.J. Influence of ferrite rolling on grain size and texture in annealed low C ad IF steels. *ISIJ Int.* **1997**, *37*, 706–714. [CrossRef]
25. Mehtonen, S.; Karjalainen, L.P.; Porter, D. Hot deformation behavior and microstructure evolution of a stabilized high-Cr ferritic stainless steel. *Mater. Sci.* **2013**, *571*, 1–12. [CrossRef]
26. Mirzadeh, H.; Najafizadeh, A. Hot deformation and dynamic recrystallization of 17-4 PH stainless steel. *ISIJ Int.* **2013**, *53*, 680–689. [CrossRef]
27. Oliveira, T.R.; Montheillet, F. Effect of Niobium and Titanium on the Dynamic Recrystallization during Hot Deformation of Stabilized Ferritic Stainless Steels. *Mater. Sci. Forum* **2004**, *467–470*, 1229–1236. [CrossRef]

28. Sellars, C.M.; Whiteman, J.A. Recrystallization and grain growth in hot rolling. *Met. Sci.* **1979**, *13*, 187–194. [CrossRef]
29. Raabe, D.; Lücke, K. The Role of Textures in Ferritic Stainless Steels. International Symposium on Strip Casting, Hot and Cold Working of Stainless Steels. *Quebec* **1993**, 221.
30. Jensen, D.J. Growth rates and misorientation relationships between growing nuclei/grains and the surrounding deformed matrix during recrystallization. *Acta Metall. Mater.* **1995**, *43*, 4117–4129. [CrossRef]

© 2020 by the authors. Licensee MDPI, Basel, Switzerland. This article is an open access article distributed under the terms and conditions of the Creative Commons Attribution (CC BY) license (http://creativecommons.org/licenses/by/4.0/).

Article

Casting and Constitutive Hot Flow Behavior of Medium-Mn Automotive Steel with Nb as Microalloying

Perla Julieta Cerda Vázquez [1], José Sergio Pacheco-Cedeño [2], Mitsuo Osvaldo Ramos-Azpeitia [3], Pedro Garnica-González [4], Vicente Garibay-Febles [5], Joel Moreno-Palmerin [6], José de Jesús Cruz-Rivera [1] and José Luis Hernández-Rivera [7,*]

1. Instituto de Metalurgia, Universidad Autónoma de San Luis Potosí, Sierra Leona 550, Lomas 2a Sección, San Luis Potosí C.P. 78210, Mexico; iq.perlavazquez@outlook.com (P.J.C.V.); jdjcr35@uaslp.mx (J.d.J.C.-R.)
2. Escuela de Ingeniería y Ciencias Región Centro, Tecnológico de Monterrey campus Morelia, Av. Montaña Monarca 1340, Michoacán C.P. 58350, Mexico; sergiopachecocedeno@gmail.com
3. Facultad de Ingeniería-Universidad Autónoma de San Luis Potosí, Dr. Manuel Nava 8, Zona Universitaria, San Luis Potosí C.P. 78290, Mexico; mitsuo.ramos@uaslp.mx
4. División de Estudios de Posgrado, Tecnológico Nacional de México Campus Instituto Tecnológico de Morelia, Av. Tecnológico 1500, Michoacán C.P. 58120, Mexico; pgarnica@itmorelia.edu.mx
5. Instituto Mexicano del Petróleo, Laboratorio de Microscopia Electrónica de Ultra Alta Resolución, Eje Central Lázaro Cárdenas 132, San Bartolo Atepehuacan C.P. 07730, Mexico; vgaribay@imp.mx
6. Departamento de Ingeniería en Minas, Metalurgia y Geología, Universidad de Guanajuato, Lascuráin de Retana No. 5, Col. Centro, Guanajuato C.P. 36000, Mexico; jmoreno@ugto.mx
7. CONACYT-Instituto de Metalurgia, Universidad Autónoma de San Luis Potosí, Sierra Leona 550, Lomas 2a Sección, San Luis Potosí C.P. 78210, Mexico
* Correspondence: luis.rivera@uaslp.mx or jlhri10@yahoo.com.mx

Received: 31 December 2019; Accepted: 28 January 2020; Published: 1 February 2020

Abstract: A novel medium-Mn steel microstructure with 0.1 wt.% Nb was designed using Thermo-Calc and JMatPro thermodynamic simulation software. The pseudo-binary equilibrium phase diagram and time–temperature transformation (TTT) and continuous cooling transformation (CCT) diagrams were simulated in order to analyze the evolution of equilibrium phases during solidification and homogenization heat treatment. Subsequently, the steel was cast in a vacuum induction furnace with the composition selected from simulations. The specimens were heat-treated at 1200 °C and water-quenched. The results of the simulations were compared to the experimental results. The microstructure was characterized using optical microscopy (OM) and scanning electron microscopy (SEM). We found that the as-cast microstructure consisted mainly of a mixture of martensite, ferrite, and a low amount of austenite, while the microstructure in the homogenization condition corresponded to martensite and retained austenite, which was verified by X-ray diffraction tests. In order to design further production stages of the steel, the homogenized samples were subjected to hot compression testing to determine their plastic flow behavior, employing deformation rates of 0.083 and 0.83 s^{-1}, and temperatures of 800 and 950 °C.

Keywords: advanced high-strength steels (AHSS); medium-Mn steel; phase equilibrium

1. Introduction

The use of advanced high-strength steels (AHSS) in automotive parts has increased in recent years as they have been designed to meet the requirements of greater resistance while retaining their performance and conformability [1]. The AHSS are classified into three generations, according to their mechanical properties, mainly considering their ultimate tensile strength (UTS) and elongation

(El.%) [2]. The first generation of AHSS reaches a UTS of more than 600 MPa, but with a ductility below 20%, limiting its formability. The second generation of AHSS reaches excellent formability combined with resistance above 700 MPa and a ductility higher than 50%, but its high alloying elements content (>17 wt.%) increases the cost and hinders its automotive application [3]. The requirements to increase both resistance (UTS > 1200 MPa) and ductility (>30%) and reduce the alloying elements content and the associated cost have aroused interest in the development of the third generation of AHSS.

One type of steel that covers the combination of properties attributable to the third generation is the so-called "medium-Mn steel" [4]. This steel has a concentration of Mn in the range of 4–12 wt.% [5]. This steel is processed by intercritical annealing between the Ac_1 and Ac_3 intercritical temperatures, which is also named austenite-reverted-transformation (ART) annealing. As a consequence of ART annealing, the microstructure consists mainly of a large fraction of retained austenite (γ_R), ultrafine ferrite, and a small amount of martensite, which are responsible for the excellent mechanical properties that are attributed to medium-Mn steel. ART annealing is critical to obtain the retained austenite because the temperature and time strongly affect its volume fraction and mechanical stability [6–8], which triggers the phenomenon of twinning-induced plasticity (TWIP) and/or deformation transformation-induced plasticity (TRIP) during tensile deformation, improving the formability of AHSS [9,10]. Recently, Dong-Woo and Sung-Joon summarized some representative mechanical properties and austenite volume fractions of medium-Mn steels reported in the literature [11]. The ultimate tensile strength (UTS), total elongation (T. El), and austenite volume fraction (γ_R) are in the range of 808-1409 MPa, 15–58%, and 8–78%, respectively.

Although tensile properties such as tensile strength (UTS) and percentage of elongation (El.%) of the medium-Mn steels are extraordinary, an enhancement in the mechanical properties of these steels is still sought. In this sense, the use of micro-alloying elements such as Nb, V, Mo, and Ti has been recently used for this purpose [12–14]. However, the number of published studies on the effect of micro-alloying elements in medium-Mn steels is currently limited.

In addition, the deformation behavior of this new steel grade at high temperature has not been extensively studied. On an industrial scale, hot deformation processes such as rolling are carried out in austenitic and austenitic–ferritic phase regions in the interval between 800 and 1100 °C. For this reason, it is important to understand the microstructure evolution and plastic flow behavior around these temperatures. This will aid the design and optimize hot rolling process parameters such as the rolling temperature range, load requirements, strain rate, and pass reduction schedule. Nevertheless, previous studies have been mainly focused on the hot flow behavior of medium-Mn steel without micro-alloying elements [15–17].

Moreover, computer simulations have become indispensable for efficiently designing and developing these steels with specific properties for automotive applications. In this sense, it is essential to use simulated diagrams that allow for the study of the thermodynamic stability of the phases and their transformations, thus selecting the optimal route for heat treatment to achieve a given set of properties. Consequently, the simulated diagrams obtained through software such as Thermo-Calc® and JMatPro® have been widely used for this purpose [18,19].

Therefore, the objective of this investigation is to analyze the stable phases at room temperature in as-cast and homogenized states by thermodynamic simulation software in conjunction with microstructural characterization and the influence of deformation temperatures (850 and 900 °C) and strain rates (0.08 and 0.8 s^{-1}) on the hot flow behavior of medium-Mn steel micro-alloyed with Nb. For this purpose, Fe–C phase diagrams and property diagrams were calculated, and the predicted microstructure was qualitatively validated by experimental methods. Thereafter, hot compression tests were done to test whether Nb would exhibit a refining effect in the microstructure of this steel, as has been reported previously in high-strength low-alloy (HSLA) steels. This model will be used in the simulation of further stages of the hot working processing of this steel.

2. Materials and Methods

Thermodynamic calculation was performed to analyze the equilibrium phases and the corresponding property diagram of medium-Mn steel by using the TCFE9 database of the Thermo-Calc software v2018b (Thermo-Calc Software Inc., Solna, Sweden) [20,21] for the following nominal composition: Fe-0.14C-1.5Al-1.7Si-6.5Mn-0.1Nb (wt.%). Afterward, the medium-Mn steels were produced in a vacuum induction melting furnace. Rectangular bars of 100 mm × 100 mm and 250 mm in length were obtained after a casting and solidification process. Subsequently, square samples of cast steel of 30 mm × 30 mm with a thickness of 5 mm were homogenized in a tube furnace at 1200 °C for 2 h followed by water quenching. Thereafter, the Rockwell hardness of the as-cast and homogenized samples was measured. The average value was calculated from 10 indentations. The homogenized steel samples were machined into cylindrical pieces 10 mm high and 5 mm in diameter to carry out the hot compression tests. These tests were performed in the universal testing machine Instron model 1100 of 60 tons of capacity (Instron corporation, Norwood, MA, USA). Hot compression tests were performed with strain rates of 0.083 and 0.83 s^{-1} and temperatures of 800 and 950 °C, using TiO_2 as a lubricant.

The metallographic analysis was used to characterize the as-cast and homogenized alloy. Optical microscopy (OM) and scanning electron microscopy (SEM) were used to observe the microstructures of the steel. The samples for optical and electron microscopy were color-etched as follows: Pre-etching for 1–2 s in Nital solution followed by etching for 20 s in sodium metabisulfite solution. The microstructural examinations were performed on an Olympus GX51-inverted reflected light microscope (Olympus Corporation, Tokyo, Japan) and a JEOL-6610LV scanning electron microscope at 30 kV (JEOL Ltd., Tokyo, Japan).

The phases present in the as-cast and homogenized conditions were determined by X-ray diffraction using the diffractometer Rigaku D/Max2200 (Rigaku Corporation, Tokyo, Japan) employing a Cu-Kα radiation source (λ = 1.5405 Å). The scanning range and step size were 30–100°, 1 s/step, and 0.02°, respectively. The MAUD software v2.92 (Luca Lutterotti, University of Trento, Italy).

Join institution (based on the Rietveld refinement method) was used for quantitative analysis by using the PDF-2/Release 2010 database. According to the powder diffraction file (PDF) database of the International Centre for Diffraction Data (ICDD), the PDF card number of austenite and ferrite phases was 00-052-0513 and 00-006-0696, respectively.

JMatPro software7.0.0 (Sente Software Ltd., Guildford, UK) was used to calculate the time–temperature transformation (TTT) and continuous cooling transformation (CCT) diagrams, to explain the evolution of the phases as a function of both time and temperature parameters during solidification and homogenization processes. It was necessary to evaluate the previous austenite grain size and the chemical composition in the as-cast state in order to build these diagrams. The CCT diagram was calculated at 1400 °C because, at that temperature, the steel had completely solidified while the TTT diagram was calculated at 1200 °C, because the homogenization heat treatment was performed at that temperature.

3. Results and Discussion

3.1. Thermodynamic Calculations

In order to predict the microstructure of the as-cast medium-Mn steel alloyed with Nb, its pseudo-binary phase diagram was simulated using Thermo-Calc software and is plotted in Figure 1a. The diagram predicts that the delta ferrite (δ) phase is formed from the liquid when the temperature decreases to around 1460 °C. As the cooling progresses, the solidification is completed at a temperature of 1400 °C, and below this temperature, the delta ferrite (δ) phase begins to transform into the austenite (γ) phase. Around 1300 °C, the transformation of the δ to γ phase is completed and the formation of the NbC phase begins. The formation of NbC ends when the temperature decreases to about 840 °C. Starting from this temperature and until it reaches about 620 °C, a part of the γ phase dissolves to form ferrite (α), resulting in a phase-field where γ + α + NbC coexists. Below 620 °C, the transformation of γ to cementite occurs. Around 460 °C, the cementite is transformed into $M_{23}C_6$ carbides and at 340 °C, the remaining γ is transformed into the β-Mn phase. Then, in the range from 200 °C to room temperature, the stable phases are α + $M_{23}C_6$ + NbC + β-Mn. In the range from 200 °C to room temperature, the β-Mn phase transforms to α-Mn. Finally, the stable phases at room temperature are α + $M_{23}C_6$ + NbC + α-Mn.

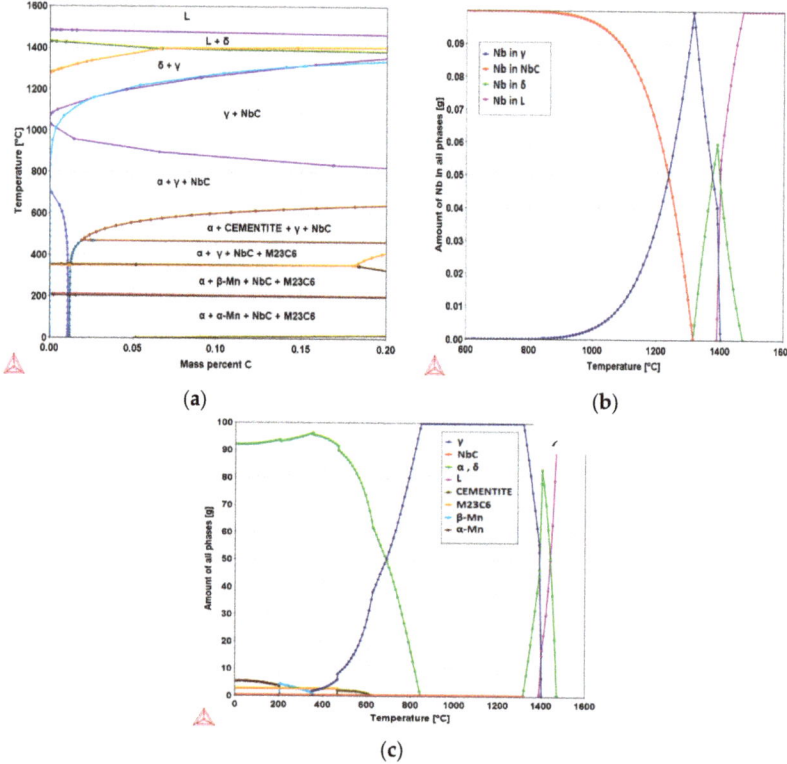

Figure 1. Simulated diagrams obtained by using Thermo-Calc software: (**a**) Pseudo-binary phase diagram, (**b**) amount of niobium in the phases as a function of temperature, and (**c**) amount of phases as a function of temperature.

In Figure 1b, it is predicted that in the stable region of austenite, from 840 to 1380 °C, Nb is present as a solute or as an NbC phase. The presence of Nb in both states affects the high-temperature thermomechanical processing in the austenite region. Solute Nb retards austenite recrystallization and grain growth after deformation through the solute drag effect. In addition, solute Nb delays the transformation kinetics from γ to α. On the other hand, the NbC phase can serve as additional ferrite nucleation sites and increases the rate of transformation from γ to α.

Furthermore, the temperatures Ac_1 and Ac_3 (intercritical temperatures calculated by Thermo-Calc) are 620 and 840 °C, respectively, and the temperature at which the volume fraction of austenite and ferrite is equal is 714 °C, as shown in Figure 1c. Defining these temperatures will serve to optimize the further processing by the application of the intercritical annealing process as the annealing temperature and time determine the microstructure, specifically, the volume fraction of retained austenite ($γ_R$) and, in turn, the mechanical properties of the steel. However, it is important to note that some equilibrium phases at lower temperatures might not form in the as-cast microstructure due to a cooling rate, unlike the thermodynamic equilibrium experiment during solidification. In order to deepen the understanding regarding the solidification and homogenization according to the cooling rate, JmatPro software was employed and the CCT and TTT transformation diagrams were built (Figure 2).

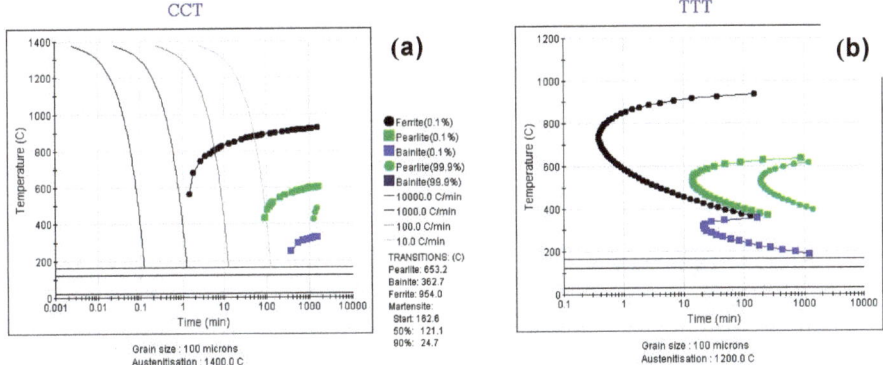

Figure 2. Diagrams simulated by using JMatPro software: (**a**) Transformation diagram in continuous cooling (CCT) and (**b**) time–temperature transformation diagram (TTT).

The cooling rate during solidification was close to 10 °C/min, which was measured using a thermocouple attached to the ingot mold. As shown in Figure 2a, this cooling rate crossed the start of ferrite formation and shows that neither perlite nor bainite could form in the sample. Therefore, according to the CCT diagram, it is expected that the microstructure in the as-cast state only consisted of martensite and ferrite. On the other hand, as shown in Figure 2b, at approximately 750 °C and 30 s, ferrite formation started, so the microstructure after the homogenization heat treatment is expected to be formed only by martensite and probably some retained austenite as the cross-section of steel samples was smaller than 5 mm and the quenching process was done in less than 15 s.

3.2. Microstructural Characterization of As-Cast and Homogenized Samples

X-ray diffractograms measured in the as-cast and homogenization condition are shown in Figure 3a,b. It can be noted that, in the first condition, there are several strong peaks from ferrite/martensite phases that formed during solidification. Weak diffraction signals are observed from the austenite phase in the corresponding diffractogram. On the other hand, it is evident that the austenite peak intensity increased after the steel was homogenized and quenched, while the intensity of the peaks from the ferrite/martensite phases decreased slightly. According to the thermodynamic calculation shown in Figure 2b, the microstructure was composed of martensite after homogenization and quenching. However, due to the segregation present in the microstructure, even after homogenization, there are regions in which the austenite phase was retained.

A quantitative analysis was performed using MAUD software, which is based on the Rietveld refining method [22]. It is shown that the austenite content in the as-cast condition was 2% (sigma = 1.49, Rwp (%) = 49.03), while this amount increased to 20% (sigma = 1.27, Rwp (%) = 39.03), after homogenization and quenching. Although austenite is a softer phase compared to martensite, the increment in the amount of the former phase after homogenization and water quenching was not reflected in the hardness of the steel, which was 5.03 GPa (50 HRc) in the as-cast condition and 5.83 GPa (55 HRc) after homogenization and quenching. This increment in hardening can be attributed to the dissolution and re-distribution of all alloying elements in the austenite matrix during the homogenization treatment, as it is shown in Figure 4, which was calculated by Thermo-Calc simulations. Consequently, solid solution strengthening had a higher effect in martensite formed from the homogenization compared to the same phase in the as-cast state as Mn, Al, and Si were completely dissolved at 1200 °C.

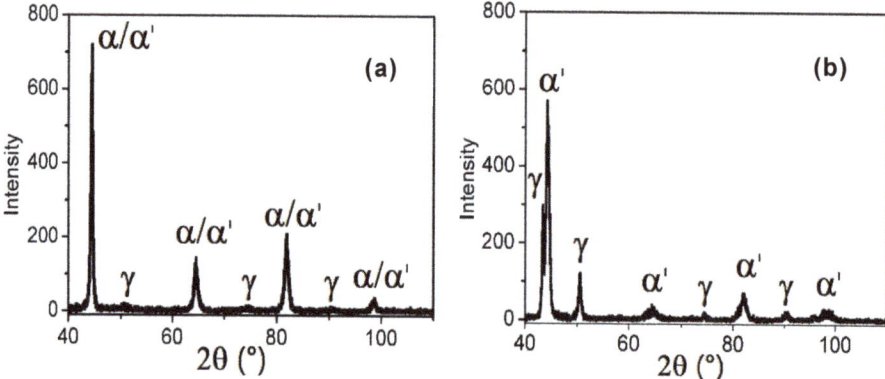

Figure 3. X-ray diffractograms of (a) as-cast and (b) homogenized samples.

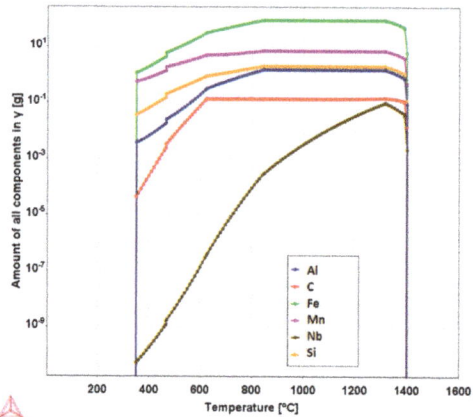

Figure 4. Distribution of alloying elements in the austenite matrix as a function of temperature.

The microstructure was characterized before and after homogenization heat treatment. Figure 5a,b and Figure 5c,d are optical and scanning electron microscope images showing the typical microstructure of the as-cast condition. Figure 5a,b show that the microstructure was composed of martensite with some off-white ferrite in the grain boundaries. Figure 5c shows that the morphology of this martensite was thin laths. Inside the laths, some carbide precipitates were observed (Figure 5d); presumably, these carbides can be of type $Mn_{23}C_6$ or cementite according to Thermo-Calc predictions (Figure 1b). Recently, the former type was observed in 0.086%C–8.05%Mn–0.138%Si–0.0215%Al–Fe (wt.%) in the as-cast condition [19] and the latter has been reported to occur in GX12CrMoVNbN9-1 (GP91) cast steel [23]. When there is carbide precipitation in martensite laths, usually as small bars or plates morphologies, the phenomenon has been reported in the literature as auto-tempered martensite [24,25].

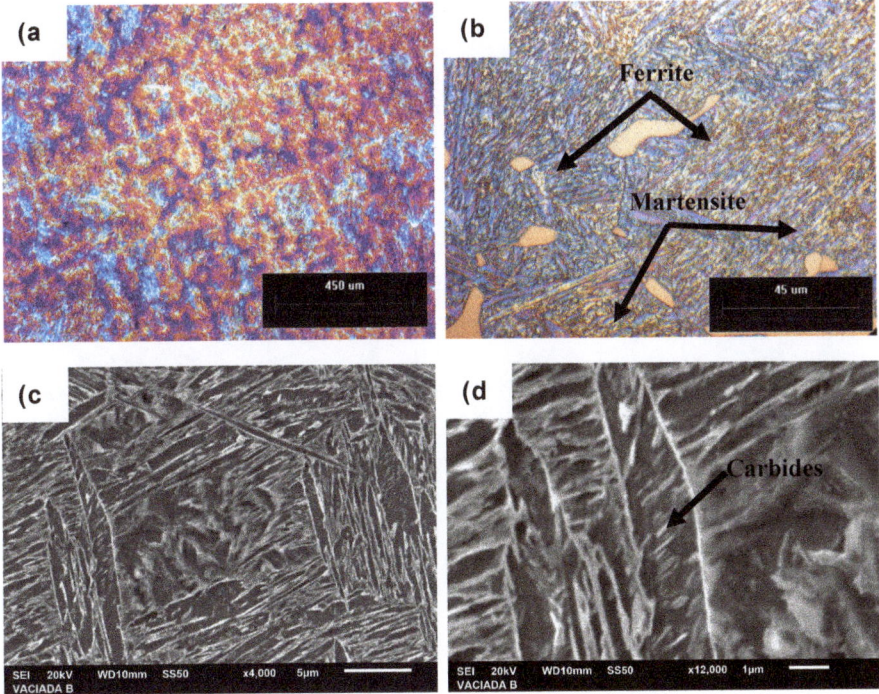

Figure 5. (**a,b**) Optical and (**c,d**) scanning electron microscope images showing the microstructure of as-cast condition.

At the same magnification, in Figure 6a,b and Figure 6c,d the microstructure of the steel is shown after homogenization and water quenching. Figure 6a,b show that the microstructure consisted of lath martensite in addition to retained austenite. On the other hand, there is no ferrite evidence in the microstructure. The absence of this phase is the result of the cooling rate (water quenching), considering that it was high enough to avoid the formation of this phase in agreement with the CCT diagram. The above results matched well with the diagrams simulated via JMatPro® and X-ray diffraction tests. Likewise, in Figure 6c, martensite laths are observed, but the presence of carbides inside the laths is not evident. As the homogenization was performed at 1200 °C, this temperature allowed M23C6 and/or cementite formed in the as-cast state to dissolve in the austenite, resulting in a greater amount of carbon available to incorporate in solid solution into this phase. On the other hand, in Figure 6d,e, the presence of NbC carbides in the austenite, previous grain boundaries, and their corresponding energy-dispersive X-ray (EDX) spectrum can be seen. The carbide evidence coincides with Thermo-Calc simulations, as according to Figure 1b, at 1200 °C, the Nb is both in the form of carbide and solid solution in the austenite matrix.

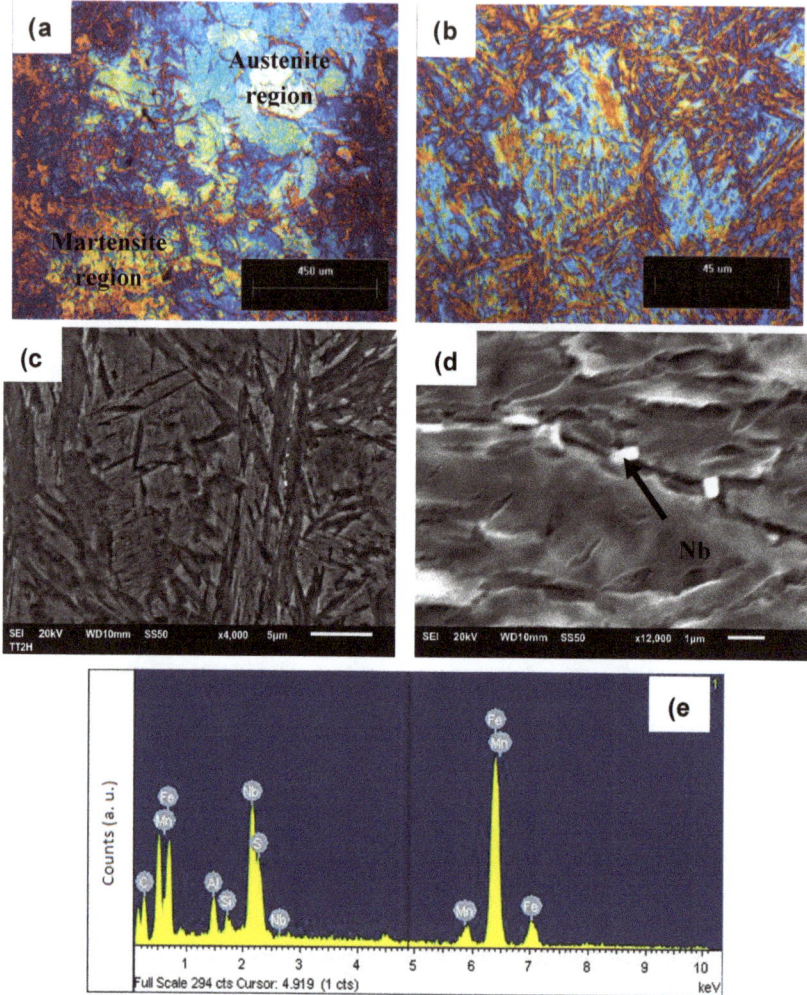

Figure 6. (**a**,**b**) Optical and (**c**,**d**) scanning electron microscope images showing the microstructure of the homogenized condition; (**e**) energy-dispersive X-ray (EDX) spectra of Nb carbides.

Figure 7 shows the results of the characterization of inclusions of the steel in the as-cast condition by means of the energy-dispersive X-ray (EDX) technique. The measurements confirmed the presence of coarse particles of aluminum nitrides (AlN) (Figure 7a) and small particles of manganese sulfides (MnS) (Figure 7b).

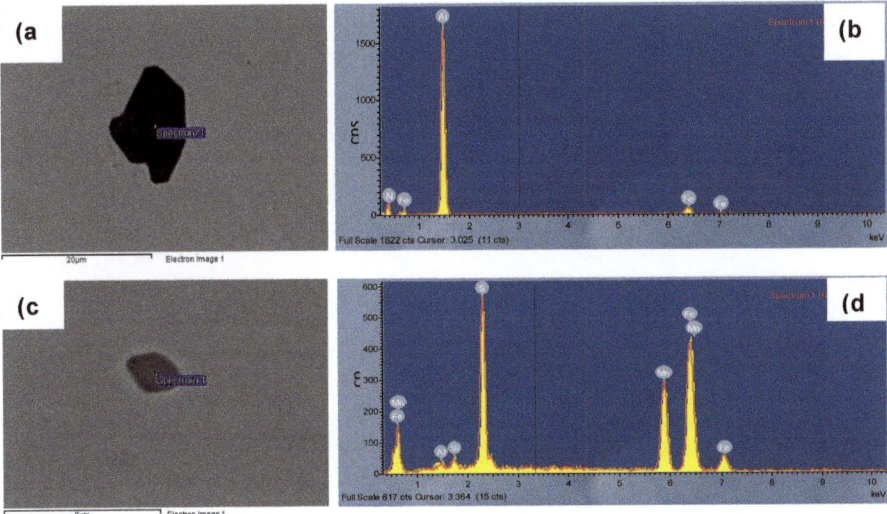

Figure 7. Micrographs of SEM and energy-dispersive X-ray (EDX) spectra of the inclusions present in the steel in casting condition. (**a**) AlN, (**c**) MnS, and (**b**,**d**) EDX spectra of inclusions.

3.3. The High-Temperature Flow Stress Behavior

As mentioned earlier, hot deformation is carried out on an industrial scale in the temperature range from 800 to 1100 °C. For this reason, hot compressions were carried out considering both γ + NbC and $\gamma + \alpha$ + NbC phase regions to observe the effect of these phases on the hot plastic flow behavior. Therefore, Figure 8 shows the high-temperature flow curves of the medium-Mn steels with 0.1% Nb at two different deformation temperatures (800 °C and 950 °C) and two strain rates (0.083 and 0.83 s^{-1}). All the curves show a similar behavior where the flow stress increases with decreasing deformation temperature and increasing strain rate. In addition, the curves show a steady-state flow stress after they reach the highest stress. This behavior is characteristic of the softening process called dynamic recovery, which may occur during deformation at high temperatures. The main reason for this behavior is that a higher strain rate gives rise to a higher dislocation density, which will promote a higher stored energy and flow stress, while a higher deformation temperature weakens the resistance to dislocation motion, consequently reducing work hardening. These results are consistent with previous research about the effect of Nb on hot plastic flow behavior [26,27]. On the other hand, the results do not match at low strain rates [26,27], because, in contrast to the aforementioned investigations, we did not observe dynamic recrystallization evidence (DRX).

This disagreement may be associated with the presence of Nb, because, as predicted by Thermo-Calc calculations (Figures 1b and 4), in the stable region of austenite (840–1380 °C), this micro-alloying element is present both in the form of carbides and as Nb in solution in austenite. According to Thermo-Calc, the amount of Nb as solute and as NbC at 800 °C is close to 0.2 and 0.8 wt.%, respectively, whereas at 950 °C, it is approximately 0.4 and 0.6 wt.%, respectively.

Xiao et al. [28] suggested that the Nb in solution has a greater inhibiting effect on austenite dynamic recrystallization than NbC during hot deformation. However, it should be mentioned that, at present, there are very few published studies regarding the effect of niobium in medium-manganese steels during its thermomechanical processing [29].

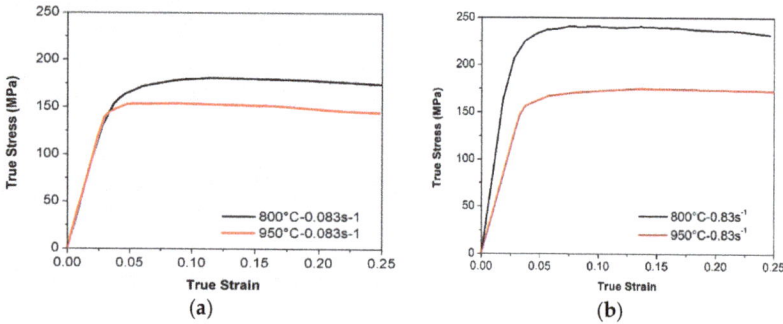

Figure 8. Hot flow curves of medium-Mn steel with 0.1% Nb analyzed at two temperatures (800 and 950 °C); and constant strain rates of (a) 0.083 and (b) 0.83 s^{-1}.

The modified Johnson–Cook model was selected to develop the hot constitutive equation of the analyzed steel [30]. This constitutive equation will aid further simulation by the finite element of the thermomechanical forming processes to which the material will be subjected. The modified Johnson–Cook model was proposed by Lin et al. [30]:

$$\sigma = \left(A_1 + B_1\varepsilon + B_2\varepsilon^2\right)\left(1 + C_1 \ln \dot{\varepsilon}^*\right)\exp\left[\left(\lambda_1 + \lambda_2 \ln \dot{\varepsilon}^*\right)\left(T - T_{ref}\right)\right], \quad (1)$$

where A_1, B_1, B_2, C_1, λ_1, λ_2 are material constants, σ is the equivalent stress, ε is the equivalent plastic strain, and $\dot{\varepsilon}^* = \frac{\dot{\varepsilon}}{\dot{\varepsilon}_{ref}}$ is the dimensionless strain rate. ε_{ref} and T_{ref} are the reference strain and reference deformation temperature, respectively. In this experiment, the reference strain and reference temperature were 0.83 s^{-1} and 800 °C, respectively, to determine the material constants in Equation (1).

To determine the A_1, B_1, B_2 constants, the reference temperature T_{ref} = 800 °C and reference deformation ε_{ref} = 0.83 s^{-1} were substituted into Equation (1), which transforms to

$$\sigma = A_1 + B_1\varepsilon + B_2\varepsilon^2. \quad (2)$$

Substituting stress and strain data under the deformation conditions, a $\sigma \sim \varepsilon$ curve was drawn, two-order polynomial fitting was conducted, and the values of A_1, B_1, and B_2 were determined to be 232.92, 131.87, and −551.35 MPa, respectively.

The constant C_1 was determined by substituting the reference temperature T_{ref} = 800 °C into Equation (1), which transforms to

$$\sigma = \left(A_1 + B_1\varepsilon + B_2\varepsilon^2\right)\left(1 + C_1 \ln \dot{\varepsilon}^*\right). \quad (3)$$

Rearranging Equation (3):

$$\frac{\sigma}{A_1 + B_1\varepsilon + B_2\varepsilon^2} = 1 + C_1 \ln \dot{\varepsilon}^*. \quad (4)$$

At two different strain rates of 0.083 and 0.83 s^{-1}, the stress values corresponding to eight strain values of 0.06, 0.07, 0.08, 0.09, 0.10, 0.15, 0.20, and 0.25 were selected. A total of 16 groups of data were gained. Substituting these data and A_1, B_1, and B_2 into Equation (4), a $\frac{\sigma}{A_1+B_1\varepsilon+B_2\varepsilon^2} \sim \ln \dot{\varepsilon}^*$ curve was drawn, and it was linearly fitted, obtaining $C_1 = 0.1208$.

Then, to determine λ_1 and λ_2 constants, the equation was rearranged in the following form:

$$\frac{\sigma}{\left(A_1 + B_1\varepsilon + B_2\varepsilon^2\right)\left(1 + C_1 \ln \dot{\varepsilon}^*\right)} = \exp\left[\left(\lambda_1 + \lambda_2 \ln \dot{\varepsilon}^*\right)\left(T - T_{ref}\right)\right]. \quad (5)$$

Natural logarithms were taken on both sides of Equation (5), obtaining

$$\ln\left\{\frac{\sigma}{(A_1 + B_1\varepsilon + B_2\varepsilon^2)(1 + C_1 \ln \dot{\varepsilon}^*)}\right\} = \left[(\lambda_1 + \lambda_2 \ln \dot{\varepsilon}^*)(T - T_{ref})\right]. \tag{6}$$

At two different strain rates of 0.083 and 0.83 s^{-1} and two different deformation temperatures of 800 and 900 °C, stress values corresponding to the eight above-mentioned strains were taken. A total of 32 groups of data were obtained. Substituting these data and A_1, B_1, B_2, and C_1 into Equation (6), $\ln\left\{\frac{\sigma}{(A_1+B_1\varepsilon+B_2\varepsilon^2)(1+C_1 \ln \dot{\varepsilon}^*)}\right\} \sim (T - T_{ref})$ curves were drawn under the two strain rates, and a linear fitting was done. The values of $\lambda_1 + \lambda_2 \ln \dot{\varepsilon}^*$ under strain rates of 0.083 and 0.83 s^{-1} were determined to be −0.0022 and −0.0010, respectively.

Finally, a $\lambda_1 + \lambda_2 \ln \dot{\varepsilon}^* \sim \ln \dot{\varepsilon}^*$ curve was drawn and fit to a line, obtaining $\lambda_1 = -0.0022$ and $\lambda_2 = -0.0005$ from this fitting procedure.

Therefore, the constitutive equation that relates the stress σ, strain ε, deformation rate $\dot{\varepsilon}$, and deformation temperature T was established according to the modified Johnson–Cook model:

$$\sigma = \left(232.92 + 131.87\,\varepsilon - 551.35\,\varepsilon^2\right)\left(1 + 0.1208 \ln \frac{\dot{\varepsilon}}{\dot{\varepsilon}_{ref}}\right)\exp\left[\left(-0.0022 - 0.0005 \ln \frac{\dot{\varepsilon}}{\dot{\varepsilon}_{ref}}\right)(T - T_{ref})\right]. \tag{7}$$

The comparison between experimental flow stress values and predicted values by the modified Johnson–Cook model is shown in Figure 9a,b. It can be observed that the predicted and experimental values displayed minor deviations; however, it has acceptable relative accuracy, which was demonstrated by the calculation of the correlation coefficient (R) and mean absolute relative error (MARE).

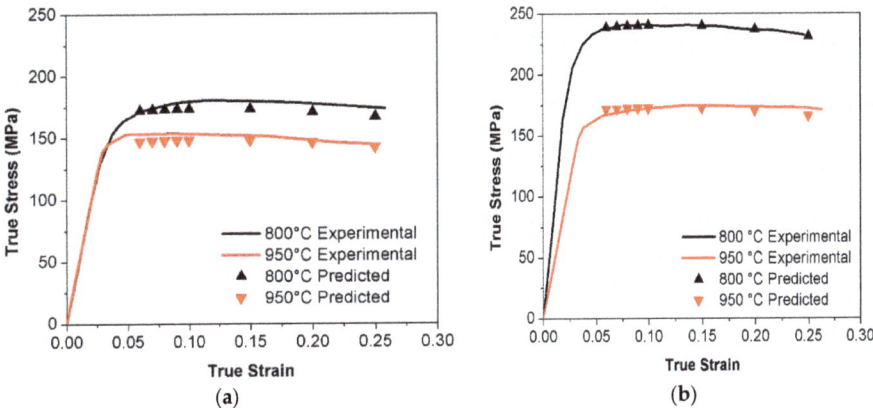

Figure 9. Comparison between experimental and predicted values by the modified Johnson–Cook model at two temperatures (800 and 950 °C); and constant strain rates of (a) 0.083 and (b) 0.83 s^{-1}.

The correlation coefficient (R) and mean absolute relative error (MARE) were used to provide information on the accuracy and effectiveness of the linear relationship between experimental and predicted values and measuring the predictability of a numerical model, respectively, as it is shown on Figure 10. The correlation coefficient was 0.9962 and the corresponding mean absolute relative error was 1.43%. Finally, it is important to point out that this model can only be used effectively for the simulation of the mechanical behavior of medium-Mn steel with 0.1 wt.% Nb in the range of the strain rate and deformation temperature conditions under which it was established [30].

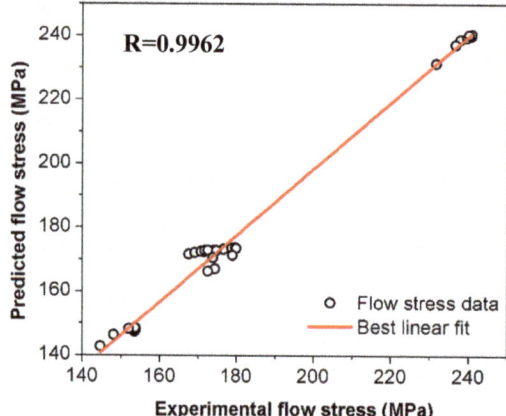

Figure 10. The proximity of experimental and predicted values by the modified Johnson–Cook model.

4. Conclusions

(1) We found that the microstructure of the medium-Mn steel with 0.1 wt.% Nb consisted of lath martensite, ferrite, and small retained austenite in the as-cast condition, whereas, in the homogenized condition, the microstructure was lath martensite and retained austenite. The volume fraction of retained austenite in the as-cast and homogenized conditions was 2 and 20%, respectively.

(2) The thermodynamic simulation of the equilibrium diagram allowed the selection of homogenization parameters and the predicted precipitation of carbides of the type $M_{23}C_6$ and/or NbC. Experimental observations confirmed the existence of these carbides in the homogenized condition along the previous austenite grain boundaries.

(3) We found that Nb delayed recrystallization during hot plastic deformation as the stress–strain curves did not show any peak stress and only showed a steady-state flow stress value after reaching the highest stress. These results were associated with the NbC precipitates, as well as Nb in solid solution.

(4) The constitutive equation established by the modified Johnson–Cook model and its corresponding parameters were calculated by fitting the experimental data, and the accuracy of the constitutive equation obtained was verified by using the correlation coefficient (R) and mean absolute relative error (MARE), which were 0.9962 and 1.42%, respectively.

Author Contributions: Formal analysis, J.M.-P.; methodology, P.G.-G. and J.L.H.-R.; resources, M.O.R.-A., V.G.-F. and J.d.J.C.-R.; Supervision, J.L.H.-R.; Validation, J.S.P.-C.; writing—original draft, P.J.C.V.; Writing—review & editing, V.G.-F. and J.L.H.-R. All authors have read and agreed to the published version of the manuscript.

Funding: This research received no external funding.

Acknowledgments: The authors appreciate the invaluable technical support provided by Rosalina Tovar, Alfredo Nuñez, Francisco Nuñez, Jesus Flores Sandoval, and Claudia Guadalupe Elias Alfaro from Instituto de Metalurgia. JLHR acknowledges CONACYT for supporting the Institutional Program "CÁTEDRAS CONACYT 2014" under the project "Fortalecimiento de las líneas de Investigación del Instituto de Metalurgia" number 2198.

Conflicts of Interest: The authors declare no conflict of interest.

References

1. Abey, A. Metallic Material Trends in the North American Light Vehicle. In Proceedings of the Great Designs in Steel Seminars, Livonia, MI, USA, 13 May 2015.
2. Singh, S.; Nanda, T. A Review: Production of Third Generation Advance High Strength Steels. *IJSRD* **2014**, *2*, 388–392.
3. Cai, M.; Di, H. Advanced High Strength Steels and Their Processes. In *Rolling of Advanced High Strength Steels: Theory, Simulation and Practice*, 1st ed.; Zhao, J., Jiang, Z., Eds.; CRC Press: Boca Raton, FL, USA, 2017; pp. 1–25.
4. Fonstein, N. Evolution of Strength of Automotive Steels to Meet Customer Challenges. In *Advanced High Strength Sheet Steels*; Springer: Berlin, Germany, 2015; pp. 1–14.
5. Fonstein, N. Candidates for the Third Generation: Medium Mn Steels. In *Advanced High Strength Sheet Steels*; Springer: Berlin, Germany, 2015; pp. 297–325.
6. Shao, C.; Hui, W.; Zhang, Y.; Zhao, X.; Weng, Y. Microstructure and mechanical properties of hot-rolled medium-Mn steel containing 3% aluminum. *Mater. Sci. Eng. A* **2017**, *682*, 45–53. [CrossRef]
7. Bansal, G.K.; Madhukar, D.A.; Chandan, A.K.; Ashok, K.; Mandal, G.K.; Sirvastava, V.C. On the intercritical annealing parameters and ensuing mechanical properties of low-carbon medium-Mn steel. *Mater. Sci. Eng. A* **2018**, *733*, 246–256. [CrossRef]
8. Li, Z.C.; Ding, H.; Misra, R.D.K.; Cai, Z.H. Deformation behavior in cold-rolled medium-manganese TRIP steel and effect of pre-strain on the Lüders bands. *Mater. Sci. Eng. A* **2017**, *679*, 230–239. [CrossRef]
9. Lee, S.; De Cooman, B.C. Tensile Behavior of Intercritically Annealed Ultra-Fine Grained 8% Mn Multi-Phase Steel. *Steel Res. Int.* **2015**, *86*, 1170–1178. [CrossRef]
10. Latypov, M.I.; Shin, S.; Cooman, B.C.D.; Kim, H.S. Micromechanical finite element analysis of strain partitioning in multiphase medium manganese TWIP + TRIP steel. *Acta Mater.* **2016**, *108*, 219–228. [CrossRef]
11. Suh, D.-W.; Kim, S.-J. Medium Mn transformation-induced plasticity steels: Recent progress and challenges. *Scr. Mater.* **2017**, *126*, 63–67. [CrossRef]
12. Haijun Pan, H.D.M.C. Microstructural evolution and precipitation behavior of the warm-rolled medium Mn steels containing Nb or Nb-Mo during intercritical annealing. *Mater. Sci. Eng. A* **2018**, *736*, 375–382.
13. Hu, J.; Du, L.-X.; Dong, Y.; Meng, Q.-W.; Misra, R.D.K. Effect of Ti variation on microstructure evolution and mechanical properties of low carbon medium Mn heavy plate steel. *Mater. Charact.* **2019**, *152*, 21–35. [CrossRef]
14. Lee, D.; Kim, J.-K.; Lee, S.; Lee, K.; Cooman, B.C.D. Microstructures and mechanical properties of Ti and Mo micro-alloyed medium Mn steel. *Mater. Sci. Eng. A* **2017**, *706*, 1–14. [CrossRef]
15. Yan, N.; Di, H.-S.; Huang, H.-Q.; Misra, R.D.K.; Deng, Y.-G. Hot Deformation Behavior and Processing Mapps of Medium Manganese TRIP Steel. *Acta Metall. Sin. (Engl. Lett.)* **2019**, *32*, 1021–1031. [CrossRef]
16. Klinkenberg, C.; Varghese, A.; Heering, C.; Lamukhina, O.; Grafe, U.; Tokmakov, K. 3rd Generation AHSS by Thin Slab Technology. *Mater. Sci. Forum* **2018**, *941*, 627–632. [CrossRef]
17. Steineder, K.; Dikovits, M.; Beal, C.; Sommitsch, C.; Krizan, D.; Schneider, R. Hot deformation behavior of a 3rd generation advenced high strength steel with a Medium-Mn content. *Mater. Sci. Forum* **2015**, *651–653*, 120–125.
18. Na, H.-S.; Kim, B.-S.; Lee, S.-S.; Kang, C.Y. Thermodynamic Alloy Design of High Strength and Toughness in 300 mm Thick Pressure Vessel Wall of 1.25Cr-0.5Mo Steel. *Metals* **2018**, *8*, 70. [CrossRef]
19. Magalhaes, A.; Moutinho, I.; Oliveira, I.; Ferreira, A.; Alves, D.; Santos, D.B. Ultrafinegrained Microstructure in a Medium Manganese Steel after Warm Rolling without Intercritical Annealing. *ISIJ Int.* **2017**, *57*, 1121–1128. [CrossRef]
20. Djurovic, D.; Hallstedt, B.; Appen, J.; Dronskowski, R. Thermodynamic assessment of the Fe–Mn–C system. *Calphad* **2011**, *35*, 479–491. [CrossRef]
21. Zheng, W.-S.; Lu, X.-G.; He, Y.-L.; Li, L. Thermodynamic modeling of Fe-C-Mn-Si alloys. *J. Iron Steel Res. Int.* **2017**, *24*, 190–197. [CrossRef]
22. Young, R.A. *The Rietveld Method*; Oxford University Press: Oxford, UK, 1993; p. 298.
23. Golański, G. Mechanical properties of GX12CrMoVNbN91 (GP91) cast steel after different heat treatments. *Mater. Sci.* **2012**, *48*, 384–391. [CrossRef]

24. Zajac, S.; Schwinn, V.; Tacke, K.-H. Characterisation and Quantification of Complex Bainitic Microstructures in High and Ultra-High Strength Linepipe Steels. *Mater. Sci. Forum* **2005**, *500–501*, 387–394. [CrossRef]
25. Bhadeshia, H.K.D.H.; Honeycombe, S.R. 9-The Tempering of Martensite. In *Steels*, 3rd ed.; Bhadeshia, H.K.D.H., Honeycombe, S.R., Eds.; Butterworth-Heinemann: Oxford, UK, 2006; pp. 183–208.
26. Wei, H.-L.; Liu, G.-Q. Effect of Nb and C on the hot flow behavior of Nb microalloyed steels. *Mater. Des.* **2014**, *56*, 437–444. [CrossRef]
27. Bao, S.; Zhao, G.; Yu, C.; Chang, Q.; Ye, C.; Mao, X. Recrystallization behavior of a Nb-microalloyed steel during hot compression. *Appl. Math. Model.* **2011**, *35*, 3268–3275. [CrossRef]
28. Xiao, F.-R.; Cao, Y.-B.; Qiao, G.-Y.; Zhang, X.-B.; Liao, B. Effect of Nb Solute and NbC Precipitates on Dynamic or Static Recrystallization in Nb Steels. *J. Iron Steel Res. Int.* **2012**, *19*, 52–56. [CrossRef]
29. Speer, J.G.; Araujo, A.L.; Matlock, D.K.; Moor, E. Nb-Microalloying in Next-Generation Flat-Rolled Steels: An Overview. *Mater. Sci. Forum* **2016**, *879*, 1834–1840. [CrossRef]
30. He, A.; Xie, G.; Zhang, H.; Wang, X. A Comparative Study on Johnson–Cook, Modified Johnson–Cook and Arrhenius-Type Constitutive Models to Predict the High Temperature Flow Stress in 20CrMo Alloy Steel. *Mater. Des.* **2013**, *52*, 677–685. [CrossRef]

© 2020 by the authors. Licensee MDPI, Basel, Switzerland. This article is an open access article distributed under the terms and conditions of the Creative Commons Attribution (CC BY) license (http://creativecommons.org/licenses/by/4.0/).

Article

Tracing Microalloy Precipitation in Nb-Ti HSLA Steel during Austenite Conditioning

Johannes Webel [1,2,*], **Adrian Herges** [1], **Dominik Britz** [1,2], **Eric Detemple** [3], **Volker Flaxa** [4], **Hardy Mohrbacher** [5,6] **and Frank Mücklich** [1,2]

[1] Department of Materials Science, Saarland University, Campus D3.3, 66123 Saarbrücken, Germany; adherges@gmx.de (A.H.); d.britz@mx.uni-saarland.de (D.B.); muecke@matsci.uni-sb.de (F.M.)
[2] Materials Engineering Center Saarland, Campus D3.3, 66123 Saarbrücken, Germany
[3] AG der Dillinger Hüttenwerke, 66763 Dillingen, Germany; eric.detemple@dillinger.biz
[4] Engineering Office and Consultant, 38229 Salzgitter, Germany; volker.flaxa@gmail.com
[5] Department of Materials Engineering, KU Leuven, 3001 Leuven, Belgium; hm@niobelcon.net
[6] NiobelCon bvba, 2970 Schilde, Belgium
* Correspondence: j.webel@mx.uni-saarland.de; Tel.: +49-681-302-70518

Received: 23 January 2020; Accepted: 9 February 2020; Published: 12 February 2020

Abstract: The microalloying with niobium (Nb) and titanium (Ti) is standardly applied in low carbon steel high-strength low-alloy (HSLA) steels and enables austenite conditioning during thermo-mechanical controlled processing (TMCP), which results in pronounced grain refinement in the finished steel. In that respect, it is important to better understand the precipitation kinetics as well as the precipitation sequence in a typical Nb-Ti-microalloyed steel. Various characterization methods were utilized in this study for tracing microalloy precipitation after simulating different austenite TMCP conditions in a Gleeble thermo-mechanical simulator. Atom probe tomography (APT), scanning transmission electron microscopy in a focused ion beam equipped scanning electron microscope (STEM-on-FIB), and electrical resistivity measurements provided complementary information on the precipitation status and were correlated with each other. It was demonstrated that accurate electrical resistivity measurements of the bulk steel could monitor the general consumption of solute microalloys (Nb) during hot working and were further complemented by APT measurements of the steel matrix. Precipitates that had formed during cooling or isothermal holding could be distinguished from strain-induced precipitates by corroborating STEM measurements with APT results, because APT specifically allowed obtaining detailed information about the chemical composition of precipitates as well as the elemental distribution. The current paper highlights the complementarity of these methods and shows first results within the framework of a larger study on strain-induced precipitation.

Keywords: niobium-titanium microalloyed steel; electrical resistivity; atom probe tomography; scanning electron microscopy

1. Introduction

The development of weldable low-carbon steels with high mechanical strength and good toughness is the basis for many modern applications in the structural, energy, and automotive sectors. Key to this steel development is the use of microalloying in combination with thermo-mechanical controlled processing (TMCP) [1–4]. Such high-strength low-alloyed (HSLA) steels are being produced as strip and plate products covering a wide range of thicknesses and yield strength levels up to 700 MPa. The involved strengthening mechanisms are in first place grain refinement followed by precipitation and dislocation strengthening. Grain refinement in combination with low carbon content is particularly effective in lowering the ductile-to-brittle transition temperature (DBTT) and increasing the ductile plateau toughness.

Microalloying elements, in particular niobium and titanium, play a major role in achieving these strengthening mechanisms. Grain refinement in the final product starts with preventing excessive austenite coarsening during soaking treatment. Therefore, particles insoluble at that high temperature (>1200 °C), typically consisting of TiN, are required [5]. Niobium ideally is brought into solution by the soaking treatment. It is the most efficient microalloying element for suppressing recrystallization during rolling at lower austenite temperatures [6]. The obstruction of recrystallization is caused by solute drag of niobium atoms segregated to the austenite grain boundary, as well as by grain boundary pinning after strain-induced precipitation of small Nb(C,N) particles [5,7]. Any niobium left in solid solution after austenite conditioning has further metallurgical effects that contribute to strengthening. Solute niobium delays the austenite-to-ferrite transformation provoking under-cooling and thus further refining grain structure, or in combination with accelerated cooling, promoting transformation into bainite [8]. Depending on the finishing conditions, solute niobium can also precipitate as ultra-fine particles in ferrite, enhancing strength according to the Orowan-Ashby mechanism [9]. Therefore, it is of great importance to predict and verify the solute and precipitate status of microalloying elements along the entire process route.

In Nb-Ti dual microalloyed steels, a considerable fraction of the nominal titanium addition precipitates at the late stages of solidification caused by segregation of titanium and nitrogen to the residual melt. Any titanium that has not precipitated during solidification does so at temperatures below 1200 °C, even in undeformed austenite as detailed by Kunze et al. [10]. At higher austenite temperatures, these precipitates nucleate on dislocations, yielding particle rows or tapes where the particle diameters are in the range of 20 to 50 nm. Below 1000 °C, supersaturation becomes so high that nucleation of TiN can also take place in the matrix resulting in randomly distributed particles with sizes below 15 nm [10].

Niobium carbide has quite a good solubility in low-carbon steels. For industrially relevant carbon and nitrogen levels, more than 0.1 wt. % Nb can be dissolved at 1200 °C [11]. The niobium solubility diminishes with decreasing temperature to a value of around 0.01 wt. % at 900 °C. Despite this, sufficient supersaturation allowing spontaneous precipitation of NbC in the matrix is not being reached. However, heterogeneous nucleation of NbC on pre-existing TiN particles is possible. It has been found that up to 0.02 wt. % niobium can be trapped in this way [12,13]. The amount of solute niobium effective for retarding recrystallization is therefore reduced [5,7,14]. This can be overcome by increasing the amount of niobium added to the steel. Higher niobium additions are also motivated by the increase of the temperature of non-recrystallization (T_{NR}) enabling higher finish rolling temperatures and thus, yielding better efficiencies in the mill [15,16].

The precipitation of niobium at lower austenite temperatures, where solubility is sufficiently reduced and supersaturation occurs, is greatly facilitated by introducing deformation [11]. This is shown schematically in Figure 1. The precipitation kinetics reaches a maximum at temperatures between 950 and 900 °C [17,18]. Rolling in that temperature range induces simultaneous precipitation of NbC thereby strongly obstructing the recrystallization of deformed austenite grains. It is also obvious that rolling schedules adopted by hot strip mills result in a much lower amount of niobium precipitation than plate mill rolling schedules.

The niobium left in solute solution after rolling is available for precipitation during or after the phase transformation from austenite to ferrite. These NbC precipitates can nucleate as a consequence of partitioning and local supersaturation in the vicinity of the moving phase front, known as interphase precipitation [19,20]. Spontaneous precipitation after phase transformation can occur by nucleation on dislocations or by a mechanism of replacing iron by niobium atoms in previously formed nano-cementite particles as described by Hin et al. [21]. The kinetics of spontaneous niobium precipitation in ferrite is also relatively slow under typical finishing conditions for plate and strip products (Figure 2) [22]. It is apparent that even under hot finishing conditions, niobium precipitation will not be complete. Colder finishing conditions promoting bainite formation largely prevent niobium precipitation. Secondary heat treatment such as tempering, however, allows nearly complete precipitation [23].

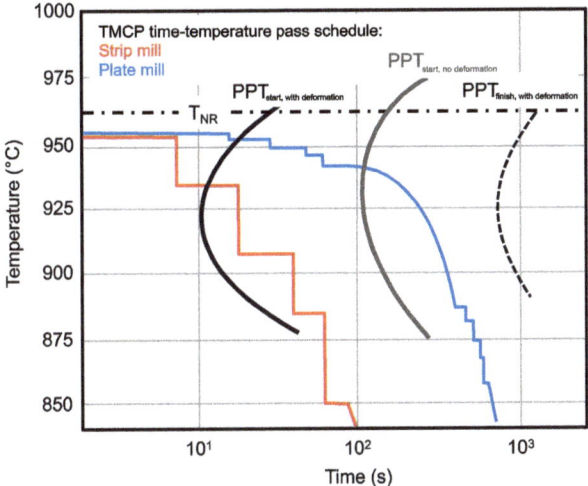

Figure 1. Schematic of rolling pass schedules in the thermo-mechanical controlled processing (TMCP) regime for strip and plate mills with respect to T_{NR} (dashed line) and precipitation (PPT) kinetics.

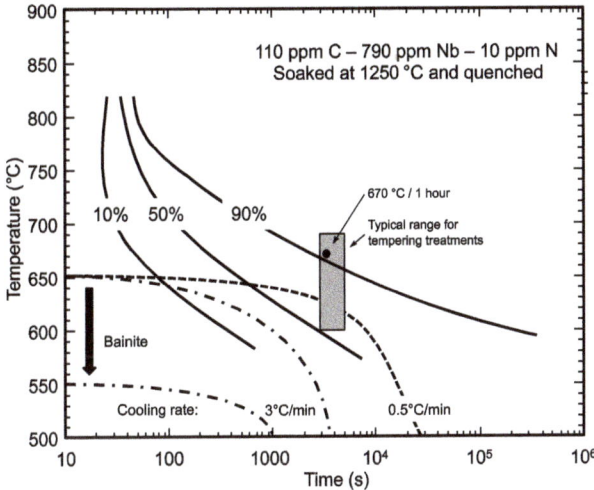

Figure 2. Kinetics of spontaneous NbC precipitation in ferrite in an experimental high-Nb steel after solution annealing and quenching [22]; typical cooling curves and tempering conditions are indicated.

The optimum exploitation of niobium with regard to the aforementioned physical metallurgical effects demands a precise knowledge of the solute niobium status before and after hot deformation. Numerous studies exist on how to calculate the behavior of microalloying elements using thermodynamics, which was recently summarized [24]. Several techniques that either directly or indirectly measure the niobium in solution exist. Early work of LeBon et al. [25] used hardness testing to quantify the amount of precipitated niobium based on the assumption of a higher hardness being due to less niobium in solution. However, such measurements are liable to misinterpretation caused by competing effects on the hardness.

Electrical resistivity measurements of alloys rely on the principle that solute precipitant elements are consumed while precipitation advances. As the iron matrix of the steel is more and more depleted of elements participating in precipitation, the resistivity of the steel drops because solute elements are stronger scattering centers for electrons in comparison to precipitates. Simoneau et al. [26] measured the electrical resistivity of niobium-microalloyed steels isothermally held at 900 °C for different times and found a good correlation with different precipitation stages. Park et al. [27] estimated the dissolution temperature of a Nb-microalloyed steel to be the point where the electrical resistivity reached a plateau and verified it by using transmission electron microscopy (TEM) replica. Jung et al. [28,29] expanded this technique to the more complex Nb-Ti-V system to investigate the precipitation behavior in thermo-mechanically processed microalloyed steels with and without deformation in the austenite field. By applying the lever rule to the resistivity curve, they could show that the precipitation mole fraction of pre-strained samples increases earlier with respect to unstrained samples. These reported results rely on the fact that the upper and lower limit of the electrical resistivity are tied to the total dissolution or precipitation of niobium, respectively. However, for (Nb,Ti)(C,N) steels, the maximum amount of niobium in solution is generally less than the nominal amount due to the above-mentioned reasons. Therefore, in that case, the electrical resistivity curves need to be calibrated on the actual amount of niobium in solution.

One common technique to assess the amount of solute niobium is the extraction of precipitates by dissolving the metallic matrix in acid. Subsequently, filtering the precipitates from the solution and measurement of both filter and filtrate yields the niobium precipitated and in solid solution, respectively [16,30]. This method ensures a very elegant and statistically significant measurement. However, it cannot be excluded that small precipitates either pass through the filter or partially dissolve during the extraction and thereby bias the analysis of solute niobium concentration.

A feasible alternative is given by atom probe tomography (APT), which detects all elements with the same high sensitivity and a spatial resolution in the sub-nanometer range [31]. Therefore, the technique applies especially well to investigation of segregation lines or nanometer-sized precipitates. Furthermore, APT can also measure the chemical composition of the matrix, excluding even the smallest precipitates or clusters inside the composition calculation [32], contrary to other spectroscopy techniques. APT was used to characterize (Nb,Ti)(C,N) particles formed during hot-rolling of microalloyed HLSA steels [33–37]. Nöhrer et al. investigated the influence of deformation level [33], while Kostryzhev and Pereloma et al. investigated the influence of the deformation temperature [34,35] on the precipitation of (Nb,Ti)(C,N) in low-alloyed steels. Due to the small analysis volume of APT measurements, quantitative data of the precipitation number density or size distribution must be treated with caution. (S)TEM characterization, which is often used to obtain counting statistics, is liable to misinterpretation because of the user-dependent production of thin foils or carbon replica films. In addition, depending on the user, there might be a precipitate size range that will be not detected by either APT or TEM. This might be also due to the loss of smallest precipitates during excessive etching when producing the replica, as shown in the case of Kostryzhev et al. [35].

Accordingly, the correlation and calibration of electrical resistivity measurements on the APT analysis of precipitates and matrix offers a high potential for obtaining meaningful quantitative data for the precipitation kinetics of (Nb,Ti)(C,N). This can help to establish a guideline for the heat treatments during steel processing.

In this work, APT was used to detect and differentiate precipitates that form during various TMCP stages, in combination with scanning transmission electron microscopy (STEM-on-FIB). Furthermore, the amount of solute niobium measured by APT for different processing stages was correlated with the electrical resistivity method. It was shown that both, deformation level and temperature during hot working, could be distinguished based on electrical resistivity. Finally, the composition of precipitates formed at lower austenite temperature as well as after transformation to ferrite was characterized with APT.

2. Materials and Methods

The current study was conducted on a laboratory melt of a Nb and Ti dual-microalloyed steel according to the composition as listed in Table 1. Similar compositions are in industrial use for advanced linepipe or structural steel grades either produced as plate or hot-rolled strip. The weight percent ratio of titanium and nitrogen in this analysis has a value of 3.4 corresponding almost exactly to an atomic Ti:N ratio of 1:1.

Table 1. Steel composition.

Element	C	Si	Mn	P	Cr	Mo	Ni	Cu	Al	N	Nb	Ti
wt. %	0.042	0.32	1.69	0.012	0.036	0.021	0.195	0.215	0.033	0.005	0.085	0.017

Cylinders of as-cast material were cut into a dimension of 16 mm length and 8 mm diameter. The sample cylinders were then thermo-mechanically processed in a Gleeble 3800 simulator (Dynamic Systems Inc., Poestenkill, NY, USA), using a hot-compression module with compressed helium gas as quenching medium. A schematic of the different heat-treatment simulations is shown in Figure 3. It consisted of an austenitizing stage at 1200 °C for 10 min, followed by cooling down to the deformation temperatures of 950 °C, 900 °C, and 850 °C, respectively, with a rate of 5 °C/s. Double-hit compression tests ($\varepsilon = 0.3 + 0.3$) with variable inter-pass times (ranging from 2 to 100 s) were performed to determine the non-recrystallization temperature (T_{NR}) and to reveal the softening stasis. Single-hit compression tests ($\varepsilon = 0.3$ or $\varepsilon = 0.6$) and subsequent isothermal holding periods ranging from 2 to 6000 s followed by quenching were performed to characterize strain-induced precipitation. Additional samples were directly quenched from temperatures of 1200 °C and 950 °C without deformation, whereas one sample was held isothermally at 670 °C for 1 h after cooling down from 1200 °C, and subsequently quenched. These treatments aimed at characterizing the precipitation status before austenite deformation as well as after the austenite-to-ferrite transformation.

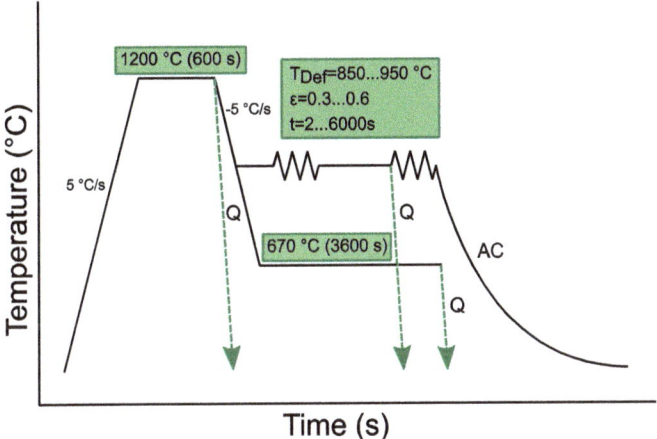

Figure 3. Schematic of the simulated heat treatments. Samples were characterized as-quenched (Q) from austenitization temperature or cooled down to deformation temperature T_{Def} = 950, 900, and 850 °C with varying strain ε, holding times t, and quenching/second compression to measure the softening, and air cooling (AC). In addition, one sample was cooled down from austenitization temperature, aged at 670 °C for 1 h, and quenched afterwards.

For each TMCP-stage described above, cylindrical samples were machined for electrical resistivity measurements by cutting cylinders of length 10 mm and diameter 5.6 mm from the thermo-mechanically deformed samples. The rather short length is due to constraints originating from the maximum cylinder length which could be treated in the Gleeble compression module without buckling. The setup for measuring the electrical resistivity was a 4-point measurement, as described in early works of Simoneau et al. [26]. The voltage drop within fixed potential points on the sample is proportional to the internal resistivity. Oxides or contaminations with organic residue will greatly increase the resistivity and produce erroneous values. Therefore, before measuring, the samples were further ground with 1200 grid sandpaper and cleaned with isopropyl alcohol to establish the most optimal surface conditions. The temperature was maintained at 23 °C for all measurements.

For selected samples, cylinders were also cut along the length and the samples were prepared for metallographic inspection by light-optical microscopy, carbon replica technique and APT. Polishing with 6, 3, and 1 µm diamond suspension and a final oxide polishing using silica slurry (0.05 µm) was conducted as final surface preparation steps. For all investigations, sample sections in the half-length and quarter-width were investigated because at this position the local strain was calculated to be approximately equal to the nominal strain during compression. Metallography for assessing the prior austenite grain structure was conducted using hot aqueous supersaturated picric acid and images were taken with a light-optical microscope Leica DM6000 (Leica Microsystems, Wetzlar, Germany). The prior austenite grain sizes where analyzed by computational image analysis in the software package AxioVision SE64 Rel. 4.9.1 (Carl Zeiss Microscopy GmbH, Jena, Germany) by applying simple gray-level thresholding and watershed segmentation. Carbon replica were prepared by using a 2 vol. % Nital etching after carbon coating and investigated in STEM-on-FIB (Helios Nanolab^{TM} 600, Thermo Fisher Inc., Waltham, MA, USA, formerly FEI Company, Hillsboro, OR, USA). In the same setup, electron dispersive spectroscopy (EDS) was conducted to obtain a qualitative chemical analysis of selected precipitates. For this, each precipitate was measured with 50,000 X-ray photon counts.

APT was done using a LEAP 3000X HR (Cameca SAS, Gennevilliers, France, USA) with voltage pulsing mode. APT analyzes specimen volumes of about 80 nm × 80 nm × 200 nm (typically tens of millions of atoms). The instrumental details are described elsewhere [31]. The specimens were prepared using the conventional lift-out technique in the FIB-SEM dual station (Helios Nanolab^{TM} 600, Thermo Fisher Inc., Waltham, MA, USA, formerly FEI Company, Hillsboro, OR, USA) [38]. A final low voltage milling at 2 kV was performed to minimize the gallium-induced damage. APT measurements were done at 15% pulse fraction, temperature of around 60 K, pressure lower than 1.33×10^{-8} Pa, frequency of 200 kHz and evaporation rate set at 5 atoms per 1000 pulses. All APT data reconstruction was done within the software IVAS 3.6.14 (Cameca SAS, Gennevilliers, France, USA). The composition measurements between detected precipitates and the steel matrix were adequately distinguished. Precipitates were analyzed using a constant iso-concentration surface with 1 at. % Nb, which gave best visual representation of the precipitate surfaces. Small clusters were analyzed using the maximum separation method [39], whereas the steel matrix composition was measured using precipitation/cluster or grain boundary/segregation-free volume of at least 1,000,000 atoms.

Lastly, the thermo-kinetic software MatCalc v 6.0 (Vienna, Austria, https://matcalc.at/) was used for a computational simulation of the precipitation density of different deformation temperatures and grades.

3. Results

Based on double-hit compression tests in the Gleeble apparatus and evaluation of respective softening behavior (Figure 4a), significant softening retardation was measured below 975 °C. The softening of double-compressed samples was recorded for constant temperatures between 1050 and 850 °C and after holding for 15 s. It equaled 20% at 963 °C and was defined as the T_{NR} of the laboratory steel [3]. Representative microstructures for all deformation temperatures were recorded in light-optical microscopy after etching to reveal the prior austenite structure (Figure 4c). The mean prior austenite grain (PAG) equivalent diameter decreased and the number of grains (edge grains in micrographs excluded) increased with longer inter-pass times between compression hits (Figure 4b and Table 2).

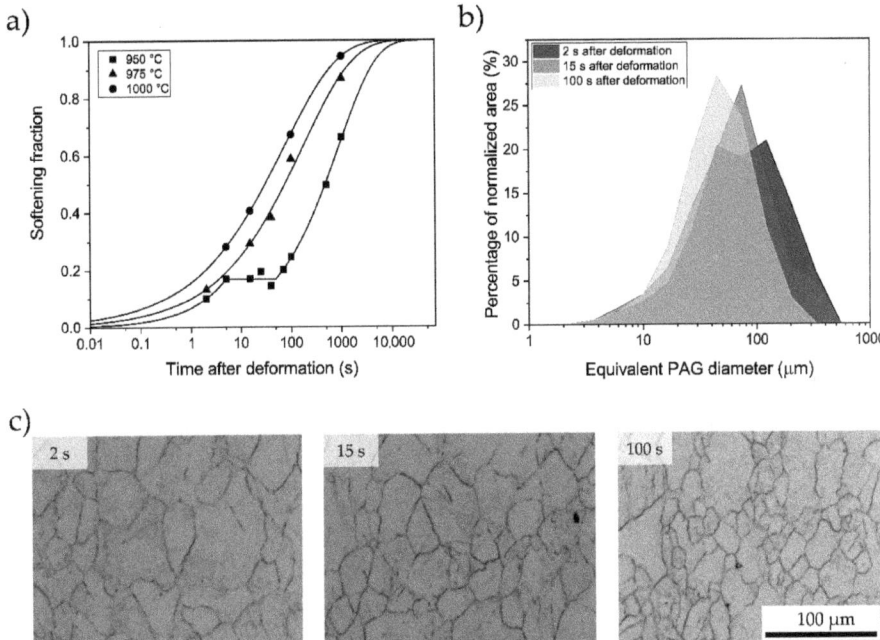

Figure 4. (a) Softening behavior of austenite after Gleeble double-hit deformation recorded for different deformation temperatures, (b) austenite grain size distribution as a function of the inter-pass time between two deformation hits derived from (c) light-optical metallography of the respective prior austenite microstructures after etching.

Table 2. Mean prior austenite grain (PAG) equivalent diameter and grain count in a 500 × 500 µm² area.

Inter-Pass Time	Mean PAG Equivalent Grain Diameter (µm)	Number of Grains
2 s	27.35	4144
15 s	24.33	5121
100 s	22.59	5634

3.1. Electrical Resistivity Measurements

The electrical resistivity behavior as a function of the time elapsed after deformation is shown in Figure 5 for the temperatures 850, 900, and 950 °C. For all temperatures, there was an almost steady decrease of resistivity after deformation which slowed over time (note the logarithmic time axis). Directly after the deformation, the resistivity was around 272 and 270 nΩm for samples deformed at 850 and 900 °C, respectively, and decreased to 266 and 264 nΩm after 6000 s. The resistivity of the samples deformed at 900 °C was mostly lower than that for 850 °C, having an offset of roughly 1 nΩm. For 950 °C, after 1200 s, the resistivity dropped below the respective values of 850 °C and 900 °C after being larger for shorter holding times. The resistivity of the sample deformed at 900 °C and held for 1200 s increased compared to other values and was reproduced in many different measurements. When the deformation was increased to $\varepsilon = 0.6$, compared to $\varepsilon = 0.3$, the resistivity was lower even for short holding times, and then dropped finally to resistivity values of roughly 259 nΩm. Just above this value, with 261 nΩm, lay the electrical resistivity of the sample that was isothermally held at 670 °C for 1 h.

For these measurements to provide a quantitative understanding of the evolution of precipitation, calibration on the solute niobium was carried out. This was achieved by using APT, as shown in Section 3.3.

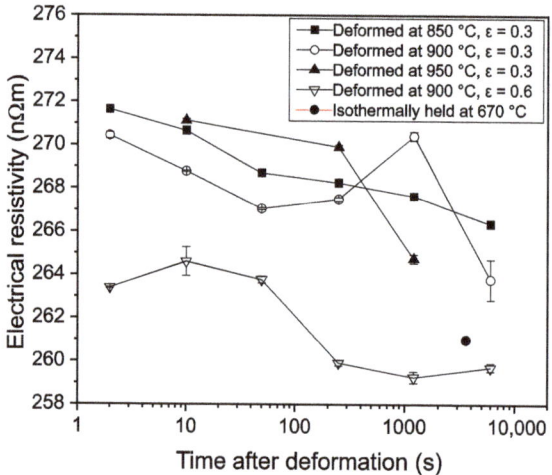

Figure 5. Evolution of electrical resistivity with the time after deformation for samples quenched from deformation at 850 °C, 900 °C, and 950 °C with $\varepsilon = 0.3$ and $\varepsilon = 0.6$ and after quenching from isothermal holding at 670 °C for 1 h.

3.2. Precipitate Characterization by STEM

Precipitation at different treatment stages was investigated by producing carbon replica films and analyzing them using STEM-on-FIB. Samples directly quenched from 1200 °C revealed precipitates that did not dissolve during the soaking treatment, whereas those quenched from 950 °C indicated whether additional precipitates have formed during the cooling phase from soaking temperature. Analyzing the same condition after deformation by a strain of $\varepsilon = 0.6$ and a holding time of 200 and 1200 s revealed newly formed strain-induced precipitates.

Figure 6a shows the precipitate population of a sample cooled down to 950 °C after soaking at 1200 °C, followed by quenching. One can observe particles decorating the prior austenite grain boundaries (indicated by white arrows), particle clusters inside the austenite grain and rather randomly distributed particles. The stage after soaking at 1200 °C comprised aligned clusters of cube-shaped particles having sizes of up to 80 nm that had not dissolved. Energy dispersive x-ray spectroscopy (EDS) identified the particle composition as being Nb-rich TiN (Figure 6b). Al, Si, and Cu elemental signals had their origin from the STEM setup. The particle clusters likely originated from casting and had not dissolved during soaking. A fraction of smaller cube-shaped particles with a size of around 12 nm, appeared to be rather pure TiN (Figure 6c). In addition, these particles must have existed before the soaking treatment. Similar-sized particles are found carrying a Nb-rich layer on two opposing sides (Figure 6d) that seemed to have nucleated on pre-existing TiN cubes. For part of these compound particles, the Nb-rich layer had grown into a larger dimension (Figure 6e). However, the Nb-rich layer appeared to grow only along one direction on opposite sides of the TiN core. Finally, bean-shaped particles were found showing niobium and titanium peaks in the EDS spectrum, yet not comprising a core-shell morphology (Figure 6f). All these particle species were also found in the various hot-deformed samples.

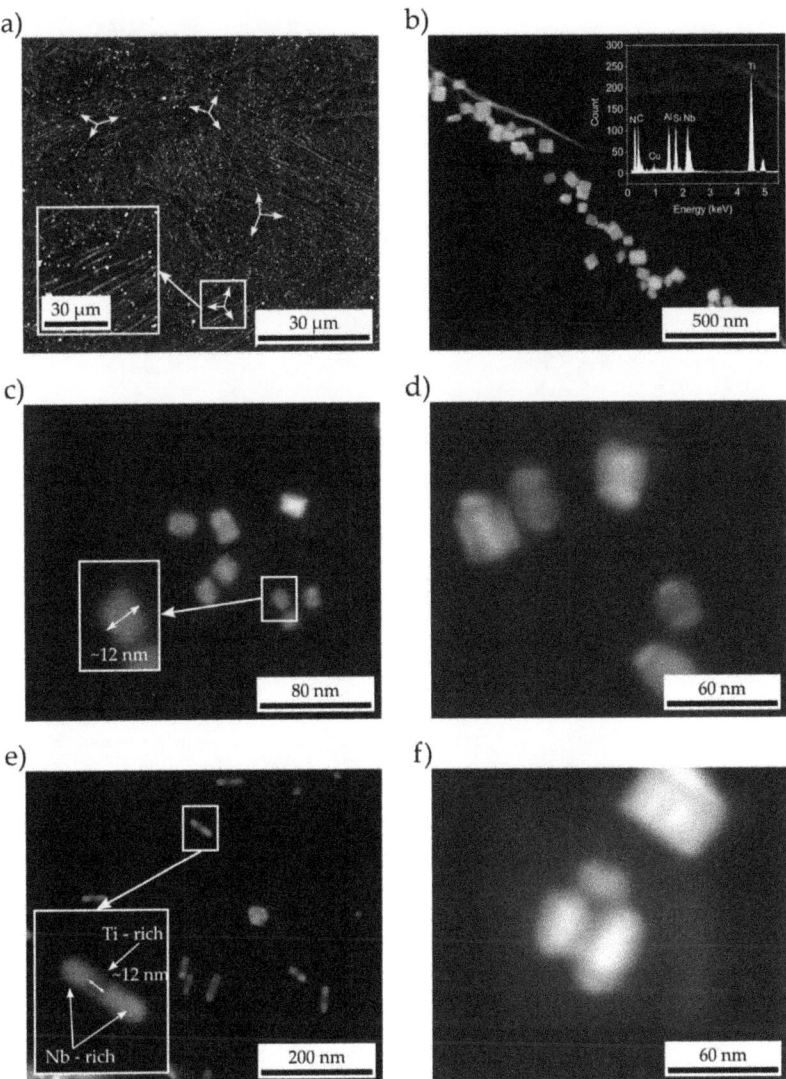

Figure 6. High Angular Diffraction Dark Field (HADDF)-STEM images of carbon replica films comprising precipitates present after quenching from 1200 °C and 950 °C. (**a**) Grain boundaries decorated by line-like precipitation of (**b**) large (approx. 80 nm), cube-shaped precipitates which EDS identified as Nb-rich TiN. (**c**) Homogeneously distributed in the grain interior were TiN of smaller size (<15 nm) as well as (**d**) TiN with heterogeneously nucleated Nb-rich caps, which (**e**) also could be larger. (**f**) Nb- or Ti-rich Nb (or Ti)(C,N) were found with a homogeneous atom distribution.

After applying deformation, the precipitate population observed by STEM was strongly augmented by a large amount of small precipitates which were homogeneously distributed inside the grains. Figure 7 shows their size and distribution exemplarily on the sample deformed at 950 °C with a strain of $\varepsilon = 0.3$. The size of precipitates is in the range of 5 to 20 nm depending on the holding period after the compression hit, which was 200 and 1200 s, respectively. EDS shows that the precipitates are free or nearly free of titanium even after prolonged holding times.

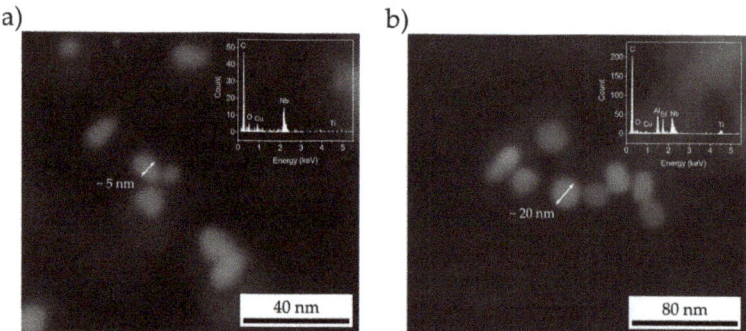

Figure 7. Dark Field (DF)-STEM image and EDS spectra depict precipitates that had formed after hot deformation at 950 °C, ε = 0.3, and holding for (**a**) 200 or (**b**) 1200 s.

3.3. APT Analysis of Precipitates and Solute Niobium Dtatus

Along with the precise analysis of precipitates, the precipitate-free material volume bore valuable information as it allowed the determination of the amount of niobium dissolved in the steel matrix, or, in other words, the characterization of the solute depletion after precipitation in the neighborhood. Figure 8 compares the decreasing mass fraction of solute niobium measured by APT in the steel matrix with the electrical resistivity as a function of the time elapsed after deformation (950 °C, ε = 0.3, and 670 °C isothermally held for 1 h). For the deformed sample, the APT analysis revealed an initial solute niobium content of 0.061 ± 0.009 wt. %, whereas the steel nominally contained 0.085 wt. % niobium (Table 1). This indicates that the difference of around 0.024 wt. % niobium had precipitated before the deformation treatment. For the applied temperature-deformation condition, solute niobium was depleted to 0.019 ± 0.004 wt. % after a holding for 1200 s. Thus, approximately 70% of the initially dissolved niobium precipitated as deformation-induced particles. In the case of the sample isothermally held at 670 °C for 1 h, the concentration of solute niobium is 0.014 ± 0.003 wt. % which is 16.5% of nominal concentration or, taking the measured initial amount in the deformed sample, 23%. In the precipitation/segregation-free areas, the other carbide forming elements—titanium, carbon and nitrogen—were not detected in the mass spectra of the APT measurements, because they were either below the detection limit (the background of the APT measurements was around 10 atoms), or the respective peaks in the mass spectrum had an overlap with a much higher concentrated element (e.g., nitrogen totally fell into the peaks of silicon, which has a much higher concentration and therefore quantification of nitrogen would have produced values with high uncertainty).

Overlaying the data from electrical resistivity measurements showed a good correlation with the APT data (Figure 8). In addition to the matrix measurements, homogeneous precipitates were analyzed by APT in samples deformed at a lower austenite temperature (850 °C, ε = 0.6, quenched after holding for 200 s), which had a bean-shaped morphology (Figure 9). Their morphology best resembled the morphology of precipitates shown in Figure 6f, indicating that these precipitates also existed in the samples that were directly quenched from high austenite temperature without deformation. Their existence is rather interesting, since the understanding of particles having a heterogeneous TiN-core-Nb(C,N)-shell morphology is quite established. As visible in Figure 9, the particle predominantly consisted of carbon and niobium as well as smaller fractions of titanium and nitrogen. The elemental distribution in this particle was homogeneous and its composition is given in Table 3. The amount of niobium and titanium was nearly the same as the amount of carbon and nitrogen, so that the particle can be considered being a stoichiometric (Nb,Ti)(C,N) precipitate.

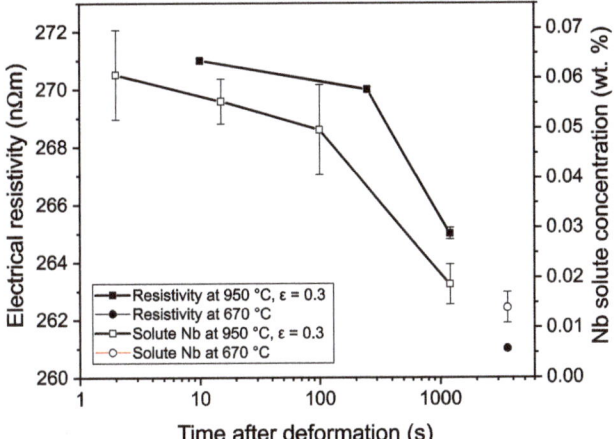

Figure 8. Atom probe tomography (APT) evaluation of solute Nb in the steel matrix and correlation with electrical resistivity measurements for samples deformed at 950 °C, $\varepsilon = 0.3$, and isothermally held at 670 °C for 1 h.

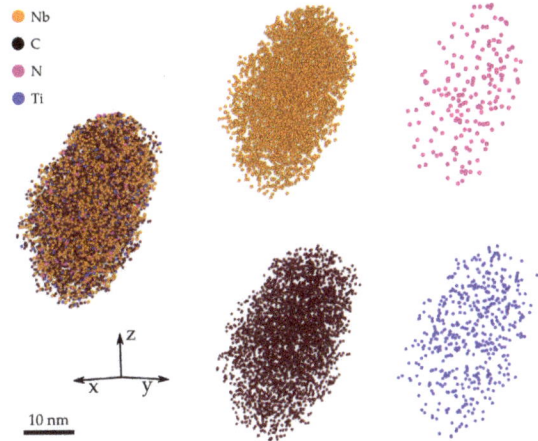

Figure 9. APT analysis of a bean-shaped particle similar to those in Figure 6f, but measured in the sample deformed at a lower austenite temperature and higher strain (850 °C, $\varepsilon = 0.6$), and quenched after holding for 200 s. Niobium, carbon, titanium, and nitrogen atoms are displayed.

Table 3. Elemental composition of APT-measured, bean-shaped particle displayed in Figure 9.

Element	Nb	C	Ti	N
Atom count	4582	4400	480	177
at. %	47.5	45.6	5.0	1.7

As shown before, after applying strain, the precipitate population observed by STEM was strongly expanded by a large number of small particles (Figure 7). In the sample deformed with a higher strain of $\varepsilon = 0.6$ at 850 °C, APT could detect small precipitates which are represented in Figure 10a. Table 4 shows that carbon and niobium were indeed the dominant elements in these precipitates, while only a small fraction of nitrogen and practically no titanium was found. The APT mass spectra peaks of $^{12}C^{2+}$ and $^{12}C^+/^{12}C_2^{2+}$ both contained significantly more atoms than the combined niobium atoms in the respective peaks at 23.3 Da (^{93}Nb^{4+}), 30.9 Da (^{93}Nb^{3+}), 46.5 Da (^{93}Nb^{2+}), and the (^{93}Nb^{14}N)$^{3+}$ (Figure 10b). The peak at 24 Da corresponded either to carbon or titanium. Since no Ti^{2+} isotopes at 23, 23.5, 24.5, and 25 Da were present, the peak at 24 Da was assigned to ^{12}C$_2{}^+$. Although it is not possible to assure that all the counts at this peak correspond unambiguously to carbon, these counts represent only 5% of the particle composition. Adding up all carbon counts, the Nb/C ratio was only 1.2:2. Yet, if the nitrogen counts are considered as well, the ratio of niobium to interstitial atoms approaches 1:2. Consequently, the chemistry of the precipitates was considered to be sub-stoichiometric (Table 4).

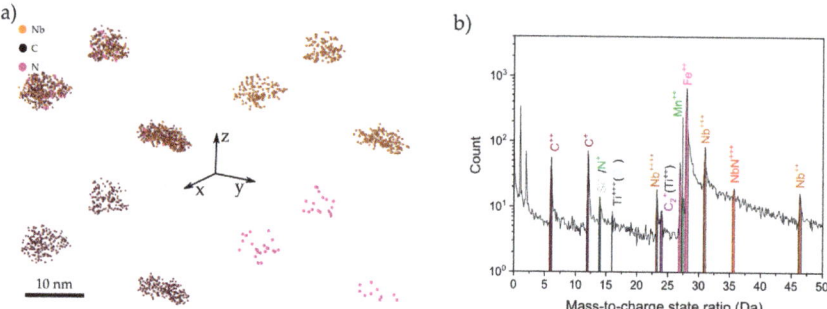

Figure 10. APT analysis of precipitates formed after deformation at a lower austenite temperature and higher strain (850 °C, ε = 0.6) and holding for 200 s. (a) Atom distribution map with niobium, carbon and nitrogen depicted; (b) mass-to-charge state ratio and the detected elemental peaks.

Table 4. Elemental composition of precipitates measured by APT in Figure 10.

Element	Nb	C	Ti	N
Atom count	410	688	-	74
at. %	35.0	58.7	-	6.3

In samples deformed at 950 °C and quenched after short holding times up to 1200 s, carbon clusters were observed as shown in Figure 11b, whereas for niobium, the atom map (Figure 11a) indicated no obvious clustering in any of the cases (for ten or more measurements for each holding time 2, 15, 100, and 1200 s, analogous to the APT analysis shown in Figure 8). To rule out early stages of clustering, the distribution of the carbon and niobium atoms in the APT reconstructions was compared to a random distribution (frequency distribution analysis in IVAS software, binning 200 ions; Figure 11e). It can be seen that in the particular case of the atom maps of Figure 11a,b, only carbon was exhibiting a small deviation of 0.5 to 1 at. % from random distribution for a small number of clusters (Figure 11b), while niobium appeared to be randomly distributed throughout the volume (Figure 11a). No precipitates as detected in STEM (Figure 7) were detected by APT.

Finally, APT analysis was performed on the sample that was isothermally held at 670 °C after soaking at 1200 °C and cooling without deformation. It is reasonable to assume that the precipitation in this sample will be similar to the sample cooled to 850 °C and subsequently quenched. In absence of strain-induced precipitation, a substantial amount of solute niobium is thus still present at the austenite-to-ferrite transformation temperature, when the niobium solubility drastically decreases. Figure 11c,d depict two atom reconstructions showing only niobium and carbon from samples after the isothermal holding period. The atom map in Figure 11c contained short-range clustering of niobium and carbon with sizes of 0.5 to 2 nm (after the maximum separation algorithm [39] with the clustering atom species niobium and carbon using the parameters d_{max} = 1.2 nm, N_{min} = 10 atoms, and an erosion step of 1.2 nm). The nearest-neighbor distribution showed a deviation from randomness (Figure 11f) for both carbon and niobium. In another ferrite grain of the same sample (Figure 11d), precipitates of typically less than 10 nm size have developed and the matrix was depleted from niobium although a small residual amount was still in solution. The precipitates were aligned in arrays and comprised an aspect ratio not equal to 1 and that was more pronounced for the larger particles. This alignment of particles is a typical feature of interphase precipitation, which occurs during the austenite-to-ferrite transformation [19,20,40]. For areas exhibiting interphase precipitation like in Figure 11d, a solute amount of niobium of 0.018 ± 0.002 wt. % was measured in the surrounding matrix, whereas for areas containing clusters such as in Figure 11c, the solute amount could be only estimated because of the diffuse boundary between NbC clusters and the steel matrix. In an exaggerated generous definition of a cluster—using a minimum cluster size of only 4 atoms—the remaining volume still contained 0.005 ± 0.002 wt. % of niobium according to peak decomposition. This result corresponds well to the data provided in Figure 2, where 90% of niobium is calculated to be precipitated after a similar heat treatment of a comparable steel. In the as-cast state of the steel melt used in this work, interphase precipitation analogous to Figure 11d was found, but with the difference, that the matrix was almost completely depleted of niobium, having only 0.002 ± 0.002 wt. % in solution.

The compositions of both cases in Figure 11c,d are given in Table 5. For the cluster composition, clusters larger than 20 atoms were regarded, which would mean a count of roughly 60 atoms (for the reported detector efficiency of 37% this equals, for example, a round nano-precipitate of one atomic layer and 2 nm diameter, as found by Breen et al. [41]). The precipitates and clusters had very similar compositions and were also stoichiometric like the homogenous precipitate from the soaked and quenched sample (Figure 9).

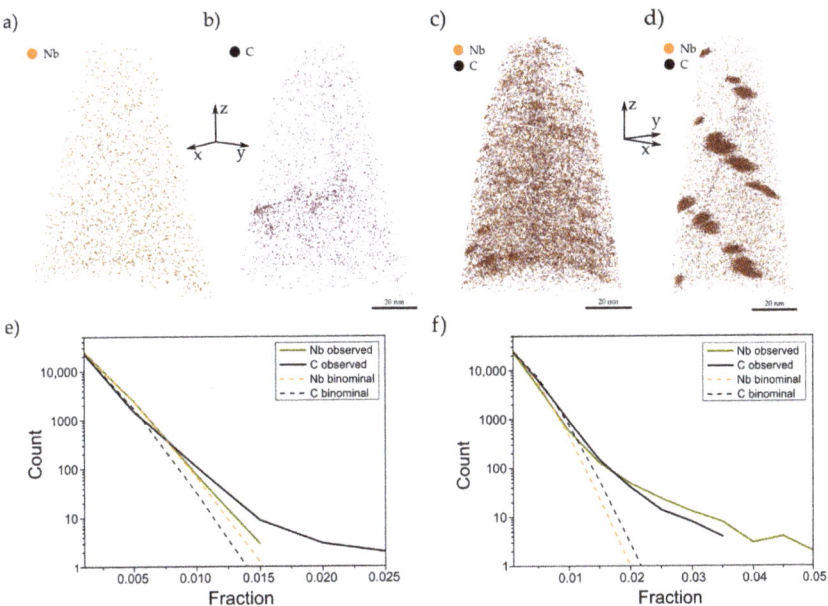

Figure 11. (**a,b,e**) APT analysis of a sample deformed at 950 °C and quenched after a short holding time of 2 s. (**a**) Niobium atom map, (**b**) carbon atom map, and (**e**) nearest-neighbor distribution of niobium and carbon observed in atom reconstruction (solid lines) vs. random distribution (dashed lines). (**c,d,f**) Quenching after isothermal holding at 670 °C for 1 h on samples soaked at 1200 °C led to (**c**) pronounced clustering or (**d**) interphase precipitation. (**f**) Nearest-neighbor distribution analysis of the dataset reconstructed in (**c**) shows strong deviation of the observed niobium and carbon distribution (solid lines) from a random distribution (dashed lines).

Table 5. Elemental composition of interphase precipitates (IP) and clusters of samples which went through the austenite-to-ferrite transformation at 670 °C for 1 h (Figure 11c,d).

Element	Nb	C	Ti	N
IP (Figure 11d)	-	-	-	-
Atom count	752	846	54	7
at. %	45.3	51.0	3.3	0.4
Clusters (Figure 11c)	-	-	-	-
Atom count	7708	8566	405	190
at. %	45.7	50.8	2.4	1.1

4. Discussion

4.1. Precipitation Behavior before TMCP Treatment

The types of precipitates observed and their evolution with simulated processing are generally in agreement with the results of other publications in literature [5,10,11]. Cube-shaped TiN particles typically containing a smaller fraction of niobium were already present at high austenite temperature (>1200 °C). The spatial arrangement of these particles and their proximity suggest that these had been formed in residual liquid pools during the late solidification phase. Titanium, niobium, nitrogen, and carbon all strongly segregated to the liquid phase as the amount of solidified delta-ferrite increased. The local supersaturation with titanium and nitrogen locally became so high that spontaneous precipitation occurred. The severity of segregation was controlled by the distribution coefficient of the respective element and might have resulted in local supersaturation exceeding the solubility limit. The segregation ratios, c_L/c_0, of the relevant elements titanium, niobium, nitrogen, and carbon were calculated as a function of the remaining liquid volume fraction, f_L, with progressing solidification using Scheil's equation [42] and published distribution coefficients ($k_{Ti} = 0.4$, $k_{Nb} = 0.3$, $k_N = 0.28$, and $k_C = 0.2$, compiled by Morita and Tanaka [43]) according to:

$$\frac{c_L}{c_0} = f_L^{k-1} \qquad (1)$$

Figure 12b demonstrates the segregation ratio of the four elements as a function of the remaining liquid volume fraction. According to Kunze et al. [10], TiN can precipitate in liquid steel if the supersaturation product of both elements exceeds the limit solubility by a factor of approximately 10. For steels having near-stoichiometric titanium and nitrogen contents as considered in the present study, this condition is fulfilled when the volume fraction of liquid phase is becoming less than 5 vol. %. Niobium has a higher segregation ratio in the residual liquid phase than titanium. Nevertheless, supersaturation levels necessary for spontaneous precipitation typically are not reached. However, type II dendritic precipitates can nucleate on titanium-rich nitride particles at the very final stage of the inter-dendritic liquid, or alternatively, on delta-ferrite phase boundaries as described in detail by Chen et al. [44]. The precipitates have a cored structure and are unstable at higher austenite temperature, where niobium carbide precipitates re-dissolve, leaving behind a string of small insoluble titanium-rich particles.

Upon full solidification, the significantly unequal distribution of solute microalloys between the center of a secondary dendrite and the last liquid periphery is quickly diminished by fast diffusion of these atoms in the delta-ferrite (Figure 12c). Short secondary dendrite arm spacings exhibit a quite equalized concentration profile at the transformation to austenite.

TiN is practically insoluble in austenite so that any remaining solute titanium is expected to precipitate upon further down-cooling. Such particles nucleate at dislocations and austenite grain boundaries. Considering the degree of solute titanium depletion and the diffusion range, these particles are expected to have a relatively small size in the range of 10 to 20 nm.

At the simulated reheating temperature of 1200 °C, niobium for the current alloy composition has a solubility of 0.27 wt. % (based on a commonly accepted solubility product log [Nb] × [C] = −7900/T + 3.42 [45]). An estimation of the dissolution kinetics predicts that particle sizes in the order of 1 µm could dissolve within the 10 min of soaking treatment. Therefore, it is concluded that the NbC caps are growing onto TiN cubes during down-cooling from soaking temperature to a lower austenite temperature. The fact that the NbC caps grow unidirectionally to opposite sides, suggests that the TiN particle is located at a dislocation or grain boundary, providing fast diffusion of niobium and carbon in this direction.

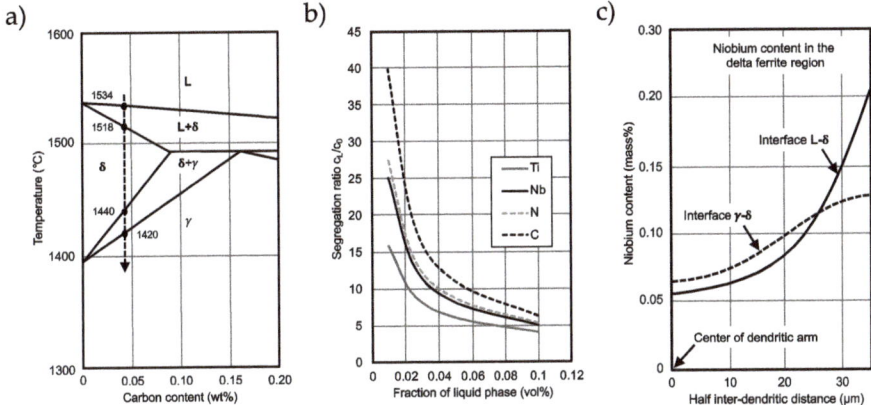

Figure 12. (a) Solidification scenario in the delta-ferrite range of the binary Fe-C diagram (equilibrium transformation temperatures for carbon content of 0.04 wt. %); (b) calculated segregation ratios as function of the residual liquid share; (c) homogenization of unequal niobium distribution after solidification by fast diffusion in delta-ferrite as a function of the secondary dendrite arm spacing.

According to EDS measurements in Figure 6b, the precipitates that form above the deformation temperature incorporated a considerable amount of niobium. The APT measurements of the steel matrix (for the samples 950 °C, $\varepsilon = 0.3$) confirmed this (Figure 8), as the concentrations lay approximately at 0.06 wt. % even 2 s after deformation, whereas the nominal bulk composition was around 0.085 wt. %. This means that more than one quarter of the added niobium had already precipitated, prior precipitated on TiN or TiCN, and as a consequence lost for later strain-induced precipitation or precipitation strengthening. This was also found by Hegetschweiler et al. [14], who used particle extraction methods. On the contrary, in niobium microalloyed steels without titanium, all niobium can be brought back into solution, as was shown by an APT study by Palmiere et al. [32].

4.2. Precipitation Behavior during TMCP Treatment

The electrical resistivity method allowed to follow the evolution of strain-induced precipitation. The drop of resistivity can be explained by the nucleation and/or growth of precipitates and the subsequent consumption of solute niobium, and consequently, less scattering of electrons. According to the results of the different temperatures in Figure 5, the samples deformed at 900 °C had the lowest resistivity in general, which also corresponds to the well-accepted fact that precipitation both commences and completes much earlier in that temperature range (Figure 2). According to the resistivity data, the consumption of niobium for precipitation at 900 °C was multiple times faster than for 850 °C. However, the difference between samples of varying deformation temperatures but constant deformation grade ($\varepsilon = 0.3$) was relatively low compared with an increased deformation (900 °C, $\varepsilon = 0.6$). As deformation induces many dislocations, the dominating effect for solute niobium and carbon consumption was due to strain-induced precipitation. The strong decrease could have been either due to a higher nucleation rate or an increased niobium diffusion along the dislocation core (pipe diffusion), both consuming niobium and carbon faster. In principle, the deformation degree seemed to have the dominating effect over deformation temperature, which was also found by Jung et al. [29]. The cooling rate from austenitizing temperature down to the deformation temperature was 5 °C/s and therefore the difference in total TMCP time for the different temperatures was only 10 s between each other. Consequently, the influence of precipitates formed or grown during cooling on the difference of resistivity should be negligible in this discussion. In Figure 8, the falling niobium concentration in the steel matrix, although only analyzed in detail for the sample deformed at 950 °C, correlated well with the downward trend of the electrical resistivity with increasing holding time after deformation. Theoretically, the other carbide-forming elements—titanium, carbon, and nitrogen—also increase the resistivity when in solid solution, however, their quantification in the steel matrix was not possible for the investigated steel composition. For the counts in the mass spectrum belonging to titanium (Figure 10b), the matrix spectrum neither had visible peaks at 16 Da, nor at 24 Da. With the background of roughly 10 atoms, this means that for all measurements the solute titanium amount was lower than 5 ppm (for illustration, the amount of 0.002 wt. % solute niobium which was measured in the as-cast sample with visible peaks equaled 200 atoms in an APT measurement of 10,000,000 atoms which was well above the background noise). Since even at the austenite-to-ferrite transformation temperature titanium could be found in precipitates, a small amount must have still been in solution. Likewise, no carbon (with the exception of the carbon atmospheres, and occasional lath boundaries or retained austenite islands) was detected in the matrix. For nitrogen, the quantification is additionally complicated by the overlap of the nitrogen peak at 14 Da with silicon, that had a much larger abundance in the sampled steel. As it will be discussed further on, when addressing potential errors of the resistivity measurements, carbon and nitrogen influences might actually be ruled out for quenched samples.

The large family of precipitates that typically form during TMCP of Nb-Ti-microalloyed HSLA steels makes it challenging to fully distinguish them relying solely on (S)TEM. Adding APT to the characterization toolbox brought about the ability to characterize the chemistry of single precipitates with the highest resolving power available to date and without interfering with the precipitate environment, i.e., steel matrix or neighbor precipitates. For example, the bean-shaped precipitates with homogeneous niobium and titanium atom distribution in Figure 9 had a size and morphology that could be identified as both the STEM-characterized precipitates present at soaking (Figure 6f), or the ones detected after deformation (Figure 7). However, APT reveals the composition of the precipitate to contain also small fractions of titanium (Table 3), whereas the careful comparison of precipitates after deformation and austenite-to-ferrite transformation (Figure 10, Table 4, and Figure 11c,d, Table 5, respectively) showed that strain-induced precipitation contains no titanium with regard to the detection limits of the APT. Characterization of the precipitate in Figure 9 on the basis of (S)TEM could erroneously identify it as the product of deformation, which would overestimate the growth kinetics of strain-induced precipitates for the investigated deformation temperature (note that this Ti-containing precipitate was found after a holding time of 200 s, whereas even at 950 °C, only after 1200 s can strain-induced precipitates reach such a size (Figure 7)). Despite a very precise chemical quantification of the big and small (Nb,Ti)(C,N) precipitates, small size deviations are possible due to APT-related artifacts. The so-called local magnification effect [46] appears when precipitates have a very different evaporation field compared to the matrix. In this special case, (Nb,Ti)(C,N) showed a high evaporation field of around 55–60 V/nm for pure NbC. In the steel matrix with an evaporation field of 33 V/nm, the reconstructed precipitates would therefore appear larger than they were. Although a direct size comparison of APT and STEM-measured precipitates was not possible, the medium-sized and homogeneous (Nb,Ti)(C,N) precipitates found in STEM for samples quenched from the soaking temperature (Figure 6f) fell in a similar range as the APT-measured precipitate in Figure 9.

A more plausible explanation for the origin of the precipitate in Figure 9 is, that it had formed during the casting. If these precipitates did not (re-)dissolve during reheating, they would still be present in the samples quenched from a higher austenite deformation temperature after austenitizing at 1200 °C (Figure 6f), as well as in all other deformed samples. As it was shown for the Ti-containing small precipitates in the samples transformed isothermally into ferrite (Figure 11,d), titanium was still in solution at the austenite-to-ferrite transformation temperature. The bigger (Nb,Ti)(C,N) precipitates might therefore be the outcome of the coarsening of clusters such as displayed in Figure 11c. Preliminary work on the co-microalloying of niobium and titanium considered the higher temperature stability of (Nb,Ti)(C,N) precipitates to be the reason for their presence even after soaking at high austenite temperatures [5,7].

When applying deformation, STEM images showed a sharp increase in very small precipitates which grew with the holding time (Figure 7). Likewise, APT results depicted small precipitates in the sample deformed with $\varepsilon = 0.6$ strain at 850 °C and found them to be free or nearly free of titanium but with an over-stoichiometric amount of carbon (Figure 10 and Table 4).

In the sample deformed at 950 °C with $\varepsilon = 0.3$ strain, no precipitates were detected by APT although their existence and coarsening are visible in STEM (Figure 7). MatCalc simulations were conducted (Figure 13) to investigate the density of precipitates that could be expected after deformation at higher and lower austenite temperatures and with varying deformation. According to simulation, the precipitation density increased by a factor of 10^3 when lowering the deformation temperature from 950 °C to 850 °C and increasing total strain from $\varepsilon = 0.3$ to $\varepsilon = 0.6$. The number density of $10^{17}/m^3$ in the case of 950 °C, $\varepsilon = 0.3$ would approximately correspond to the detection of one particle every 8000 measurements, assuming homogeneous particle spacing and an average volume of one APT measurement with a diameter of 80 nm at the base and a length of 200 nm. On the contrary, for 850 °C, $\varepsilon = 0.6$, an average of one detection every six measurements was predicted. This lies in the range of the experimental findings for these cases. Furthermore, compared to the influence of the deformation temperature, the decrease in electrical resistivity was more pronounced when raising the deformation from $\varepsilon = 0.3$ to $\varepsilon = 0.6$, as shown for the samples deformed at 900 °C (Figure 5). This was also well reflected by the increased detectability of NbC in APT in higher-deformed samples. Together with the APT measurements, it is concluded that the measured effect stems from strain-induced precipitation. The strain-induced NbC had a non-stoichiometric composition with a carbon content being more than 1.5 times larger than the niobium concentration. A previous investigation using APT had discussed the hypothesis of a deficiency of carbon detection in carbides [47], which is contrary to the findings of the present work. In addition, if these precipitates are strain-induced, then it is conceivable that the carbon surplus came from its segregation along dislocations which had preceded the precipitation formation. Carbon atmospheres as possible precursors were detected in almost every APT measurement of the samples deformed at 950 °C, and many of them showed line-like features like in Figure 11b. The carbon enrichments are therefore assumed to be Cottrell atmospheres and niobium, which has a slow diffusion rate compared to carbon, diffused during a later stage to form precipitates. When deformation increased, more dislocations were produced, facilitating the diffusion along the dislocation core. Therefore, the particle detection statistics during APT experiments strengthen the assumption that the precipitates found were indeed induced by strain. The existence of metastable NbC precipitates was already discussed earlier by Danoix et al. [48] and the detected Nb(C,N) in the sample subjected to a higher deformation provided evidence of their existence.

Kostryzhev et al. [35] proposed early stages of precipitation in the form of clusters (in the range of 30 to 60 atoms) during austenite deformation. No evidence of such a clustering was found in this work. The atom maps of deformed and then quenched samples did not show any visible clustering of niobium or titanium except the already discussed precipitates in the sample 850 °C, $\varepsilon = 0.6$. In Figure 11, the atom distribution of the sample deformed at 950 °C, $\varepsilon = 0.3$ and the sample held at 670 °C for 1 h were compared. In the latter case, Figure 11f displays a marked deviation of niobium and carbon distribution from random, which is also visibly clustered in the atom map of Figure 11c. In contrast, sample 950 °C, $\varepsilon = 0.3$ had no visible niobium clustering (Figure 11a), while the carbon clustering in Figure 11b shows a possible Cottrell atmosphere as discussed above. It is therefore concluded that strain-induced precipitation during deformation in the higher austenite temperature regime is challenging to trace with APT alone due to the low particle density in low-deformed austenite. An ongoing project is developing a methodology that aims at increasing the precipitate yield in APT measurements by their extraction and re-encapsulation in a suitable matrix material. It is envisioned that by this methodology, precipitates of any heat treatment can be measured with sufficient yield.

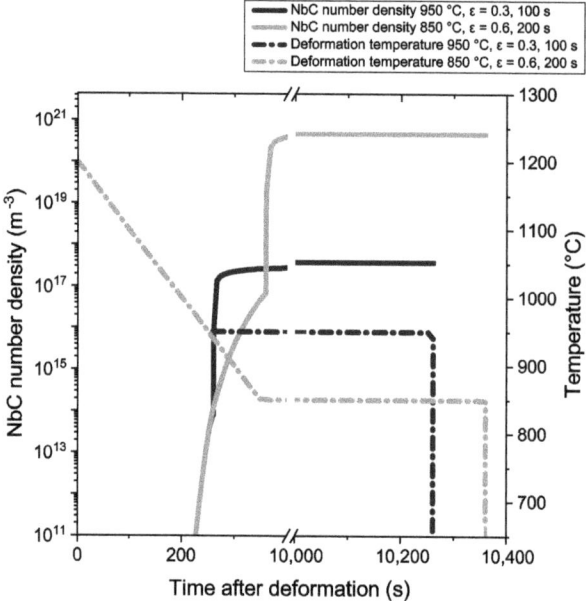

Figure 13. MatCalc simulation of the precipitate number density evolution over time for varying compression with a total deformation of $\varepsilon = 0.3$ at 950 °C and $\varepsilon = 0.6$ at 850 °C.

The correlation of APT results with the electrical resistivity proved that bulk measurements can successfully describe the precipitation sequence occurring at the atomic scale. As specific resistance measurements were quite challenging for the investigated samples, the sources of error need to be addressed as well. The electrical resistivity relies on the precise measurements of the current that flows through the sample, the voltage drop over the sample, and the precise knowledge of the relative position of the potential points. The latter aspect was especially important, as for the currently used Gleeble setup, the hot-deformed samples were relatively short to prevent buckling. The total increment of resistivity decrease from dissolved to fully precipitated accounts only for approximately 10% of the total resistivity range. The distance of the potential is proportional to the measured resistance, therefore a misplacement of several hundreds of microns in the present setup introduces large errors of the calculated resistivity. Future experiments with larger samples may help to reduce the error in the measurement of the distance of the potential points. Another aspect worth mentioning is the assumption, that the other elements in the sampled steel do not change their atomic distribution in the different investigated heat treatment stages. This approximation might be correct for most elements, such as manganese, copper, silicon, and nickel, which did not take part in the precipitation process according to APT results. However, long-range chemical inhomogeneities which stem from casting of the ingots could produce variations in the resistivity irrespective of the solute niobium consumption during precipitation. Additionally, pores in the sample would decrease the cross section for the passing current and increase the measured electrical resistance. All this could be the reason for unexpected resistivity values as was the case for the sample 900 °C, $\varepsilon = 0.3$ and held for 1200 s before quenching (Figure 5). Regarding carbon and nitrogen, which actively took part in the precipitation process, tempering at low temperature was conducted in prior work to bring all solute carbon and nitrogen to lath or grain boundaries [26,28,29]. However, more recent experimental data suggest that even under heat treatment conditions severely obstructing diffusion—such as quenching from the austenite field to room temperature—there is sufficient time for interstitial atoms to diffuse to bainite or martensite lath boundaries or prior austenite grain boundaries in large quantity [49,50], or, as in some cases, to dislocations (Figure 11b). Therefore, the matrix is depleted of those elements, which was also confirmed by APT measurements in the present work. Consequently, their influence on changes of the electrical resistivity was assumed to be negligible.

4.3. Precipitation during Austenite-to-Ferrite Transformation

When isothermally held at 670 °C for 1 h, APT measured a high density of (Nb,Ti)(C,N) precipitates. In some ferrite grains, very fine clusters with sizes even below 1 nm were found, whereas in other grains, an ordered arrangement of bigger (5–10 nm) precipitates typical for interphase precipitation appeared. These two phenomena coexisted in the same heat treatment as evidenced in Figure 11c,d. During transformation, the interphase precipitation proceeded over the moving ferrite grain front into the austenite and when a certain level of supersaturation was reached, the solute niobium precipitated along the whole front. However, some grains might have transformed too quickly for interphase precipitation to follow, or the grain orientation was not favorable [20]. Therefore, instead of interphase precipitation, a Nb supersaturation of the freshly formed ferrite led to a high nucleation density resulting in the fine-scale clustering/precipitation visible in Figure 11c.

The interphase precipitates had a similar composition as the medium-sized (Nb,Ti)(C,N) (Tables 3 and 5, respectively). Interphase precipitation was also observed in the as-cast state, likewise, containing titanium. As these small (Nb,Ti)(C,N) precipitates were not present in the hot-deformed and quenched samples, they must have either totally dissolved again during soaking or coarsened to larger particles, similar to the one shown in Figure 9. The Gibbs-Thompson effect predicts solubilities almost twice as high for small precipitates compared to larger ones as found in the samples quenched from austenitization temperature [51]. Therefore, smaller (Nb,Ti)(C,N) precipitates could have completely re-dissolved, whereas bigger precipitates did not dissolve during re-heating. The residual solute amount of niobium after the isothermal holding phase was measured to be approximately 0.014 ± 0.003 wt. %. Considering that 0.061 ± 0.009 wt. % niobium was re-dissolved after soaking and was therefore available for precipitation during isothermal holding, the fraction of precipitation was approximately 77%. This is close to the value that would be expected according to the kinetics shown in Figure 2 and was also well reflected in the electrical resistivity measurements.

5. Conclusions

Multi-scale characterization methods were used to quantify the precipitation status of niobium during austenite conditioning in a typical Nb-Ti microalloyed high-strength low-alloy steel, namely electrical resistivity, STEM and APT measurements. These methods provided complementary information on nucleation, growth, spatial distribution, and chemical composition of Nb(C,N) precipitates.

The reported results showed that the electrical resistivity method can efficiently monitor the progress of precipitation for different TMCP conditions. Measuring the average amount of solute niobium in the steel matrix by APT allowed to calibrate the electrical resistivity method. Such measurements also demonstrated that, in Nb-Ti dual microalloyed steel, part of the niobium is bound in mixed Ti,Nb(C,N) particles that do not dissolve during typical soaking treatment.

The chemical composition and spatial arrangement of such Nb-rich TiN particles indicated that both, Nb and Ti, had reached a high segregation level in inter-dendritic liquid pools existing just before full solidification.

APT allowed a detailed chemical characterization of particles as compared to electron microscopy. It was demonstrated that precipitates being present before the deformation stage can be distinguished from strain-induced precipitates, thus, preventing erroneous estimations of growth kinetics. In this context, current typical restrictions on the detectability and precise size estimation of precipitates will be addressed in future work through an extraction and re-encapsulation methodology.

The observed core-shell structures of TiN cubes with NbC caps were assumed to form after reheating by re-precipitation of niobium onto insoluble TiN particles. The fact that the caps did not form on all cube faces allowed to conclude that niobium is supplied by preferred diffusion paths such as a grain boundary or dislocation.

Precipitates that formed at lower austenite temperature or during transformation to ferrite had a chemical composition dominated by niobium and carbon. It was found that these particles scavenged the small quantity of titanium and nitrogen atoms that remained in solution. The titanium and nitrogen atoms are randomly incorporated within the NbC particle.

APT was able to show the formation of atom clusters in the nucleation stage of particles. During TMCP treatment applied in this study, only weak clustering of carbon could be detected, yet no significant clustering of niobium was seen. However, after isothermal holding in the ferrite phase, the formation of niobium-carbon clusters was evident.

With regard to strain-induced precipitation, all characterization methods used in this study confirmed that the degree of deformation is dominating over temperature with regard to precipitate nucleation during TMCP treatment, which is in agreement with established knowledge.

Author Contributions: Conceptualization, J.W., E.D. and H.M.; methodology and experimental validation, J.W., D.B. and A.H.; Formal analysis and data curation, J.W.; visualization, J.W. and H.M.; writing—original draft preparation, J.W. and H.M.; writing—review and editing, H.M., A.H., D.B., E.D., V.F. and F.M.; supervision, project administration, D.B., H.M., V.F., and F.M.; funding acquisition, D.B., J.W. and F.M. All authors have read and agreed to the published version of the manuscript.

Funding: The authors express their gratitude to CBMM for providing the financial funding of this work. Furthermore, J.W. wishes to acknowledge the EFRE funds (C/4-EFRE-13/2009/Br) of the European Commission. The Atom Probe was financed by the DFG and the Federal State Government of Saarland (INST 256/298-1 FUGG).

Acknowledgments: The authors also thank Comtes FHT for the fruitful discussions.

Conflicts of Interest: The authors declare no conflict of interest.

References

1. Dutta, B.; Palmiere, E.J. Effect of prestrain and deformation temperature on the recrystallization behavior of steels microalloyed with niobium. *Metall. Mater. Trans. A* **2003**, *34*, 1237–1247. [CrossRef]
2. Dutta, B.; Sellars, C.M. Effect of composition and process variables on Nb(C,N) precipitation in niobium microalloyed austenite. *Mater. Sci. Technol.* **1987**, *3*, 197–206. [CrossRef]
3. Deardo, A.J. Niobium in modern steels. *Int. Mater. Rev.* **2003**, *48*, 371–402. [CrossRef]
4. Morrison, W.B. Microalloy steels-the beginning. *Mater. Sci. Technol.* **2009**, *25*, 1066–1073. [CrossRef]
5. Hong, S.G.; Kang, K.B.; Park, C.G. Strain-induced precipitation of NbC in Nb and Nb-Ti microalloyed HSLA steels. *Scr. Mater.* **2002**, *46*, 163–168. [CrossRef]
6. Cuddy, L.J.; Raley, J.C. Austenite grain coarsening in microalloyed steels. *Metall. Trans. A* **1983**, *14*, 1989–1995. [CrossRef]
7. Gong, P.; Palmiere, E.J.; Rainforth, W.M. Dissolution and precipitation behaviour in steels microalloyed with niobium during thermomechanical processing. *Acta Mater.* **2015**, *97*, 392–403. [CrossRef]
8. Jia, T.; Militzer, M. The Effect of Solute Nb on the Austenite-to-Ferrite Transformation. *Metall. Mater. Trans. A* **2014**, *46*, 614–621. [CrossRef]
9. Altuna, M.A.; Iza-Mendia, A.; Gutiérrez, I. Precipitation of Nb in ferrite after austenite conditioning. Part II: Strengthening contribution in high-strength low-alloy (HSLA) steels. *Metall. Mater. Trans. A* **2012**, *43*, 4571–4586. [CrossRef]
10. Kunze, J.; Mickel, C.; Backmann, G.; Beyer, B.; Reibold, M.; Klinkenberg, C. Precipitation of titanium nitride in low-alloyed steel during cooling and deformation. *Steel Res.* **1997**, *68*, 441–449. [CrossRef]
11. Klinkenberg, C.; Hulka, K.; Bleck, W. Niobium Carbide Precipitation in Microalloyed Steel. *Steel Res. Int.* **2004**, *75*, 744–752. [CrossRef]

12. Zhou, C.; Priestner, R. The evolution of precipitates in Nb-Ti microalloyed steels during solidification and post-solidification cooling. *ISIJ Int.* **1996**, *36*, 1397–1405. [CrossRef]
13. Craven, A.J.; He, K.; Garvie, L.A.J.; Baker, T.N. Complex heterogeneous precipitation in titanium–niobium microalloyed Al-killed HSLA steels—I.(Ti,Nb)(C,N) particles. *Acta Mater.* **2000**, *48*, 3857–3868. [CrossRef]
14. Hegetschweiler, A.; Borovinskaya, O.; Staudt, T.; Kraus, T. Single-particle mass spectrometry of titanium and niobium carbonitride precipitates in steels. *Anal. Chem.* **2019**, *91*, 943–950. [CrossRef]
15. Miao, C.L.; Shang, C.J.; Zhang, G.D.; Subramanian, S.V. Recrystallization and strain accumulation behaviors of high Nb-bearing line pipe steel in plate and strip rolling. *Mater. Sci. Eng. A* **2010**, *527*, 4985–4992. [CrossRef]
16. Cao, Y.; Xiao, F.; Qiao, G.; Huang, C.; Zhang, X.; Wu, Z.; Liao, B. Strain-induced precipitation and softening behaviors of high Nb microalloyed steels. *Mater. Sci. Eng. A* **2012**, *552*, 502–513. [CrossRef]
17. Akben, M.G.; Bacroix, B.; Jonas, J.J. Effect of vanadium and molybdenum addition on high temperature recovery, recrystallization and precipitation behavior of niobium-based microalloyed steels. *Acta Metall.* **1983**, *31*, 161–174. [CrossRef]
18. Watanabe, H.; Smith, Y.E.; Pehlke, R.D. Precipitation kinetics of niobium carbonitride in austenite of high-strength low-alloy steels. In *The Hot Deformation of Austenite*; TMS-AIME: New York, NY, USA, 1977; pp. 140–168.
19. Herman, J.C.; Donnay, B.; Leroy, V. Precipitation kinetics of microalloying additions during hot-rolling of HSLA steels. *ISIJ Int.* **1992**, *32*, 779–785. [CrossRef]
20. Okamoto, R.; Borgenstam, A.; Ågren, J. Interphase precipitation in niobium-microalloyed steels. *Acta Mater.* **2010**, *58*, 4783–4790. [CrossRef]
21. Hin, C.; Bréchet, Y.; Maugis, P.; Soisson, F. Kinetics of heterogeneous dislocation precipitation of NbC in alpha-iron. *Acta Mater.* **2008**, *56*, 5535–5543. [CrossRef]
22. Perrard, F.; Deschamps, A.; Bley, F.; Donnadieu, P.; Maugis, P. A small-angle neutron scattering study of fine-scale NbC precipitation kinetics in the α-Fe–Nb–C system. *J. Appl. Crystallogr.* **2006**, *39*, 473–482. [CrossRef]
23. Huang, B.M.; Yang, J.R.; Yen, H.W.; Hsu, C.H.; Huang, C.Y.; Mohrbacher, H. Secondary hardened bainite. *Mater. Sci. Technol.* **2014**, *30*, 1014–1023. [CrossRef]
24. Costa e Silva, A. Challenges and opportunities in thermodynamic and kinetic modeling microalloyed HSLA steels using computational thermodynamics. *Calphad Comput. Coupling Phase Diagrams Thermochem.* **2020**, *68*, 101720. [CrossRef]
25. LeBon, A.; Rofes-Vernis, J.; Rossard, C. Recristallisation et précipitation provoquées par la déformation à chaud: Cas d'un acier de construction sondable au niobium. *Mem. Soc. Rev. Met.* **1973**, *70*, 577–588.
26. Simoneau, R.; Bégin, G.; Marquis, A.H. Progress of NbCN precipitation in HSLA steels as determined by electrical resistivity measurements. *Met. Sci.* **1978**, *12*, 381–386. [CrossRef]
27. Park, J.S.; Lee, Y.K. Determination of Nb(C,N) dissolution temperature by electrical resistivity measurement in a low-carbon microalloyed steel. *Scr. Mater.* **2007**, *56*, 225–228. [CrossRef]
28. Jung, J.-G.; Bae, J.-H.; Lee, Y.-K. Quantitative evaluation of dynamic precipitation kinetics in a complex Nb-Ti-V microalloyed steel using electrical resistivity measurements. *Met. Mater. Int.* **2013**, *19*, 1159–1162. [CrossRef]
29. Jung, J.-G.; Park, J.-S.; Kim, J.; Lee, Y.-K. Carbide precipitation kinetics in austenite of a Nb–Ti–V microalloyed steel. *Mater. Sci. Eng. A* **2011**, *528*, 5529–5535. [CrossRef]
30. Rivas, A.L.; Matlock, D.K.; Speer, J.G. Quantitative analysis of Nb in solution in a microalloyed carburizing steel by electrochemical etching. *Mater. Charact.* **2008**, *59*, 571–577. [CrossRef]
31. Gault, B.; Moody, M.P.; Cairney, J.M.; Ringer, S.P. *Atom Probe Microscopy*; Springer Series in Materials Science; Springer: New York, NY, USA, 2012; Volume 160, ISBN 978-1-4614-3435-1.
32. Palmiere, E.J.; Garcia, C.I.; De Ardo, A.J. Compositional and microstructural changes which attend reheating and grain coarsening in steels containing niobium. *Metall. Mater. Trans. A* **1994**, *25*, 277–286. [CrossRef]
33. Nöhrer, M.; Mayer, W.; Primig, S.; Zamberger, S.; Kozeschnik, E.; Leitner, H. Influence of Deformation on the Precipitation Behavior of Nb(CN) in Austenite and Ferrite. *Metall. Mater. Trans. A* **2014**, *45*, 4210–4219. [CrossRef]

34. Pereloma, E.V.; Kostryzhev, A.G.; AlShahrani, A.; Zhu, C.; Cairney, J.M.; Killmore, C.R.; Ringer, S.P. Effect of austenite deformation temperature on Nb clustering and precipitation in microalloyed steel. *Scr. Mater.* **2014**, *75*, 74–77. [CrossRef]
35. Kostryzhev, A.G.; Al Shahrani, A.; Zhu, C.; Ringer, S.P.; Pereloma, E.V. Effect of deformation temperature on niobium clustering, precipitation and austenite recrystallisation in a Nb-Ti microalloyed steel. *Mater. Sci. Eng. A* **2013**, *581*, 16–25. [CrossRef]
36. Kapoor, M.; O'Malley, R.; Thompson, G.B. Atom probe tomography study of multi-microalloyed carbide and carbo-nitride precipitates and the precipitation sequence in Nb-Ti HSLA steels. *Metall. Mater. Trans. A* **2016**, *47*, 1984–1995. [CrossRef]
37. Maruyama, N.; Uemori, R.; Sugiyama, M. The role of niobium in the retardation of the early stage of austenite recovery in hot-deformed steels. *Mater. Sci. Eng. A* **1998**, *250*, 2–7. [CrossRef]
38. Thompson, K.; Lawrence, D.; Larson, D.J.; Olson, J.D.; Kelly, T.F.; Gorman, B. In situ site-specific specimen preparation for atom probe tomography. *Ultramicroscopy* **2007**, *107*, 131–139. [CrossRef]
39. Vaumousse, D.; Cerezo, A.; Warren, P.J. A procedure for quantification of precipitate microstructures from three-dimensional atom probe data. *Ultramicroscopy* **2003**, *95*, 215–221. [CrossRef]
40. Timokhina, I.B.; Hodgson, P.D.; Ringer, S.P.; Zheng, R.K.; Pereloma, E.V. Precipitate characterisation of an advanced high-strength low-alloy (HSLA) steel using atom probe tomography. *Scr. Mater.* **2007**, *56*, 601–604. [CrossRef]
41. Breen, A.J.; Xie, K.Y.; Moody, M.P.; Gault, B.; Yen, H.-W.; Wong, C.C.; Cairney, J.M.; Ringer, S.P. Resolving the morphology of niobium carbonitride nano-precipitates in steel using atom probe tomography. *Microsc. Microanal.* **2014**, *20*, 1100–1110. [CrossRef]
42. Scheil, E. Bemerkungen zur Schichtkristallbildung. *Z. Met.* **1942**, *34*, 70–72.
43. Morita, Z.; Tanaka, T. Thermodynamics of solute distributions between solid and liquid phases in iron-base ternary alloys. *Trans. Iron Steel Inst. Jpn.* **1983**, *23*, 824–833. [CrossRef]
44. Chen, Z.; Loretto, M.H.; Cochrane, R.C. Nature of large precipitates in titanium-containing HSLA steels. *Mater. Sci. Technol.* **1987**, *3*, 836–844. [CrossRef]
45. Narita, K. Physical Chemistry of the groups IVa(Ti,Zr), Va(V,Nb,Ta) and the rare earth elements in steel. *Trans. Iron Steel Inst. Jpn.* **1975**, *15*, 145–152. [CrossRef]
46. Vurpillot, F.; Bostel, A.; Blavette, D. Trajectory overlaps and local magnification in three-dimensional atom probe. *Appl. Phys. Lett.* **2000**, *76*, 3127–3129. [CrossRef]
47. Thuvander, M.; Weidow, J.; Angseryd, J.; Falk, L.K.L.; Liu, F.; Sonestedt, M.; Stiller, K.; Andrén, H.-O. Quantitative atom probe analysis of carbides. *Ultramicroscopy* **2011**, *111*, 604–608. [CrossRef]
48. Danoix, F.; Bémont, E.; Maugis, P.; Blavette, D. Atom probe tomography i. early stages of precipitation of nbc and nbn in ferritic steels. *Adv. Eng. Mater.* **2006**, *8*, 1202–1205. [CrossRef]
49. Li, Y.J.; Ponge, D.; Choi, P.; Raabe, D. Segregation of boron at prior austenite grain boundaries in a quenched martensitic steel studied by atom probe tomography. *Scr. Mater.* **2015**, *96*, 13–16. [CrossRef]
50. Li, Y.J.; Ponge, D.; Choi, P.; Raabe, D. Atomic scale investigation of non-equilibrium segregation of boron in a quenched Mo-free martensitic steel. *Ultramicroscopy* **2015**, *159*, 240–247. [CrossRef]
51. Porter, D.A.; Easterling, K.E. *Phase Transformations in Metals and Alloys*, 2nd ed.; CRC Press: Boca Raton, FL, USA, 1992; ISBN 0748757414.

© 2020 by the authors. Licensee MDPI, Basel, Switzerland. This article is an open access article distributed under the terms and conditions of the Creative Commons Attribution (CC BY) license (http://creativecommons.org/licenses/by/4.0/).

Article

Hot Strip Mill Processing Simulations on a Ti-Mo Microalloyed Steel Using Hot Torsion Testing

Caleb A. Felker *, John. G. Speer, Emmanuel De Moor and Kip O. Findley

Department of Metallurgical and Materials Engineering, Colorado School of Mines, Golden, CO 80401, USA; jspeer@mines.edu (J.G.S.); edemoor@mines.edu (E.D.M.); kfindley@mines.edu (K.O.F.)
* Correspondence: cafelker@mines.edu

Received: 8 January 2020; Accepted: 28 February 2020; Published: 3 March 2020

Abstract: Precipitation strengthened, fully ferritic microstructures in low-carbon, microalloyed steels are used in applications requiring enhanced stretch-flange formability. This work assesses the influence of thermomechanical processing on the evolution of austenite and the associated final ferritic microstructures. Hot strip mill processing simulations were performed on a low-carbon, titanium-molybdenum microalloyed steel using hot torsion testing to investigate the effects of extensive differences in austenite strain accumulation on austenite morphology and microstructural development after isothermal transformation. The gradient of imposed shear strain with respect to radial position inherent to torsion testing was utilized to explore the influence of strain on microstructural development for a given simulation, and a tangential cross-section technique was employed to quantify the amount of shear strain that accumulated within the austenite during testing. Greater austenite shear strain accumulation resulted in greater refinement of both the prior austenite and polygonal ferrite grain sizes. Further, polygonal ferrite grain diameter distributions were narrowed, and the presence of hard, secondary phase constituents was minimized, with greater amounts of austenite strain accumulation. The results indicate that extensive austenite strain accumulation before decomposition is required to achieve desirable, ferritic microstructures.

Keywords: low-carbon steel; microalloyed; hot torsion testing; prior austenite; polygonal ferrite

1. Introduction

A challenge exists in the automotive industry to develop new, hot-rolled, high-strength low-alloy (HSLA) steels offering a balance of high tensile strength and superior stretch-flange formability to reduce vehicle weight without compromising safety, performance, or manufacturability [1,2]. The steel industry has responded by developing ferritic steels strengthened with extensive nano-sized precipitation [2–4]. The single-phase ferritic matrix eliminates hard constituents and imparts superior stretch-flange formability, while its high yield and tensile strengths are derived from nano-sized precipitates. Titanium (Ti)-, niobium (Nb)-, or vanadium (V)-based microalloy systems are typically used for such HSLA steels, and molybdenum (Mo) is often added to strongly retard the precipitate coarsening rate [4,5]. Substitutional manganese (Mn) additions are also made to such HSLA steels to compensate for the low carbon levels and lower the Ar_3 transformation temperature for better refinement of the microalloy precipitation sizes [6]. However, the Mo and Mn additions can enhance the hardenability of the steel, resulting in slower austenite decomposition kinetics and hard secondary phase constituents (e.g., bainite and/or martensite) in the final microstructures. Hard constituents are undesirable because stretch-flange formability is markedly reduced due to the nucleation of voids at the interfaces between the relatively hard and soft phases [7,8]. Therefore, the ability to obtain both fine microalloy precipitates and a single-phase ferritic matrix in the final microstructure requires attention to thermomechanical processing, due to its effect on the austenite decomposition behavior [9].

Hot strip mills (HSM) are a key part of HSLA steel production. A typical, semi-continuous HSM consists of the following units: reheat furnace, roughing stand, transfer table, coilbox, finishing mill, runout table, and coiler [10]. The reheat furnace heats the slab to a suitable temperature to start hot rolling. The roughing stand is used for major reductions in slab thickness. The transfer table carries the slab, now called the transfer bar, from the roughing stand to the finishing mill. A coilbox is sometimes used to wrap the transfer bar into a coil to obtain a more uniform temperature profile [10]. The transfer bar is then delivered to the finishing mill, which is used for more precise gauge reductions. The strip is water-cooled from coolant headers along the runout table to control the final microstructure and then wrapped into a coil, which slowly cools to room temperature. Important processing steps during HSM processing to develop the desired, ferritic microstructures are controlled rolling in the finishing mill and accelerated cooling on the runout table (lowers the coiling temperature) [9].

Three types of austenite recrystallization behaviors are typical in HSM processing: static recrystallization, austenite pancaking (i.e., avoidance of recrystallization), and dynamic/metadynamic recrystallization [11]. The long interpass times and high temperatures, above the non-recrystallization temperature (T_{nr}), during rough rolling allow for nearly complete static recrystallization (SRX) to take place between rolling passes. Fine, equiaxed austenite grains are produced with negligible strain accumulation through SRX. Finish rolling is typically close to or below the T_{nr}, thus encouraging rapid strain-induced precipitation (SIP) of microalloy carbonitrides (e.g., Nb [12], Ti [13], and V [14] containing HSLA steels). These precipitates retard or even prevent the SRX of austenite grains; leading to pancaking, greater strain accumulation, and the generation of defects like dislocations and deformation bands. One of the major differences between HSM and plate mill rolling schedules is the interpass times, which are much shorter in the HSM and range from 0.2–5 s [11,15,16]. These short interpass times encourage strain accumulation and dynamic recrystallization (DRX) [11,17]. Fine, equiaxed austenite grains with negligible strain accumulation are also produced through DRX [18].

Hot torsion testing has been employed in numerous hot rolling simulations and recrystallization studies of HSLA steels for its ability to impose large amounts of strain while accurately controlling temperature, interpass time, and strain rate [9,15,18–21]. The conditioning of austenite into pancaked grains during HSM processing is important for final microstructural development. Whitley et al. [19] considered the evolution of the austenite grain morphology during hot torsion testing (shear deformation), and Figure 1 [19] shows a schematic overview of the expected austenite morphology at several stages throughout testing. The torsional axis is vertical to the page in Figure 1. Grains undergo one or both of the following processes at a given time: (i) shear deformation, thus becoming more elongated in morphology; and (ii) recrystallization, thus becoming refined and equiaxed in nature. Austenite grains are assumed to be initially equiaxed after soaking at high temperatures (Figure 1a), and they become elongated and rotated by the application of shear strain (Figure 1b) when viewed normal to the torsional axis of a cylindrical specimen [19]. If deformation occurs above the T_{nr}, SRX of the austenite is expected given sufficient interpass time (Figure 1c). However, if deformation occurs below the T_{nr}, pancaking of the austenite is expected and results in rotated grains with higher aspect ratios (Figure 1d). Austenite grains accommodate strain in this manner until there is sufficient driving force for recrystallization in the form of stored strain energy [19]. Strain-free grains can form in the regions of highest stored strain energy (e.g., grain boundaries and deformation substructure) (Figure 1e).

A metallographic technique was developed by Whitley et al. [19] to quantify the shear strain accumulation within (prior) austenite microstructures produced via hot torsion testing. Tested samples are sectioned parallel to the torsional axis and metallographically prepared to observe the prior austenitic microstructures. In this "tangential" plane cross-section, the inclination angles (θ') of prior austenite grains and other microstructural features can be measured with respect to the torsional axis and used to estimate the amount of shear strain that accumulated during testing (γ_{acc}) with the relationship

$$\gamma_{acc} = \tan(\theta'). \tag{1}$$

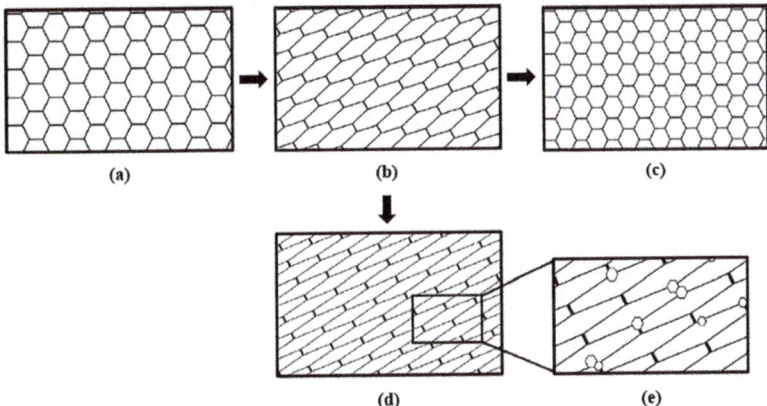

Figure 1. Schematic representation of the expected morphological evolution of austenite grains during hot torsion testing (shear deformation). The arrows indicate the progression of morphological changes. Austenite grains are (**a**) initially equiaxed and (**b**) elongate from imposed shear strain. Next, austenite grains either (**c**) recrystallize or (**d**) continue to deform in shear until (**e**) sufficient driving force is present for partial recrystallization. The torsional axis is vertical to the page. Adapted with permission from [19].

Multiple austenite deformation-recrystallization cycles during multi-pass torsion testing can result in mixed (prior) austenite microstructures displaying a distribution of inclination angles [18]. These distributions reflect the local variation in strain accumulation that can result during thermomechanical processing. Figure 2 [19] shows an example of this technique applied to a 1045 steel microalloyed with V that underwent industrial bar rolling simulations via hot torsion testing. Various microstructural features are highlighted with their corresponding inclination angles with respect to the torsional axis, which indicated:

(A) Manganese sulfide (MnS) inclusion, elongated and initially oriented parallel to the rolling direction. Since MnS inclusions do not recrystallize during thermomechanical processing, γ_{acc} represents the total shear strain imposed;
(B) Highly elongated prior austenite grain with the same inclination angle as MnS, suggesting no recrystallization during thermomechanical processing;
(C) Elongated prior austenite grain with an inclination angle less than (A) and (B), indicating some degree of recrystallization during thermomechanical processing, followed by subsequent deformation. The measured inclination angle represents the amount of shear strain that accumulated after the last recrystallization event, assuming an equiaxed grain morphology after recrystallization;
(D) Fine, equiaxed prior austenite grains that indicate recrystallization without shear strain accumulation.

The main aim of this work was to investigate differences in austenite strain accumulation before decomposition and the associated influence on (prior) austenite morphology and microstructural development after isothermal transformation for a low-carbon, Ti-Mo microalloyed steel.

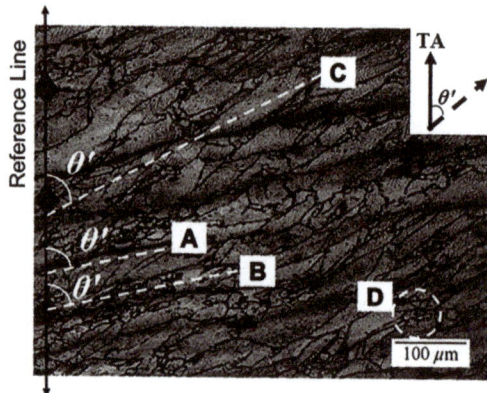

Figure 2. Light optical micrograph from the tangential plane cross-section of the prior austenite microstructure produced via industrial bar rolling simulations on a 1045 steel microalloyed with V using hot torsion testing. MnS inclusions (A) and some austenite grains (B) accumulate all imposed shear strain without recrystallizing. Other austenite grains (C) deform in shear after recrystallizing or (D) recrystallize following the final deformation pass. The torsional axis (TA) is vertical to the page, parallel to the reference line indicated. Adapted with permission from [19].

2. Materials and Methods

A low-carbon, Ti-Mo microalloyed steel was investigated, and its chemical composition is shown in Table 1. The experimental alloy was received as 16 mm thick, hot-rolled steel from Baoshan Iron & Steel Co. (Shanghai, China). Table 2 provides estimates of particular critical transformation temperatures using empirical equations found in the literature. These critical transformation temperatures were used to guide the thermomechanical processing of the experimental alloy. The following empirical equations were used: Andrews for Ac_1 and Ac_3 [22], Schacht for Ar_1 [23], Pickering for Ar_3 [24], Lee #2 for B_s [25], and Borrato for T_{nr} [26]. The M_s temperature was determined experimentally using dilatometry [27].

Table 1. Experimental Alloy Composition.

Wt pct	C	Mn	Si	Mo	Ti	Al	N	S	P	Fe
Ti-Mo	0.053	1.86	0.10	0.24	0.120	0.035	0.0036	0.0024	0.0085	Balance

Table 2. Critical Transformation Temperature Estimates for the Experimental Alloy.

°C	Ac_1	Ac_3	Ar_1	Ar_3	B_s	M_s	T_{nr}
Ti-Mo	706	875	725	871	606	362	995
Reference	[22]	[22]	[23]	[24]	[25]	Experimental	[26]

Solutionizing temperatures for relevant compounds in austenite were determined based on solubility expressions [28,29]. The calculations showed that titanium nitride (TiN) remains undissolved during solid-state processing, thus all nitrogen (N) was assumed to be removed from solid solution. The evolution of equilibrium phases as a function of temperature was predicted with Thermo-Calc® (Thermo-Calc Software, Solna, Sweden, Version 2019) using the TCFE9 database (assuming all N was already incorporated into TiN precipitates), and the results are shown in Figure 3. The MC equilibrium phase represents a mixed microalloy carbide exhibiting the NaCl (B1) crystal structure without the incorporation of N. From the determined equilibrium dissolution temperature of MC, a soaking temperature of 1250 °C was selected.

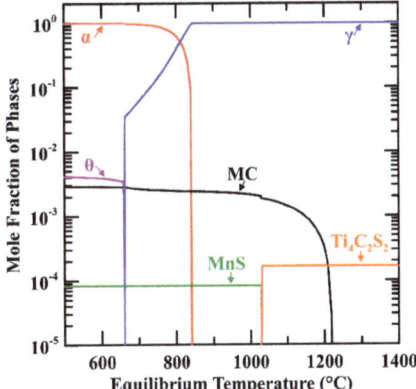

Figure 3. Evolution of equilibrium phases as a function of temperature predicted with Thermo-Calc® using the TCFE9 database. It was assumed that all N was first incorporated into TiN precipitates at much higher temperatures (greater than 1500 °C).

2.1. Hot Torsion Testing

Hot torsion testing was accomplished using a Gleeble® 3500 system equipped with the Hot Torsion Mobile Conversion Unit (Dynamic Systems Inc., Poestenkill, NY, USA). Sub-sized torsion samples were machined from the as-received material according to the schematic illustration in Figure 4, where the rolling direction was parallel to their lengths. Figure 5 shows a photograph inside the Hot Torsion Mobile Conversion Unit chamber, highlighting the setup used during testing. Both ends of the sample were restrained to keep the reduced gauge length fixed during torsion testing, and the Gleeble® 3500 was programmed to minimize axial stresses by adjusting the stroke arm. Axial stresses did not exceed ±10 Mpa during testing. Helium (He) gas was used as the quenchant for all tests and was directed from quench heads both in front and behind the sample. The torsion motor coupler within the Hot Torsion Mobile Conversion Unit was set for 20° free rotation, which allowed rapid acceleration of the torsion motor during deformation as well as a rapid reduction in torque on the sample during interpass times. Hot torsion testing was performed under the protective environment of argon (Ar) gas to minimize oxidation and decarburization near the surface of the sample. The oxygen partial pressure within the chamber was maintained under 30 ppm during testing and monitored using a PurgEye® 200 oxygen sensor (Huntingdon Fusion Techniques, Burry Port, UK).

The temperature of each sample was monitored at the mid-length of the reduced gauge section using a Metis Model MQ11 optical pyrometer (Process Sensors Corporation, Milford, MA, USA) that was calibrated at 1100 °C prior to testing. The pyrometer is not reliable below ~700 °C, so it was used to control temperature during heating, soaking, and deformation. An alternative method was required to control temperature during the accelerated cooling and isothermal holding steps due to the relatively low temperatures employed. Attaching a thermocouple to the fixed shoulder of the sample, where limited deformation occurs, and accounting for the temperature difference between the shoulder and mid-length of the reduced gauge section, proved to be an efficient method for controlling temperatures below ~700 °C. Therefore, a Type K thermocouple was spot welded to the surface of the sample roughly 0.5 mm away from the fixed shoulder (as shown in Figure 5) for testing that included isothermal holding. Each thermocouple wire was insulated with a small section of ceramic tubing to prevent short-circuiting. An isothermal holding temperature of 650 °C was planned, and preliminary testing showed that an offset value of approximately 18 °C (i.e., shoulder temperature of 632 °C) was appropriate to account for the temperature difference between the shoulder and reduced gauge section.

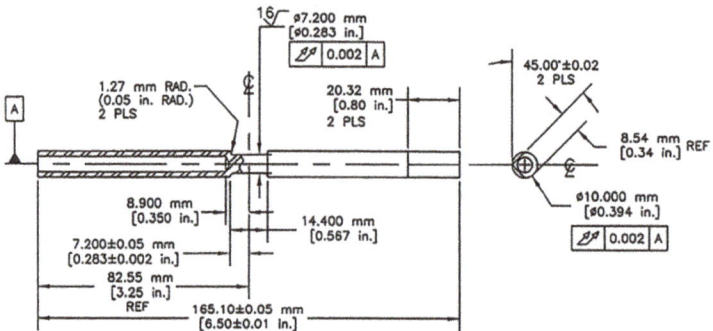

Figure 4. Schematic illustration of the sub-sized torsion samples tested with the Gleeble® 3500. Note the reduced gauge section and hollow ends. The hollow ends allow for a fairly consistent cross-sectional area throughout the entire sample and improve temperature uniformity during testing.

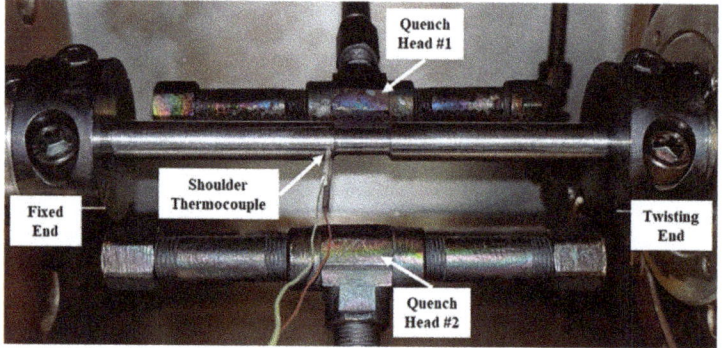

Figure 5. Photograph inside the Gleeble® 3500 Hot Torsion Mobile Conversion Unit chamber highlighting the hot torsion setup. The shoulder thermocouple was required to control temperature during accelerated cooling and isothermal holding.

Microstructural gradients within samples undergoing torsion testing require the selection of a specific radial position to determine deformation parameters since imposed shear strain (γ) varies with the radius of the reduced gauge section (r), according to

$$\gamma = \frac{r\,\varphi}{L} = \tan(\theta), \qquad (2)$$

where φ is the angle of twist (in radians), L is the reduced gauge length of the sample, and θ is the expected inclination angle with respect to the torsional axis corresponding to the imposed shear strain. The radial position used in this work to determine deformation parameters during HSM processing simulations was the "effective radius", which is positioned at 72.4 pct of the radial distance from the central axis [30]. Barraclough et al. showed that this location represents the bulk behavior for materials having a wide variety of strain rate sensitivities and/or strain hardening behaviors [30].

The following equations were used to convert the pass-by-pass shear strains and strain rates into appropriate angles of twist and twisting times for simulation purposes, as well as to convert the resulting torque to shear stress. The angle of twist for each pass was calculated according to

$$\varphi = \frac{\gamma\,L}{0.724\,r}, \qquad (3)$$

and the twisting time (t) for each pass was calculated according to

$$t = \frac{\gamma}{\dot{\gamma}}, \tag{4}$$

where $\dot{\gamma}$ is the shear strain rate. Barraclough et al. [30] developed an equation to convert torque (Γ) to shear stress (τ) considering the effective radius, which assumes pure torsion and uniform shear strain along the length of the reduced gauge section

$$\tau = \frac{3\Gamma}{2\pi[(r_2)^3 - (r_1)^3]}, \tag{5}$$

where r_2 is the outer surface radius and r_1 is the inner surface radius. Finally, the shear stress and strain values were converted to equivalent true stress (σ) and strain (ε) values by applying the Von Mises criterion

$$\sigma = \sqrt{3}\tau, \tag{6}$$

$$\varepsilon = \frac{\gamma}{\sqrt{3}}, \tag{7}$$

and used to determine the MFS for each pass according to

$$\text{MFS} = \frac{1}{\varepsilon_b - \varepsilon_a} \int_{\varepsilon_a}^{\varepsilon_b} \sigma \, d\varepsilon, \tag{8}$$

where ε_a and ε_b are the initial and final equivalent true strains per pass, respectively. The integrals were solved using analytical solutions of logarithmic regressions of the data according to

$$\int_{\varepsilon_a}^{\varepsilon_b} (a_1 \ln(\varepsilon) + b_1) d\varepsilon = [(a_1 \varepsilon) \ln(\varepsilon) - a_1 \varepsilon + b_1 \varepsilon]_{\varepsilon_a}^{\varepsilon_b}, \tag{9}$$

where a_1 and b_1 are constants. This approach typically assumes uniform constitutive mechanical properties of the material through the cross-section.

Table 3 summarizes the hot torsion testing schedule applied. The testing parameters were developed after consideration of the literature [9,11,15,18] and industrial processing [16,31]. Samples were heated at 5 °C/s to a soaking temperature of 1250 °C and held for 5 min to dissolve microalloy carbides. Rough rolling simulations consisted of four identical passes between 1240 and 1150 °C, each with relatively long interpass times to promote SRX of the austenite. Finish rolling simulations consisted of seven passes: either between 1150 and 1000 °C (designated as High T Finish) or between 1050 and 900 °C (designated as Low T Finish). Note that the designations of either High T Finish or Low T Finish simulations include the identical rough rolling simulations. These temperature ranges were selected to be above or mostly below the estimated T_{nr} of approximately 1000 °C to develop (prior) austenite microstructures with drastically different strain accumulation prior to decomposition. The roughing-to-finishing delay was 30 and 100 s for the High T Finish and Low T Finish simulations, respectively. Short interpass times are typical during HSM processing, which promote austenite pancaking and possibly DRX of the austenite later during finish rolling [11,15,16]. Overall, the amounts of true strain imparted during rough and finish rolling simulations were about 1.60 and 2.40, respectively, totaling about 4.00. The cooling rate between all passes was kept constant at 5 °C/s to ensure accurate temperature control. Additionally, relatively low target shear strain rates were utilized to ensure accuracy of the imparted shear strains. After the last finishing pass (F7), samples were either: (i) quenched as rapidly as possible (~43 °C/s) to room temperature to investigate the prior austenite grain (PAG) size and morphology, or (ii) accelerated cooled at ~30 °C/s to an isothermal holding temperature of 650 °C, held for 30 min, and finally quenched as rapidly as possible to room temperature to investigate polygonal ferrite characteristics and the presence of any secondary phase constituents.

Table 3. Hot Torsion Testing Schedule to Simulate High T Finish* and Low T Finish** Processing.

Pass No.	Temperature (°C)			True Strain	Shear Strain	Twist Angle (rad)	Shear Strain Rate (s⁻¹)	Interpass Time (s)		
R1	1240			0.40	0.80	4.42	5		20	
R2	1210			0.40	0.80	4.42	5		20	
R3	1180			0.40	0.80	4.42	5		20	
R4	1150			0.40	0.80	4.42	5	30*	or	100**
F1	1150*	or	1050**	0.50	1.00	5.52	10		8	
F2	1110*	or	1010**	0.50	1.00	5.52	10		8	
F3	1070*	or	970**	0.40	0.80	4.42	10		6	
F4	1040*	or	940**	0.40	0.80	4.42	10		4	
F5	1020*	or	920**	0.30	0.60	3.31	10		2	
F6	1010*	or	910**	0.20	0.40	2.21	10		2	
F7	1000*	or	900**	0.10	0.20	1.10	10	Quench	or	Hold

Previous research [19] has shown that a tangential orientation is best for investigating PAG morphologies and quantifying the amount of shear strain that accumulated within the microstructure for samples tested via hot torsion. This orientation is presented schematically in Figure 6. Tested samples were prepared in the tangential orientation according to the following procedure. First, the reduced gauge section was cut free from its ends on both sides of the sample. Next, the gauge section was cut in half (perpendicular to the torsional axis) to reveal the "thermal plane". Note that the thermal plane corresponds to the mid-length of the reduced gauge section, approximately where the optical pyrometer was aligned prior to testing. Then, each piece was cut in half (parallel to the torsional axis). Finally, these quartered sections were mounted in Bakelite and precision ground to the radial position of interest using measurements of chord length to reveal the tangential plane.

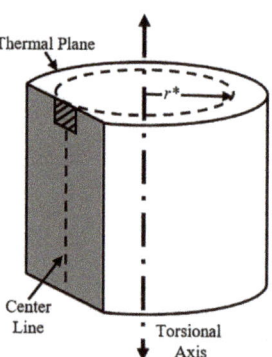

Figure 6. Schematic of the tangential orientation (evaluated at some radial position, r^*) of samples tested via hot torsion. The shaded region represents the tangential plane of interest, and the boxed region represents the approximate location of microstructural characterization.

In addition to the effective radius, two other radial positions were selected to further investigate how strain influences austenite conditioning and the final microstructures during HSM processing. Recall that shear strain varies with the radius of the reduced gauge section according to Equation (2). The target shear strain for the HSM processing simulations was 8.00, where the effective radius (0.724 radial position) was used to determine the deformation parameters. Radial positions of 0.50 and 0.90 were also selected to represent an extensive range of possible shear strain accumulation.

The approximate shear strain for the 0.50 radial position is 5.52, which might represent processing of thicker gauge material, and the approximate shear strain for the 0.90 radial position is 9.94, which might represent processing of thinner gauge material. All three radial positions were investigated using the tangential orientation for each condition produced via hot torsion testing.

2.2. Microstructural Characterization

Samples for microscopic evaluation were sectioned, mounted in Bakelite, and prepared using standard metallographic procedures. Samples were etched with either a 1 pct nital or a modified Béchet–Beaujard reagent. The 1 pct nital reagent was used to reveal ferrite grain boundaries and secondary phase constituents. The modified Béchet-Beaujard reagent was used to reveal PAG boundaries and consisted of 200 cm^3 of deionized water, 2.6 g of picric acid solids, 8 cm^3 of Teepol (wetting agent), and 2 cm^3 of hydrochloric acid. This reagent was heated to 65 °C on a temperature-controlled hot plate equipped with a thermocouple feedback and stirred with a magnetic stir rod throughout the etching process. After each interval of etching with the modified Béchet-Beaujard reagent, samples were immersed in a methanol bath, ultrasonicated, and dried with a heat gun. PAG boundaries were highlighted in black to enhance their clarity within the provided micrographs.

General imaging of microstructures was accomplished using light optical microscopy (LOM), where the micrographs were used to determine average grain size, phase area fraction, etc. LOM was performed with an Olympus Model PMG3 inverted light microscope (LECO Corporation, St. Joseph, MI, USA) with a PAXcam Model PX-CM digital camera (MIS Inc., Villa Park, IL, USA) and PAX-it! Image analysis software (MIS Inc., Villa Park, IL, USA, Version 7.8). Electron backscatter diffraction (EBSD) analysis was performed on isothermally transformed microstructures with a JSM-7000F field emission-scanning electron microscope (JEOL USA Inc., Peabody, MA, USA) to investigate polygonal ferrite grain diameter distributions. Prior to EBSD analysis, samples were metallographically prepared and vibratory polished for at least 4 h using 0.02 µm colloidal silica solution. EBSD scans were performed at an accelerating voltage of 20 keV, calibrated EBSD camera working distance of 18 mm, and step size of 0.1 µm. EBSD data were collected with a Hikari Pro detector (EDAX Inc., Mahwah, NJ, USA) using the TEAM™ software (EDAX Inc., Mahwah, NJ, USA, Version 4.5), and the datasets were analyzed with the Orientation Imaging Microscopy Analysis© software (EDAX Inc., Mahwah, NJ, USA, Version 8.1). EBSD datasets were cleaned using the following functions: Neighbor Orientation Correction (Level 3, Tolerance 5.0, Minimum Confidence Index (CI) 0.10); Grain CI Standardization (Tolerance 5.0, Minimum Size 3, Multi Row 1); and Neighbor CI Correlation (Minimum CI 0.30, Single Iteration).

Polygonal ferrite and prior austenite grain sizes were determined with the concentric circle method utilizing the ImageJ software (open-source). The intercepts of the circles with the grain boundaries were counted, and the average intercept lengths were calculated and reported as the average grain sizes. A total of 1000 or more grain boundary intercepts were counted for each condition to determine a representative grain size. Note that prior austenitic twins were observed within some PAGs, but these were not considered in the PAG size measurements. Aspect ratios of the PAGs were determined by measuring the major and minor axes of individual grains (assuming an elliptical shape) with the ImageJ software and calculating their ratios. Inclination angles of the PAGs with respect to the torsional axis were measured with the ImageJ software and used to quantify the amount of shear strain that accumulated within the (prior) austenite microstructures using the previously described metallographic technique developed by Whitley et al. [19]. A total of 100 or more PAGs were measured for each condition to determine a representative aspect ratio and inclination angle.

The ImageJ software was also used to determine the area fraction of secondary-phase constituents using an image thresholding procedure. This procedure was employed because of the distinct difference in the etching response of the polygonal ferrite (carbon depleted) and secondary phase constituents (carbon rich) with 1 pct nital. First, LOM micrographs were thresholded from grayscale images to black-and-white images using a grayscale value range that best captured the secondary phase constituents but without capturing polygonal ferrite grain boundaries. After thresholding, pixels in the desired grayscale value range were transformed to white pixels (representing the secondary phase constituents), while all other pixels were transformed to black pixels (representing polygonal ferrite). Figure 7 provides an example of the thresholding process. Finally, the ratio of the number of white or black pixels to the total number of pixels was used to determine the secondary phase constituents or polygonal ferrite area fractions, respectively. This procedure was repeated at five or more different areas of the microstructure for each condition to determine representative phase area fractions.

Samples were metallographically prepared and etched with 1 pct nital prior to Vickers microhardness testing in order to relate microhardness to microstructural features. Vickers microhardness testing was performed with a LM110 hardness tester (LECO Corporation, St. Joseph, MI, USA) according to ASTM Standard E384-17 [32]. A 5 × 5 array of indents with an indentation load of 100 g was used to determine the representative values of the polygonal ferrite in each isothermally transformed condition. Indentations immediately adjacent to secondary phase constituents (including large TiN precipitates) were disregarded. Very low indentation loads of 10 g were used to investigate the small secondary phase constituents. Note that indentation loads like 10 g can result in consistently higher microhardness values [33], so values determined with this indentation load were used for comparison only.

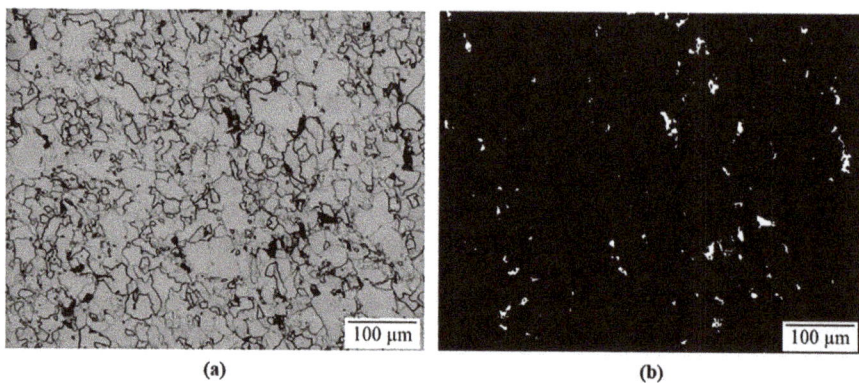

Figure 7. Example of the image thresholding process used to determine the area fraction of constituent phases in mixed microstructures: (**a**) original light optical microscopy (LOM) micrograph, where light etched regions represent polygonal ferrite and dark etched regions represent the secondary phase constituents, and (**b**) processed image, where black pixels represent polygonal ferrite and white pixels represent the secondary phase constituents.

3. Results

3.1. Strain Accumulation in the (Prior) Austenite

Figure 8 shows the MFS as a function of inverse absolute temperature for the (a) High T Finish and (b) Low T Finish HSM processing simulations. Two linear regressions are included in each figure: the first for the roughing stage and the second for the finishing stage (without including the softening regions). Note that the imparted shear strain was incrementally reduced from the F1 to F7 deformation steps (1.00 to 0.20) during finish rolling simulations, and deformation parameters were identical between High T Finish and Low T Finish simulations. The relatively low imparted shear strain for the F7 deformation step (0.20) contributed to the drop in the MFS observed for that deformation step in both simulations. Thus, the data obtained from the F7 deformation steps were not considered in the analysis. The MFS continuously increased from the F1 to F6 deformation steps for High T Finish simulations. However, a softening region was indicated following the F4 deformation step for Low T Finish simulations. This softening region was not indicated for High T Finish simulations when subjected to identical strains per pass. The rates at which MFS increased with respect to inverse absolute temperature were similar between the two rolling simulations for both the roughing and finishing stages. Low T Finish simulations were expected to cause a sharper increase in the rate at which MFS changed with respect to absolute temperature relative to High T Finish simulations, but this behavior was not observed. The lack of a sharper increase in the rate of change in MFS with respect to absolute temperature for Low T Finish simulations may be due to limited strain-induced precipitation as a result of the finish rolling start (F1) temperature being above the estimated T_{nr} coupled with the relatively short interpass times. However, the magnitudes of the MFS were greater overall for the Low T Finish simulations, since lower deformation temperatures were employed. The softening region indicated in the last passes near the end of Low T Finish simulations may suggest recrystallization of the austenite after extensive strain accumulation. Characterization of the prior austenite microstructures of samples quenched immediately after the final finishing deformation step (F7) was performed for each simulation to relate the observed PAG morphology with the amount of shear strain that accumulated during thermomechanical processing. Additionally, a separate sample underwent Low T Finish simulations but was quenched after the fourth finishing deformation step (F4) to investigate the prior austenite microstructure immediately preceding the softening region indicated in Figure 8b.

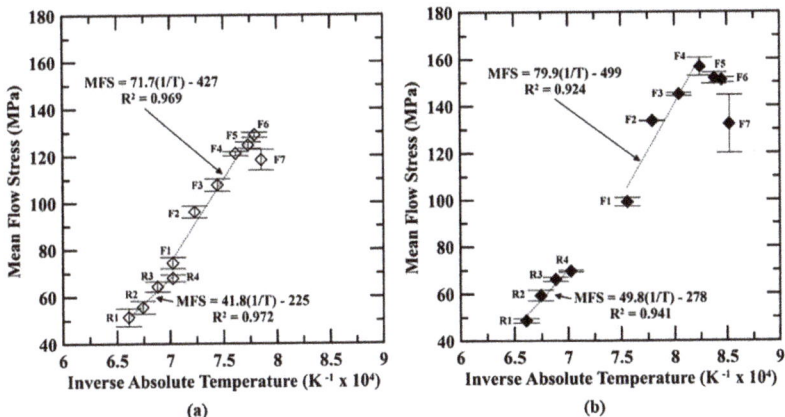

Figure 8. Summary of mean flow stress as a function of inverse absolute temperature for (**a**) High T Finish and (**b**) Low T Finish simulations via hot torsion testing. Two linear regressions are included in each figure: the first for the roughing stage (R) and the second for the finishing stage (F), excluding the softening region. The error bars represent one standard deviation.

The prior austenite microstructures produced via hot torsion testing are shown in Figure 9 after etching with a modified Béchet-Beaujard reagent. Samples that underwent High T Finish simulations are shown in Figure 9a–c, and samples that underwent Low T Finish simulations are shown in Figure 9d–f. Three radial positions are shown for each HSM processing simulation: mid-radius, 0.50 (a and d); effective radius, 0.724 (b and e); and near-surface, 0.90 (c and f). PAG sizes were significantly refined for Low T Finish simulations (roughly 6–9 μm) compared to High T Finish simulations (roughly 14–15 μm). PAGs were mostly equiaxed for High T Finish simulations, and the (prior) austenite was only slightly refined by the additional strain near the surface. However, Low T Finish simulations resulted in mixed (prior) austenite microstructures consisting of larger, pancaked, and aligned grains and smaller, equiaxed grains. PAGs were not inclined to the torsional axis (horizontal to the page) for High T Finish simulations, but PAGs were inclined to the torsional axis for Low T Finish simulations and exhibited an average inclination angle of approximately 60 ± 10°. Interestingly, the average inclination angles for the Low T Finish simulations were consistent between the three radial positions investigated. This average prior austenite inclination angle corresponds to an accumulation of roughly 1.7 shear strain (0.85 true strain) within the microstructure according to Equation (1). If no shear strain accumulates during rough rolling simulations, a total of 4.8 shear strain could accumulate during finish rolling simulations. Therefore, roughly 36 pct of the total shear strain is accumulated during Low T Finish simulations, while negligible shear strain is accumulated during High T Finish simulations. The differences in prior austenite morphology are highlighted in Figure 10, where higher magnification images are shown for the 0.724 radial position.

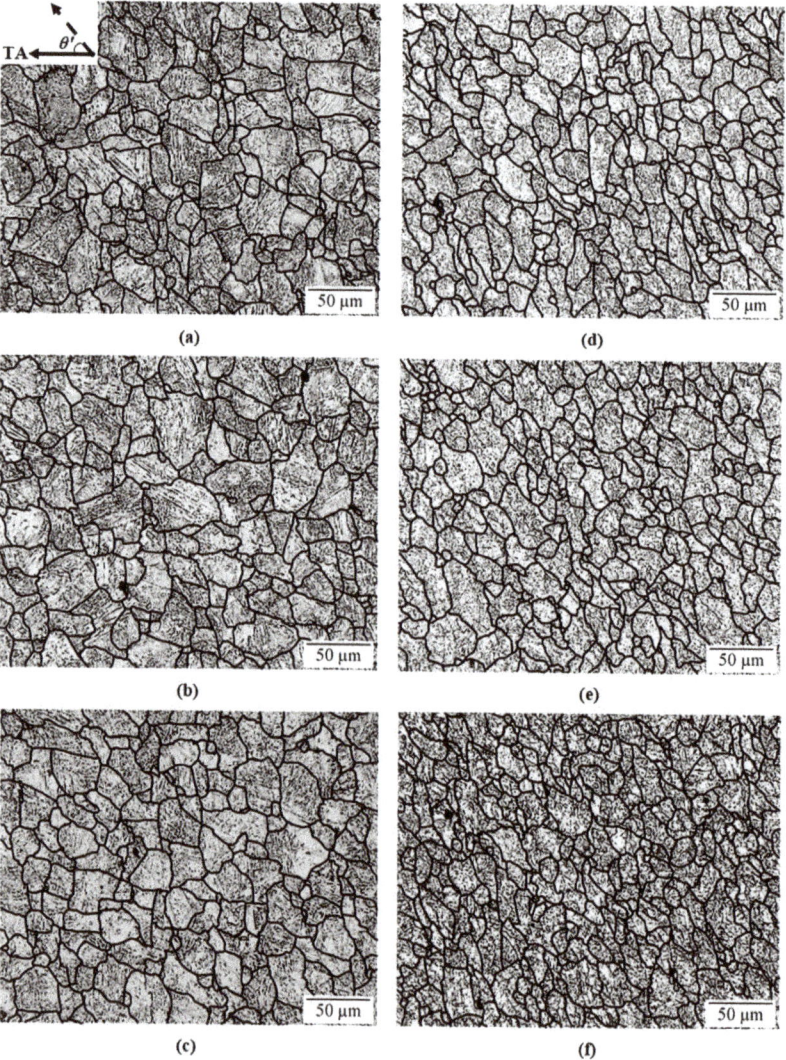

Figure 9. LOM micrographs of the prior austenite microstructures produced via (**a**–**c**) High T Finish and (**d**–**f**) Low T Finish simulations. The following radial positions are shown: (**a**) and (**d**) 0.50; (**b**) and © 0.724; and (**c**) and (**f**) 0.90. The torsional axis (TA) is horizontal to the page. Etched with a modified Béchet-Beaujard reagent, and prior austenite grain boundaries are highlighted in black.

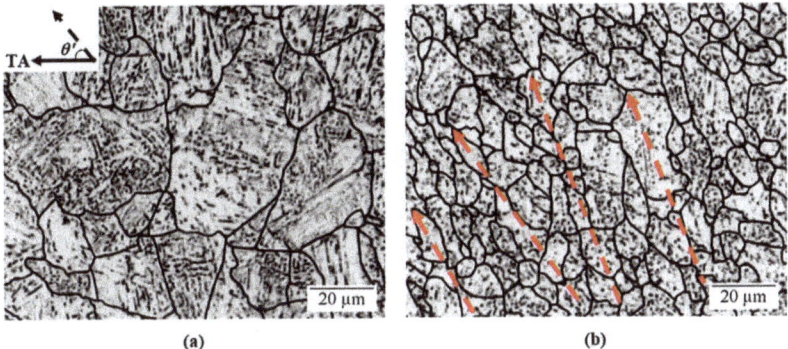

Figure 10. Higher magnification LOM micrographs of the prior austenite microstructures produced via (**a**) High T Finish and (**b**) Low T Finish simulations highlighting differences in morphology. The 0.724 radial position is shown in both. The torsional axis (TA) is horizontal to the page. Dashed lines in (**b**) show the inclination angles of various prior austenite grains. Etched with a modified Béchet-Beaujard reagent, and prior austenite grain boundaries are highlighted in black.

Figure 11 summarizes the average PAG (a) size and (b) aspect ratio with respect to radial position determined for microstructures produced via High T Finish and Low T Finish simulations. Note that as radial position increases (towards the surface), the amount of shear strain imposed during torsion testing increases according to Equation (2). PAG size decreases with increasing radial position for both High T Finish and Low T Finish simulations. Radial position had a greater influence on PAG size refinement for Low T Finish simulations compared to High T Finish simulations. PAG aspect ratios ranged between approximately 1–2 and 1–3.5 for High T Finish and Low T Finish simulations, respectively, considering all radial positions. The large range of PAG aspect ratios for Low T Finish simulations was a result of its mixed nature involving pancaked grains with higher aspect ratios along with fine, equiaxed grains with lower aspect ratios. The PAG morphology becomes more equiaxed with increasing radial position, especially for Low T Finish simulations. The micrographs provided in Figure 9d–f and data summarized in Figure 11 suggest that greater imposed shear strain results in a greater amount of fine, equiaxed (prior) austenite grains for Low T Finish simulations. The prior austenite microstructures and MFS results are consistent with the occurrence of recrystallization due to high strain accumulation.

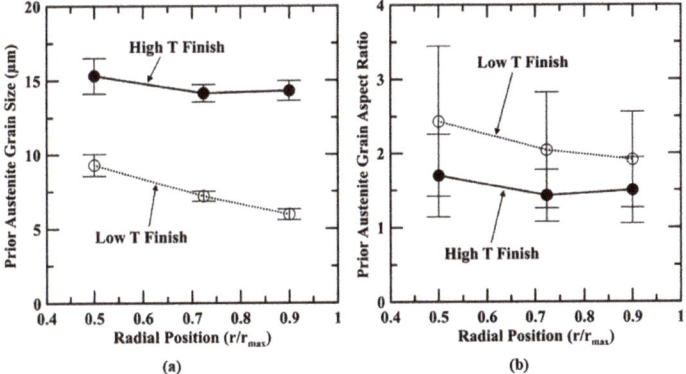

Figure 11. Prior austenite grain (**a**) size and (**b**) aspect ratio with respect to radial position determined for microstructures produced via High T Finish and Low T Finish simulations. The error bars represent one standard deviation.

The prior austenite microstructure produced via Low T Finish simulations and quenched after the fourth finishing deformation step (F4) is shown in Figure 12 after etching with a modified Béchet–Beaujard reagent. Only the 0.724 radial position is shown. PAGs were mostly pancaked and inclined to the torsional axis, but some fine, equiaxed grains can also be observed (Figure 12b). The PAG size was approximately 11 ± 1 μm, which was greater than the PAG size that resulted from quenching after the final finishing deformation step (F7). That value was approximately 7.2 ± 0.3 μm at the effective radius. The finer and more equiaxed (prior) austenitic microstructure obtained after complete Low T Finish simulations compared to quenching after the F4 step suggests that more austenite recrystallization occurred (whether statically and/or dynamically) due to the additional shear strain that was imparted during thermomechanical processing. Figure 13 shows PAG inclination angle histograms constructed for Low T Finish simulations quenched after the (a) F4 step and (b) F7 step. Data were collected at the 0.724 radial position. Interestingly, both simulations resulted in similar distributions of inclination angles, and the average inclination angles were determined to be 60° ± 8° and 59° ± 10° for interrupted and complete Low T Finish simulations, respectively. This microstructural inclination angle may correspond to a critical amount of shear strain (approximately 1.7) that must accumulate during thermomechanical processing before a significant driving force for austenite recrystallization is available.

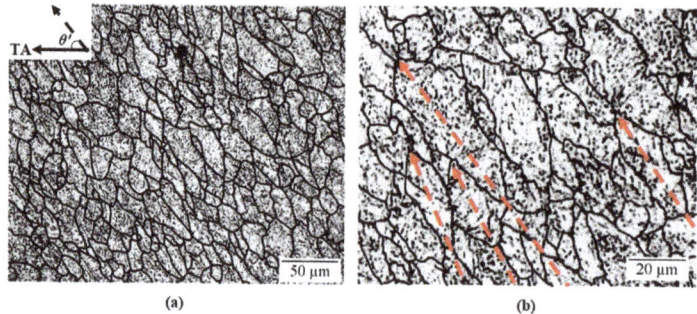

Figure 12. LOM micrographs of the prior austenite microstructures produced via Low T Finish simulations and quenched after the fourth finishing deformation step (F4): (**a**) lower and (**b**) higher magnification. The 0.724 radial position is shown in both. The torsional axis (TA) is horizontal to the page. Dashed lines in (**b**) show the inclination angles of various prior austenite grains. Etched with a modified Béchet–Beaujard reagent, and prior austenite grain boundaries are highlighted in black.

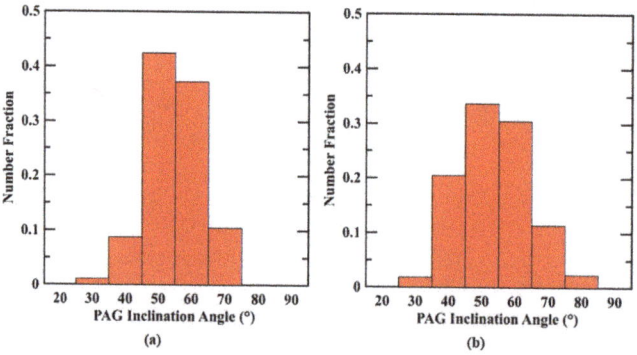

Figure 13. Prior austenite grain (PAG) inclination angle histograms for microstructures produced via Low T Finish simulations: (**a**) quenched after the fourth finishing deformation step (F4) and (**b**) quenched after the final finishing deformation step (F7). Data were obtained at the 0.724 radial position for both cases.

3.2. Isothermally Transformed Microstructures

The polygonal ferrite microstructures produced via hot torsion testing followed by isothermal holding at 650 °C for 30 min are shown in Figure 14 after a 1 pct nital etch. Samples that underwent High T Finish simulations are shown in Figure 14a–c, and samples that underwent Low T Finish simulations are shown in Figure 14d–f. Three radial positions are shown for each HSM processing simulation: mid-radius, 0.50 (a and d); effective radius, 0.724 (b and e); and near-surface, 0.90 (c and f). Table 4 summarizes the average polygonal ferrite area fraction, grain size, and Vickers microhardness for each condition. All radial positions for each HSM processing simulation exhibited a mostly ferritic microstructure (greater than 97 pct), where a small amount of secondary phase constituents was observed in some conditions. However, High T Finish simulations clearly resulted in more secondary phase constituents than Low T Finish simulations. The average Vickers microhardness for the secondary phase constituents in the High T Finish and Low T Finish simulations (both evaluated at the 0.724 radial position) was approximately 550 HV (10 g). The presence of hard, secondary phase constituents negatively impacts stretch-flange formability [7,8]. Polygonal ferrite grain sizes were significantly refined for Low T Finish simulations (roughly 2.7–3.2 µm) compared to High T Finish simulations (roughly 5.2–6.0 µm). The refinement observed for Low T Finish simulations is due to the increased boundary surface area per grain volume from the austenite pancaking and generation of defects, both of which enhance the ferrite nucleation kinetics during austenite decomposition [12]. Polygonal ferrite grain sizes were relatively consistent for all the radial positions investigated for each processing simulation. Finally, polygonal ferrite microhardness was similar for the Low T Finish conditions, due in part to the refined grain size, except for the 0.90 radial position, which exhibited a marked decrease in microhardness. Thermomechanical processing may have an important influence on microalloy precipitation within the ferrite, and this behavior is being investigated in ongoing research.

Table 4. Characteristics of Isothermally Transformed Microstructures.

Simulation Designation	Radial Position	Ferrite Area Fraction (pct)	Ferrite Grain Size (µm)	Ferrite Vickers Microhardness (100 g)
High T Finish	0.50	97.4	5.4 ± 0.3	263 ± 9
	0.724	98.7	5.2 ± 0.3	271 ± 13
	0.90	> 99	6.0 ± 0.4	256 ± 8
Low T Finish	0.50	98.8	3.2 ± 0.1	283 ± 8
	0.724	> 99	3.1 ± 0.1	284 ± 6
	0.90	> 99	2.7 ± 0.05	245 ± 11

EBSD analysis of polygonal ferrite microstructures produced via both processing simulations was performed to measure grain diameter distributions. Composite image quality (IQ) and inverse pole figure (IPF) maps for the High T Finish and Low T Finish simulations are shown in Figure 15a,b, respectively, for the 0.724 radial position. These maps provide a clear delineation of polygonal ferrite grains and highlight the substantial differences in overall morphology that resulted from the different HSM processing simulations. Figure 16 shows ferrite grain diameter histograms constructed for (a) High T Finish and (b) Low T Finish simulations, where data were collected and averaged from four 90×90 µm^2 regions. High T Finish simulations exhibited a bimodal distribution centered around 2–4 and 10–20 µm and Low T Finish simulations exhibited a unimodal distribution centered around 1–4 µm. Additionally, the average ferrite grain diameters determined by EBSD analysis were approximately 4.8 ± 0.1 and 2.7 ± 0.1 µm for the High T Finish and Low T Finish simulations, respectively. These values are similar to the values determined via LOM using the concentric circle method for the 0.724 radial position. Overall, these data suggest that extensive austenite strain accumulation before decomposition is required to achieve fine, homogeneous microstructures of polygonal ferrite and to avoid small amounts of hard, secondary-phase constituents.

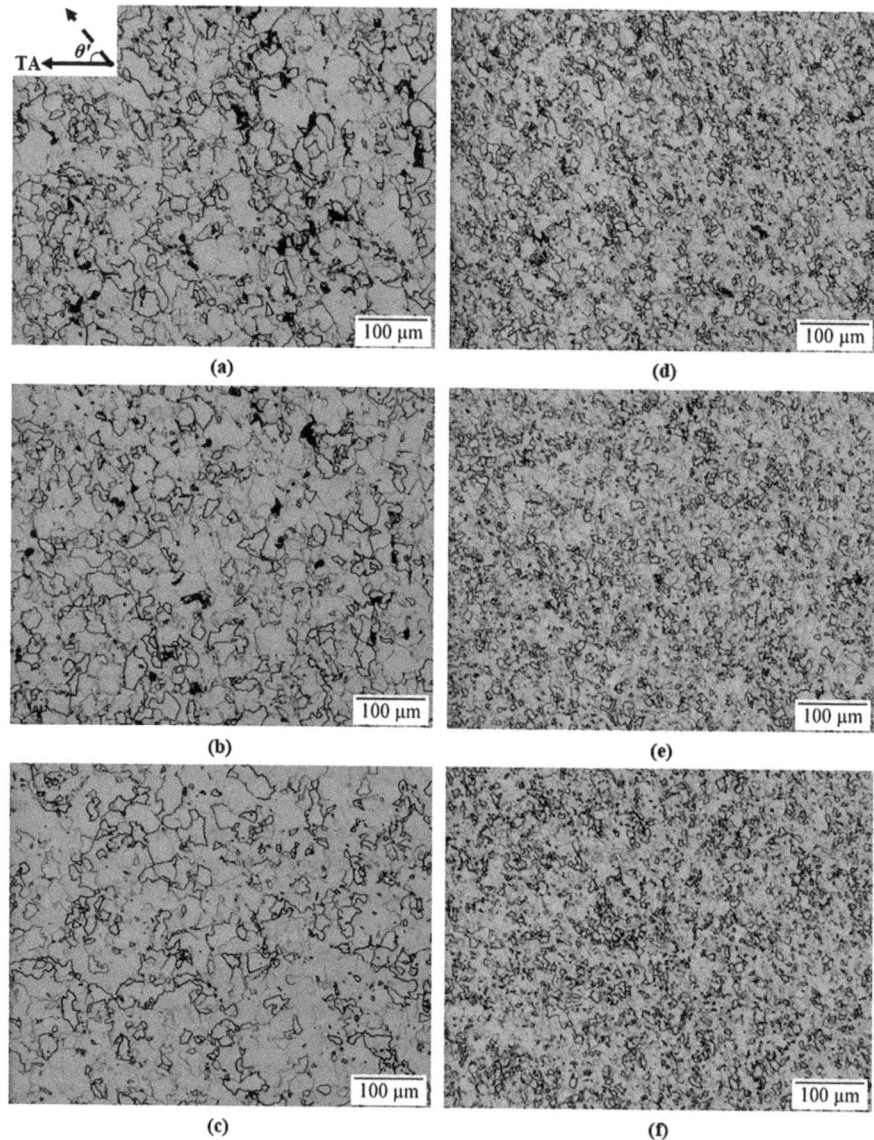

Figure 14. LOM micrographs of the polygonal ferrite microstructures produced via (**a**–**c**) High T Finish and (**d**–**f**) Low T Finish simulations followed by isothermal holding at 650 °C for 30 min. The following radial positions are shown: (**a**) and (**d**) 0.50; (**b**) and (**e**) 0.724; and (**c**) and (**f**) 0.90. The torsional axis (TA) is horizontal to the page. Etched with 1 pct nital.

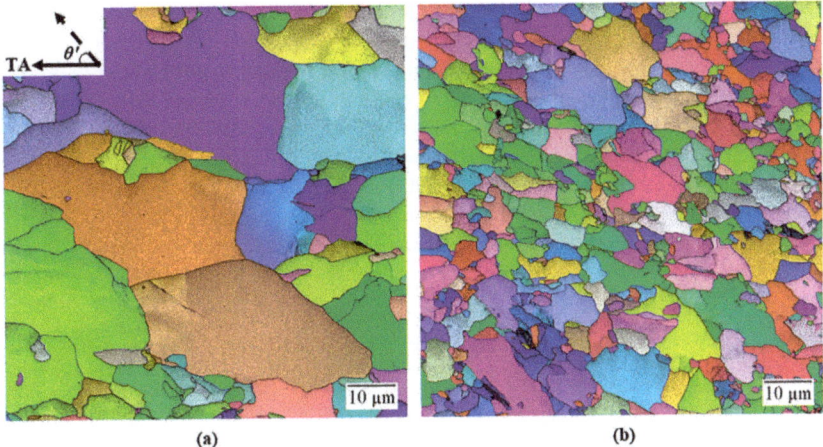

Figure 15. Composite image quality and inverse pole figure maps obtained with EBSD analysis of the polygonal ferrite microstructures produced via (**a**) High T Finish and (**b**) Low T Finish simulations followed by isothermal holding at 650 °C for 30 min. The 0.724 radial position is shown in both. The torsional axis (TA) is horizontal to the page.

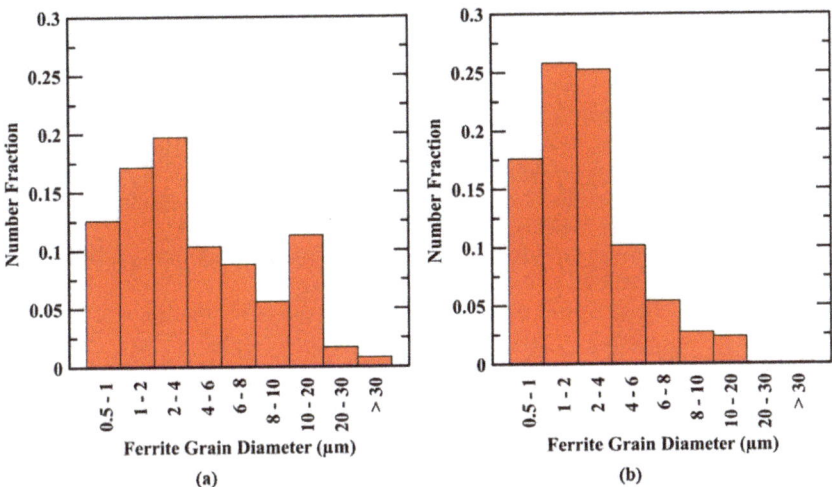

Figure 16. Ferrite grain diameter histograms obtained with EBSD analysis of the polygonal ferrite microstructures produced via (**a**) High T Finish and (**b**) Low T Finish simulations followed by isothermal holding at 650 °C for 30 min. Data were obtained at the 0.724 radial position for both.

4. Conclusions

The effects of extensive differences in austenite strain accumulation before decomposition on (prior) austenite morphology and microstructural development after isothermal transformation were investigated for a low-carbon, Ti-Mo microalloyed steel. Roughly 36 pct of the total shear strain imparted during finish rolling simulations accumulated within the prior austenitic microstructures for Low T Finish simulations, while negligible shear strain was accumulated for High T Finish simulations. The amounts of shear strain that accumulated during Low T Finish simulations were similar (roughly 1.7) at multiple radial positions subjected to different total strains, as well as for both interrupted and complete multi-pass rolling simulations. This value of accumulated shear strain may indicate a critical amount of strain that must accumulate during thermomechanical processing before a sufficient driving force for austenite recrystallization is available in this steel. Low T Finish simulations resulted in a significant refinement in PAG size compared to High T Finish simulations, and greater imposed shear strain resulted in a greater amount of fine, equiaxed austenite grains (intermixed with larger, pancaked grains) for Low T Finish simulations. Additionally, Low T Finish simulations followed by isothermal holding at 650 °C for 30 min resulted in a significant refinement in polygonal ferrite grain sizes and better homogenization compared to High T Finish simulations. Therefore, the results suggest that extensive austenite strain accumulation before decomposition is required to achieve fine, homogeneous microstructures of polygonal ferrite and to avoid small amounts of hard, secondary-phase constituents that may diminish stretch-flange formability.

Author Contributions: Conceptualization, C.A.F., J.G.S. and E.D.M.; Data curation, C.A.F.; Formal analysis, C.A.F.; Investigation, C.A.F.; Methodology, C.A.F. and J.G.S.; Project administration, J.G.S. and E.D.M.; Resources, J.G.S., E.D.M. and K.O.F.; Supervision, J.G.S. and E.D.M.; Visualization, C.A.F. and J.G.S.; Writing—original draft, C.A.F.; Writing—review and editing, J.G.S., E.D.M. and K.O.F. All authors have read and agreed to the published version of the manuscript.

Funding: This research was funded by the sponsors of the Advanced Steel Processing and Products Research Center, an NSF industry/university cooperative research center.

Acknowledgments: The authors gratefully acknowledge Huanrong Wang (Baoshan Iron & Steel Co.) for providing the experimental materials.

Conflicts of Interest: The authors declare no conflict of interest. The sponsors had no role in the execution, interpretation, or writing of the study.

References

1. Dinda, S.; DiCello, J.; Kasper, A. Using Microalloyed Steels to Reduce Weight of Automotive Parts. In Proceedings of the Microalloying 75, Washington, DC, USA, 1–3 October 1975; pp. 531–538.
2. Jha, G.; Haldar, A.; Bhaskar, M; Venugopalan, T. Development of High Strength Hot Rolled Steel Sheet for Wheel Disc Application. *Mater. Sci. Technol.* **2011**, *27*, 1131–1137. [CrossRef]
3. Kamikawa, N.; Abe, Y.; Miyamoto, G.; Funakawa, Y.; Furuhara, T. Tensile Behavior of Ti, Mo-Added Low Carbon Steels with Interphase Precipitation. *ISIJ Int.* **2014**, *54*, 212–221. [CrossRef]
4. Funakawa, Y.; Fujita, T.; Yamada, K. Metallurgical Features of NANOHITEN™ and Application to Warm Stamping. *JFE Tech. Rep.* **2013**, *18*, 74–79.
5. Okamoto, R.; Borgenstam, A.; Agren, J. Interphase Precipitation in Niobium-Microalloyed Steels. *Acta Mater.* **2010**, *58*, 4783–4790. [CrossRef]
6. Funakawa, Y.; Shiozaki, T.; Tomita, K.; Yamamoto, T.; Maeda, E. Development of High Strength Hot-rolled Sheet Steel Consisting of Ferrite and Nanometer-sized Carbides. *ISIJ Int.* **2004**, *44*, 1945–1951. [CrossRef]
7. Chun, E.; Do, H.; Kim, S.; Nam, D.; Park, Y.; Kang, N. Effect of Nanocarbides and Interphase Hardness Deviation on Stretch-Flangeability in 998 MPa Hot-Rolled Steels. *Mater. Chem. Phys.* **2013**, *140*, 307–315. [CrossRef]
8. Hu, J.; Du, L.; Wang, J.; Sun, Q. Cooling Process and Mechanical Properties Design of Hot-rolled Low Carbon High Strength Microalloyed Steel for Automotive Wheel Usage. *Mater. Des.* **2014**, *53*, 332–337. [CrossRef]
9. Nakata, N.; Militzer, M. Modelling of Microstructure Evolution during Hot Rolling of a 780 MPa High Strength Steel. *ISIJ Int.* **2005**, *45*, 82–90. [CrossRef]
10. Yildiz, S.; Forbes, J.; Huang, B.; Zhang, Y.; Wang, F.; Vaculik, V.; Dudzic, M. Dynamic Modeling and Simulation of a Hot Strip Finishing Mill. *Appl. Math. Model.* **2009**, *33*, 3208–3225. [CrossRef]
11. Siciliano, F.; Jonas, J. Mathematical Modeling of the Hot Strip Rolling of Microalloyed Nb, Multiply-alloyed Cr-Mo, and Plain C-Mn Steels. *Metall. Trans. A* **2000**, *31A*, 511–530. [CrossRef]
12. Speer, J.; Hansen, S. Austenite Recrystallization and Carbonitride Precipitation in Niobium Microalloyed Steels. *Metall. Trans. A* **1989**, *20A*, 25–38. [CrossRef]
13. Wang, Z.; Zhang, H.; Guo, C.; Liu, W.; Yang, Z.; Sun, X.; Zhang, Z.; Jiang, F. Effect of Molybdenum Addition on the Precipitation of Carbides in the Austenite Matrix of Titanium Micro-alloyed Steels. *J. Mater. Sci.* **2016**, *51*, 4996–5007. [CrossRef]

14. Baker, T. Processes, Microstructure and Properties of Vanadium Microalloyed Steels. *Mater. Sci. Technol.* **2009**, *25*, 1083–1107. [CrossRef]
15. Samuel, F.; Yue, S.; Jonas, J.; Zbinden, B. Modeling of Flow Stress and Rolling Load of a Hot Strip Mill by Torsion Testing. *ISIJ Int.* **1989**, *29*, 878–886. [CrossRef]
16. Merwin, M.; U.S. Steel Research and Technology Center, Munhall, PA, USA. Personal communication, 2018.
17. Pereda, B.; Fernandez, A.; Lopez, B.; Rodriguez-Ibabe, J. Effect of Mo on Dynamic Recrystallization Behavior of Nb-Mo Microalloyed Steels. *ISIJ Int.* **2007**, *47*, 860–868. [CrossRef]
18. Samuel, F.; Yue, S.; Jonas, J.; Barnes, K. Effect of Dynamic Recrystallization on Microstructural Evolution during Strip Rolling. *ISIJ Int.* **1990**, *30*, 216–225. [CrossRef]
19. Whitley, B.; Araujo, A.; Speer, J.; Findley, K.; Matlock, D. Analysis of Microstructure in Hot Torsion Simulation. *Mater. Perform. Charact.* **2015**, *4*, 307–321. [CrossRef]
20. Bai, D.; Yue, S.; Sun, W.; Jonas, J. Effect of Deformation Parameters on the No-Recrystallization Temperature in Nb-Bearing Steels. *Metall. Trans. A* **1993**, *24A*, 2151–2159. [CrossRef]
21. Calvo, J.; Collins, L.; Yue, S. Design of Microalloyed Steel Hot Rolling Schedules by Torsion Testing: Average Schedule vs. Real Schedule. *ISIJ Int.* **2010**, *50*, 1193–1199. [CrossRef]
22. Krauss, G. Critical Temperatures. In *Steels—Processing, Structure, and Performance*, 2nd ed.; ASM International: Materials Park, OH, USA, 2015; pp. 30–32.
23. Schacht, K.; Prahl, U.; Bleck, W. Material Models and their Capability for Process and Material Properties Design in Different Forming Processes. *Mater. Sci. Forum* **2016**, *854*, 174–182. [CrossRef]
24. Pickering, F. Steels: Metallurgical Principles. In *Encyclopedia of Materials Science and Engineering*; The MIT Press: Cambridge, MA, USA, 1986.
25. Lee, Y. Empirical Formula of Isothermal Bainite Start Temperature of Steels. *J. Mater. Sci. Lett.* **2002**, *21*, 1253–1255. [CrossRef]
26. Boratto, F.; Barbosa, R.; Yue, S.; Jonas, J. Effect of Chemical Composition on the Critical Temperatures of Microalloyed Steels. Proceedings of International Conference on Physical Metallurgy of Thermomechanical Processing of Steels and Other Metals (THERMEC), Tokyo, Japan, 6–10 June 1988; pp. 383–390.
27. Yang, H.; Bhadeshia, H. Uncertainties in Dilatometric Determination of Martensite Start Temperature. *Mater. Sci. Technol.* **2007**, *23*, 556–560. [CrossRef]
28. Krauss, G. Austenite Grain-Size Control in Microalloyed Steel. In *Steels—Processing, Structure, and Performance*, 2nd ed; ASM International: Materials Park, OH, USA, 2015; pp. 153–157.
29. Pavlina, E.; Speer, J.; Van Tyne, C. Equilibrium Solubility Products of Molybdenum Carbide and Tungsten Carbide in Iron. *Scr. Mater.* **2012**, *66*, 243–246. [CrossRef]
30. Barraclough, D.; Whittaker, H.; Nair, K.; Sellars, C. Effect of Specimen Geometry on Hot Torsion Test Results for Solid and Tubular Specimens. *J. Test. Eval.* **1973**, *1*, 220–226.
31. Fechte-Heinen, R.; thyssenkrupp Steel Europe AG, Duisburg, Germany. Personal communication, 2018.
32. *ASTM Standard E384-17: Standard Test Method for Microindentation Hardness of Materials*; ASTM International: West Conshohocken, PA, USA, 2017.
33. Brito, R.; Kestenbach, H. On the Dispersion Hardening Potential of Interphase Precipitation in Micro-Alloyed Niobium Steel. *J. Mater. Sci.* **1981**, *16*, 1257–1263. [CrossRef]

© 2020 by the authors. Licensee MDPI, Basel, Switzerland. This article is an open access article distributed under the terms and conditions of the Creative Commons Attribution (CC BY) license (http://creativecommons.org/licenses/by/4.0/).

Article

The Influence of Vanadium Additions on Isothermally Formed Bainite Microstructures in Medium Carbon Steels Containing Retained Austenite

Irina Pushkareva [1,*], Babak Shalchi-Amirkhiz [1], Sébastien Yves Pierre Allain [2], Guillaume Geandier [2], Fateh Fazeli [1], Matthew Sztanko [1] and Colin Scott [1]

1. Canmet Materials, Natural Resources Canada, 183 Longwood Road South, Hamilton, ON L8P 0A5, Canada; Babak.Shalchi_Amirkhiz@canada.ca (B.S.-A.); fateh.fazeli@canada.ca (F.F.); matthew.sztanko@canada.ca (M.S.); colin.scott@canada.ca (C.S.)
2. Institut Jean Lamour, UMR CNRS-UL 7198, 54011 Nancy, France; sebastien.allain@univ-lorraine.fr (S.Y.P.A.); guillaume.geandier@univ-lorraine.fr (G.G.)
* Correspondence: irina.pushkareva@canada.ca; Tel.: +1-905-645-0789

Received: 25 February 2020; Accepted: 16 March 2020; Published: 19 March 2020

Abstract: The influence of V additions on isothermally formed bainite in medium carbon steels containing retained austenite has been investigated using in-situ high energy X-ray diffraction (HEXRD) and ex-situ electron energy loss spectroscopy (EELS) and energy dispersive X-ray analysis (EDX) techniques in the transmission electron microscope (TEM). No significant impact of V in solid solution on the bainite transformation rate, final phase fractions or on the width of bainite laths was seen for transformations in the range 375–430 °C. No strong influence on the dislocation density could be detected, although quantitative analysis was impeded by ferrite tetragonality. A reduction in the carbon content of retained austenite C_γ that is not believed to be due to competition with VC or cementite precipitation was observed. No influence of V on the carbon supersaturation in bainitic ferrite C_b could be directly measured, although carbon mass balance calculations suggest C_b slightly increases. A beneficial refinement of blocky MA and a corresponding size effect induced enhancement in austenite stability were found at the lowest transformation temperature. Overall, V additions result in a slight increase in strength levels.

Keywords: bainite; vanadium microalloying; austenite stability; HEXRD; EELS

1. Introduction

Vanadium additions are known to be beneficial for improving the strength and/or the in-use performance of several classes of advanced high strength steel (AHSS) products. These include low carbon hot rolled bainitic steels [1], medium carbon air-cooled bainitic forging steels [2–4], intercritically annealed low and medium carbon transformation induced plasticity (TRIP) and dual phase (DP) sheets [5], medium carbon ferritic hot strips [6,7] and high carbon austenitic twinning induced plasticity (TWIP) steels [8,9]. Classically, the vanadium strengthening effect in carbon-manganese steels is attributed to a combination of microstructure refinement and precipitation hardening [10]. However, vanadium can also provide positive effects while still in solid solution. For example, in bainitic steels that are fast cooled from temperatures above Ac_3 and transformed below B_s almost all of the added vanadium remains in solid solution [3,4]. Nevertheless, for low carbon alloys it is clear that there can be a beneficial strengthening effect from vanadium due to a notable decrease in the bainite transformation temperature [1]. Further, bainite softening during coiling is reduced in a manner analogous to the well known effect of vanadium on the temper softening of martensite [11]. While the situation for low carbon (i.e., containing < 0.1 wt. % C) bainite now seems reasonably clear, few detailed studies on technologically important medium carbon bainitic alloys (0.2–0.4 wt. % C) are available. For example, in 0.25 wt. % C alloys with V additions of up to 0.4 wt. % Sourmail and co-authors [2,3] reported no significant precipitation or clustering of V and no effect on the bainite reaction for isothermal transformations in the range 375–450 °C. In fact, the dominant effect of vanadium in these alloys was a strong reduction in the critical cooling rate for pearlite formation, leading to a significant improvement in hardenability for large sections. Wang et al. [4] found that the addition of 0.13 wt. % V in bainitic forging steels was advantageous for both strength and toughness under as-forged slow air-cooled conditions, but that the benefits were reduced as the cooling rate increased.

In low carbon steels the bainitic transformation is very rapid and, depending on the composition and cooling rates applied on the hot strip mill run out table, can begin during fast cooling before the coiling temperature is reached. The situation is more complicated in the medium carbon steels considered here as the bainitic transformation is much slower and carbon partitioning will result in the presence of a second phase of carbon-stabilized retained austenite (RA). Often, silicon additions are employed to delay cementite formation in bainitic ferrite and increase the amount of carbon available for retained austenite stabilization [12]. Further, the high dislocation densities formed in fresh bainitic ferrite may accelerate vanadium carbide and nitride nucleation. In this work, a powerful combination of in-situ high energy synchrotron X-ray diffraction (HEXRD) and ex-situ electron back scattered diffraction (EBSD) and electron energy loss spectroscopy (EELS) and energy dispersive X-ray mapping (EDX) techniques to determine the effect of vanadium additions on the evolution of the microstructure before, during and after the bainitic transformation in two model alloys is applied. An important aspect of this study is that all of the experimental work including the heat treatment, in-situ diffraction, ex-situ microscopy and mechanical testing was carried out on the same samples. This was done in order to eliminate, as far as possible, the effects of heterogeneous composition and/or processing.

2. Materials and Methods

In this work the effect of vanadium on the bainitic transformation in two laboratory alloys, Ref and Ref+V was studied, using a range of in-situ diffraction and ex-situ microscopy techniques. The in-situ work was done using synchrotron X-ray diffraction to determine the evolution of the phase fractions, the average carbon content in austenite and the dislocation density in bainitic ferrite. Ex-situ optical and scanning electron microscopy (including EBSD) were applied to compare the microstructural parameters such as bainite lath size and MA island sizes. Transmission electron microscopy was also employed to characterize precipitate size and chemistry using EDX mapping and to measure the local carbon content in austenite, martensite and bainitic ferrite.

The chemical compositions of the two alloys, Ref and Ref+V, are given in Table 1. The nitrogen content was deliberately kept low in order to maximize the amount of vanadium in solution during the bainite transformation. The vanadium content was based on previous work on TRIP alloys [13] and is typical of bainitic forging applications. The steels were cast into 50 kg ingots from a single 200 kg heat in a vacuum induction melting (VIM) furnace at CanmetMATERIALS (CMAT). The chemical compositions of the solidified ingots were verified by optical emission spectroscopy (OES) and LECO combustion analysis for C and N. Hot rolling was carried out on the CMAT pilot hot strip mill. The ingots were reheated to 1220 °C, held at temperature for 3 h and then hot rolled from 75 to 15 mm in 6 passes. The final rolling pass was carried out above 900 °C and the finished plates were allowed to air cool.

Table 1. Chemical compositions of the studied steels, wt. %.

Steel	C	Mn	Si	Mo	V	Al	N
Ref	0.22	2.2	1.8	0.2	-	0.01	0.0028
Ref+V	0.22	2.2	1.8	0.2	0.15	0.01	0.0027

Hollow tube specimens with an outer diameter of 4 mm, an inner diameter of 2 mm and a length of 10 mm were directly machined from the hot rolled plates. They were cut with the cylindrical axis parallel to the rolling direction and centered at $\frac{1}{4}$ plate thickness. The tube geometry has two important advantages for this study; firstly the amount of X-ray absorption is reduced by having thin wall sections and secondly induction heating and gas cooling is much more efficient and homogeneous with hollow samples. A Bahr DIL 805 deformation dilatometer was used to determine the alloy transformation temperatures A_{c1}, and A_{c3} using a heating rate of 5 °C/s. Optical microscopy (OM), EBSD and Vickers microhardness testing were all done on wall sections obtained by cutting the tubes into two equal half cylinders and polishing the cut faces. Specimens for OM were first polished with 1 µm diamond paste, then finished with Struers OPS solution before etching with a Klemm reagent. EBSD maps were acquired from the same OM samples after repolishing. The EBSD data was acquired using an EDAX TSL system on a Nova NanoSEM 650 field emission gun scanning electron microscope (FEG-SEM). The spot spacings used were 200 nm and 80 nm. Vickers microhardness was measured using a Clemex MMT-X7B tester under 300 gf force and 10 s dwell time according to the ASTM E384 standard.

The in-situ HEXRD experiments were carried out at the Petra (Hamburg) P07 beam line operated in transmission mode under powder diffraction conditions. A high-energy monochromatic beam with 100 keV energy and 0.12 Å wavelength permitted data acquisition in the transmission mode at a frequency of $\simeq$ 10 Hz. A 2-D Perkin-Elmer charge coupled device (CCD) detector was positioned 1 m behind the sample, giving access to full Debye-Scherrer rings with a maximum 2θ angle of 12°. The X-ray beam cross section was 400 µm × 400 µm so that the analyzed volume contained at least 10,000 austenitic grains at the start of each experiment. The applied thermal cycles were controlled using another Bahr dilatometer available at the P07 beam line. Specimens were heated by induction and cooled by argon gas. The temperature was regulated and recorded by a thermocouple spot-welded at the center of the sample. The setup was the same as that used in [14].

The 2-D diffraction patterns produced during the synchrotron experiments were circularly integrated using the Fit2D software [15] after calibration using Si powder and the resulting 1-D diffractograms were analyzed with a full Rietveld refinement procedure. The diffraction peaks were fitted by pseudo-Voigt functions using the FullProf code [16] with 16 degrees of freedom (background, lattice parameters, peak shape and temperature effects) to determine the phase fractions and the lattice parameters. In all the following experiments only two phases were detected in the diffraction patterns: a face-centered cubic phase, corresponding to austenite; and a body-centered phase. The latter could correspond to either body-centered tetragonal (bct-martensite or supersaturated bct-bainitic ferrite) or body-centered cubic (bcc) ferrite. It was not possible to isolate the contribution of each based on tetragonal distortion. Therefore, during the Rietveld refinement procedures the lattices of the two phases were considered to be cubic (Fm3m for austenite and Im3m for martensite/bainite). It is also important to note that no peaks from carbides of any type (including cementite) were detected on the diffractograms.

The austenite carbon content C_γ starts to change from the nominal value during the bainitic transformation when carbon diffuses from fresh bainite into the parent austenite phase. This causes the austenite lattice parameter a_γ to increase linearly with the amount of carbon rejected from bainitic ferrite. In these trials the transformation is isothermal, so the austenite carbon content was determined by calculating the relative lattice dilatation Δa_γ between each measured lattice parameter value, a_m and a reference lattice parameter determined at the transformation temperature before any bainite has formed, a_{tr}. This difference is proportional to the austenite carbon enrichment ΔC_γ as follows [14]:

$$\Delta a_\gamma = a_m - a_{tr} = k_C \Delta C_\gamma \qquad (1)$$

The coefficient of proportionality k_C is taken to be 0.00105 nm/(at. %)C [17,18]. The absolute carbon content in austenite is then obtained by adding ΔC_γ to the nominal carbon content.

Planar TEM thin foils were prepared from the same specimens that were analyzed in the synchrotron. To do this, the tube samples were cut in half in the direction parallel to the rotation axis and one of the resulting half cylinders was gently flattened and then mechanically thinned to 100 µm. The other half cylinder was used for microscopy and microhardness testing. Standard 3 mm disks were punched out of the flattened tubes and perforated by twin-jet polishing using a Struers Tenupol 5 and Struers A8 electrolyte. The specimens were analyzed in a Technai Osiris 200 keV X-FEG (S)TEM equipped with a 4 element windowless Super-X EDX detection system and a Gatan Enfina EELS analysis system. The carbon content in martensite, bainite and retained austenite phases was measured by EELS using a standards based second difference acquisition technique developed by the authors [18]. All equilibrium and kinetic thermodynamic calculations were carried out using the MatCalc 6.02 (rel 1.003) software [19] using the mc_fe_v2.011 diffusion data base, the mc_fe_1.003 physical database and the mc_fe_2.059 thermodynamic database.

3. Results

Section 3.1 is concerned with the determination of the optimum experimental conditions. Section 3.2 contains the OM, SEM and microhardness data. Section 3.3 provides an exploitation of the in-situ synchrotron results. Complementary data obtained from TEM analyses are presented in Section 3.4.

3.1. Definition of the Thermal Cycles

The solution treatment employed here was designed as a compromise to best satisfy several mutually incompatible requirements:

- The specimens should be completely austenitized with an austenite grain size as small as possible in order to maximize the number of diffracting crystallites and minimize texture effects in the HEXRD experiments.
- Vanadium precipitates formed either during hot rolling or during the reheating ramp should be redissolved to maximize the amount of vanadium and carbon in solution before the bainite transformation.
- Any cementite formed during reheating should be dissolved to maximize the interstitial carbon content in austenite before the bainite transformation.

The equilibrium transformation temperatures Ae1 and Ae3 for Ref and Ref+V were calculated using the MatCalc 6 code and compared with experimental dilatometry data [20] in Table 2. The bainitic start temperature Bs and the martensite start temperature M_s calculated from the empirical equations of Kircaldy and Venugopalan [21] and Andrews [22] respectively are also shown.

Table 2. Experimental and calculated transformation temperatures.

Steel	A_{e1} (calc)	A_{c1} (exp)	A_{e3} (calc)	A_{c3} (exp)	T_{sol} V(CN)	B_s (calc) [21]	M_s (calc) [22]
Ref	707 °C	700 °C	831 °C	933 °C	-	423 °C	375 °C
Ref+V	713 °C	748 °C	840 °C	975 °C	1067 °C	423 °C	375 °C

As expected, the addition of vanadium tends to increase the equilibrium transformation temperatures and stabilize the ferrite phase. The most important effect is on the kinetics—during continuous heating vanadium greatly reduces the austenite nucleation and growth rate in the intercritical region, resulting in a >40 °C increase in A_{c1} and A_{c3} at a constant heating rate of 5 °C/s. This behavior has been previously observed by the authors in microalloyed dual phase (DP) steels [5]. The calculated solution temperature for V(C,N) in Ref+V is 1067 °C. In the interests of optimizing the austenite grain size for the synchrotron experiment it was decided to use the lowest possible austenitizing temperature and soaking time and tolerate the inevitable presence of a small fraction of precipitated V(C,N). This will be considered further in the discussion section. Therefore a combination of 900 °C/120 s for the austenitization parameters was chosen. It was verified by dilatometry that this was sufficient for the complete austenitization of both alloys. Not much reliable information on the effect of vanadium on B_s

and M_s exists for this class of steels. Therefore, three isothermal bainitic transformation temperatures spanning the range of the theoretical values; 375 °C, 400 °C and 430 °C were studied. The optimum cooling rate after austenitization was determined experimentally from the initial synchrotron trials. The cooling rate should be fast enough to avoid the formation of bainite before the isothermal temperature is attained, but slow enough so that there is no significant temperature undershoot at the transition point. It was found that a cooling rate of 40–50° C*s^{-1} satisfied these requirements. The experimental thermal cycles are shown in Figure 4a.

3.2. Microstructure and Mechanical Properties

The microstructure of the Ref and Ref+V alloys transformed at 375 °C can be seen after Klemm etching in Figure 1. The matrices (colored dark) are fully bainitic with second phase retained austenite and/or martensite (MA) colored light. The second phase consists of two distinct morphologies—alignments of thin films or small islands along interlath boundaries and prior austenite grain (PAG) boundaries and coarser blocky islands with a more random distribution. The area fraction and the mean size of second phase islands were determined using the ImageJ analysis software for all three transformation temperatures. The results are listed together with the corresponding synchrotron HEXRD, EBSD and TEM data in Tables 3 and 4.

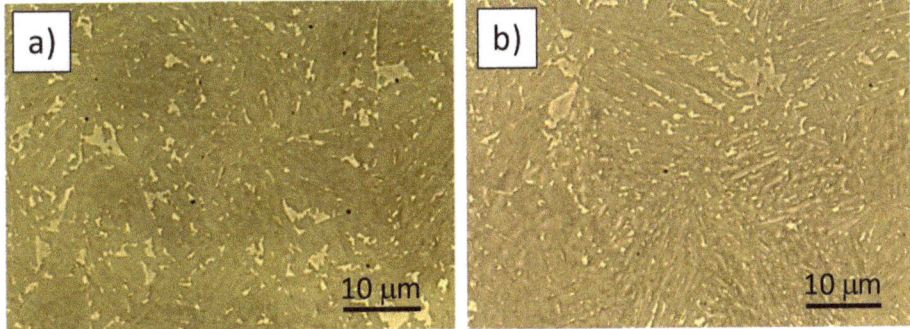

Figure 1. Optical Micrographs (Klemm etch) of Ref and Ref+V alloys isothermally transformed at 375 °C in the synchrotron (**a**) Ref and (**b**) Ref+V. The light colored phase is martensite/retained austenite.

The EBSD data from specimens transformed at 375 °C are presented in Figure 2a–f. Similar results were obtained for the 400 °C transformation (not shown). The 430 °C transformation samples were not studied. The Euler orientation maps are shown in Figure 2a,b, the size distributions determined from low angle boundaries ($\theta < 5°$) are compared in Figure 2c, the grain misorientation frequencies in Figure 2d and the image quality (IQ) maps with superimposed austenite islands (red color) in Figure 2e,f. The frequency distributions of grain boundary misorientation angles were typical of lower bainitic microstructures with dominant fractions of low misorientation ($\theta < 15°$) and high misorientation ($\theta > 50°$) boundaries. Vanadium additions clearly had little or no effect on the distributions at 375 °C. The effective grain size for low angle boundaries ($\theta < 5°$) thought to define the effective lath size [23] is reported in Table 4. An estimate of the MA fractions and mean MA island area was determined from the poorly indexed regions in the IQ maps. This information, together with the RA fraction is reported in Tables 3 and 4.

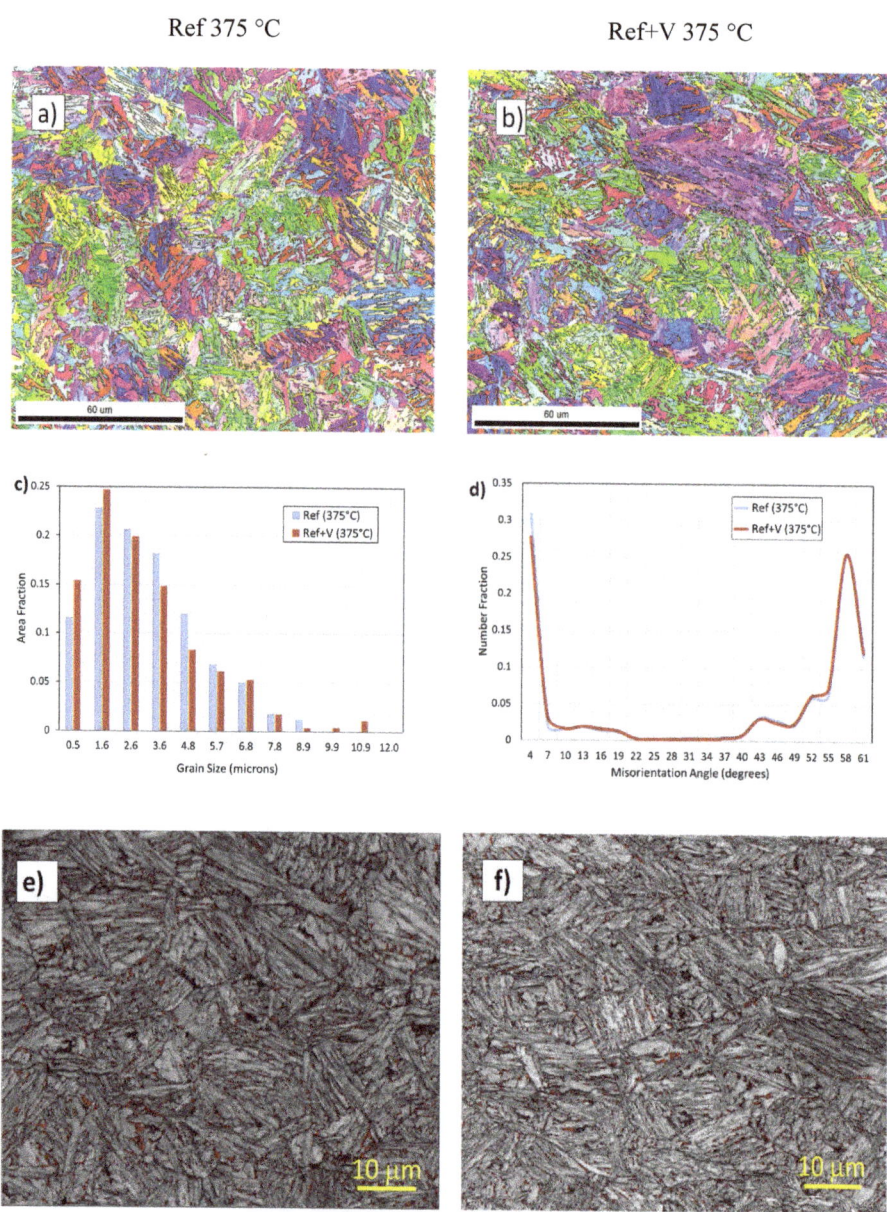

Figure 2. Electron back scattered diffraction (EBSD) data from Ref and Ref+V alloys transformed at 375 °C: (**a,b**) Euler orientation maps with high angle (θ > 15°) boundaries shown; (**c**) low angle (θ < 5°) boundary size distributions; (**d**) misorientation frequencies and (**e,f**) image quality maps with superimposed austenite islands in red.

Table 3. Phase fractions from synchrotron high energy synchrotron X-ray diffraction (HEXRD), EBSD and optical microscopy data.

Alloy	Average Cooling Rate	Transformation Temperature	B_s (5%)	Time for 95% Completion	HEXRD Final Bainite Fraction	HEXRD RA Fraction (25 °C)	HEXRD Martensite Fraction	EBSD RA Fraction	EBSD MA Fraction	OM MA Fraction
Ref	44 °C/s	375 °C	375 °C	160 s	87%	12%	0.3%	5.4%	1.5%	5.1%
Ref	50 °C/s	400 °C	400 °C	129 s	86%	14%	0.9%	5.3%	3.4%	6.9%
Ref	45 °C/s	430 °C	430 °C	117 s	78%	19%	2.5%	NA	NA	11.3%
Ref+V	42 °C/s	375 °C	381 °C	127 s	89%	11%	0.6%	4.8%	5.4%	2.3%
Ref+V	50 °C/s	400 °C	504 °C	94 s	85%	14%	0.9%	4.8%	6.8%	6.9%
Ref+V	45 °C/s	430 °C	432 °C	114 s	75%	19%	5.9%	NA	NA	10.7%

Table 4. Microstructure parameters from EBSD, TEM and OM data.

Alloy	Transformation Temperature	Vickers Hardness	EBSD Effective Grain Size q < 5°	TEM Mean Lath Width	EBSD Mean MA Area	OM Mean MA Area
Ref	375 °C	412 HV	1.01 mm	0.285 mm	0.53 mm^2	0.22 mm^2
Ref	400 °C	387 HV	0.96 mm	0.249 mm	0.60 mm^2	0.22 mm^2
Ref	430 °C	357 HV	NA	0.226 mm	NA	0.44 mm^2
Ref+V	375 °C	422 HV	0.90 mm	0.252 mm	0.38 mm^2	0.09 mm^2
Ref+V	400 °C	390 HV	1.04 mm	0.211 mm	0.49 mm^2	0.16 mm^2
Ref+V	430 °C	372 HV	NA	0.263 mm	NA	0.44 mm^2

Vickers microhardness measurements were obtained from the as-rolled plates and from the tube samples after thermal treatment in the synchrotron (Table 4, Figure 3). The hardness of the air-cooled hot strips was found to be 343 ± 15 HV for the Ref alloy and 361 ± 12 HV for Ref+V after air cooling. This is equivalent to an increase of ~+60 MPa in the YS [24]. As expected, fast cooled and isothermally transformed bainitic samples were all stronger than the as-rolled condition, with the highest hardness (422 HV) occurring for the Ref+V alloy transformed at 375 °C. Vanadium additions improved the alloy strength at all transformation temperatures, but the difference at 400 °C was very small. This could indicate that the presence of vanadium introduces different strengthening mechanisms operating at low and high transformation temperatures. The maximum vanadium-related hardness increase after isothermal transformation occurred at 430 °C (+15 HV or ~+50 MPa). The gain in strength was therefore proportionately much lower than that achieved under slow air cooling conditions.

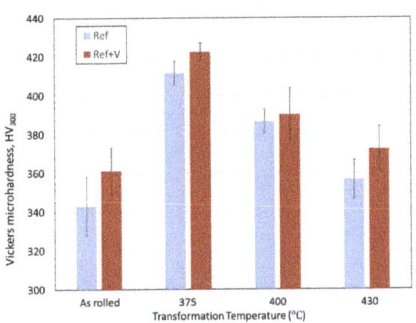

Figure 3. Vickers microhardness data from air-cooled as-rolled and isothermally transformed samples.

3.3. Synchrotron XRD—Phase Fractions, Carbon Partitioning and Dislocation Densities.

The evolution of the bcc phase fraction with time and temperature can be readily determined from the in-situ XRD data. Figure 4a shows the results obtained for the Ref and Ref+V alloys isothermally transformed at 375 °C, 400 °C and 430 °C. Time t = 0 s on the horizontal axis corresponds to the start of the fast cooling ramp after soaking at 900 °C. The isothermal section ends at t = 612 s. At this point, any unstable austenite transforms to martensite during the final cooling step. The amount of martensite formed during secondary cooling is very limited except for the Ref+V alloy transformed at 430 °C. Although the HEXRD detection limit for ferrite is excellent (0.1% volume fraction) there is a large mathematical uncertainty in the calculated volume fractions for ferrite fractions < 5% during the refinement procedure. This is partly due to the highly textured nature of the very first ferrite nuclei. In consequence only data with >5% bcc fraction is reported. The phase fractions are summarized in Table 3 (uncertainty ± 1%).

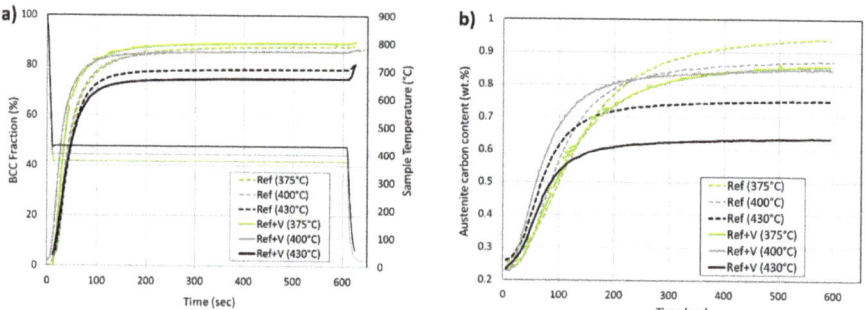

Figure 4. HEXRD data from Ref and Ref+V alloys as a function of the transformation temperature. (a) Body-centered cubic (bcc) phase fraction and (b) carbon content in austenite.

Following current theory, the total amount of bainite formed increased as the isothermal transformation temperature decreased (Figure 4a) and the RA fraction followed the opposite trend (Table 3). From the HEXRD data, the addition of vanadium had virtually no effect on the final bainite fraction after transformation at 375 °C and 400 °C and introduced a slight decrease at 430 °C. It had no influence on the final fraction of retained austenite at any transformation temperature, although it did increase the unstable austenite content at 430 °C. This slight excess austenite transformed to give 5.9% martensite fraction during final cooling, compared to only 2.5% for the Ref alloy under identical conditions. The bainite start transformation temperature, B_s, defined here as the temperature at which 5% of the final bainite fraction is detected, occurred either at or very close to the isothermal temperature for all of the trials, except for Ref+V at 400 °C. The time required to complete 95% of the bainite transformation decreased as the transformation temperature is lowered—this is expected as the thermodynamic driving force for austenite decomposition increases. The addition of vanadium appears to very slightly accelerate the transformation, but the effect is not significant.

The magnitude of the RA fraction detected by EBSD was lower than the HEXRD data by a factor of >2 and the MA fraction was much higher. Other authors have reported analogous discrepancies with similar bainitic steels [4]. Either more than 50% of all RA is contained in films and islands that are below the EBSD resolution limit (80 nm step size) or it is unstable and transforms to martensite during the specimen preparation. Nevertheless, all techniques show that the same trends are followed. For both steels, the mean MA island size decreased as the transformation temperature was lowered. However, a significant additional refinement from the vanadium addition was apparent at 375 °C where the mean EBSD MA island size in Ref+V was 0.38 µm² compared to 0.53 µm² for the Ref alloy.

The evolution of C_γ as the bainite transformation progresses can be seen in Figure 4b. Comparing Figure 4b with Figure 4a, it is interesting to note that partitioning of carbon to austenite follows a sigmoidal type kinetic that is much slower than the bainite formation rate. Indeed, at the lowest transformation temperature partitioning is still active at the end of the isothermal plateau (600 s), even although the bainite transformation stopped more than 300 s before. The final RA carbon content in both alloys (Table 5) increased monotonically with decreasing transformation temperature in agreement with the T_0 model [25].

From Figure 4b and Table 5, the addition of vanadium changed the austenite carbon enrichment, i.e., C_γ was systematically lower in Ref+V at the end of transformation, for all transformation temperatures. The difference was large at 430 °C and at 375 °C but very small at 400 °C. As all of the Ref+V samples were machined from the same plate and processed identically, there is unlikely to be any variation in the fraction of carbon trapped in vanadium carbide precipitates formed prior to the fast cooling step. Now, for a fixed transformation temperature Table 3 shows that the RA fraction is the same for both alloys, so these data suggest that vanadium is in some way restricting the amount of carbon available for partitioning to austenite. Further, the severity of the effect may have a complex temperature dependence. This could be related to carbon trapping by undissolved prior VC precipitates or the formation of new VC/V_xC_y clusters in bainitic ferrite. Alternatively, vanadium may promote faster cementite formation in bainite. It is also possible that vanadium could directly consume some of the carbon partitioned to austenite by precipitating as VC inside highly enriched austenite islands. Another theory is that solute vanadium retards recovery in bainitic ferrite leading to a higher defect (dislocation) density that in turn increases the effective carbon solubility. These points are discussed further in the next sections.

Measuring Dislocation Densities

The dislocation density in bainitic ferrite at the end of the isothermal transformation is a critical parameter in evaluating the structure/properties relationship of these steels. TEM imaging can be used to estimate dislocation densities on a microscopic scale (as long as the foil thickness is known), however the highly disorganized nature of these microstructures (see the next section) would require extensive sampling to avoid serious errors in determining the average dislocation density. HEXRD can provide an accurate measurement of global dislocation densities [26]. Normally, Williamson–Hall (W–H) or Warren–Averbach (W–A) type analyses are used to isolate the contribution to the full width half maximum (FWHM) peak broadening from microstrain and hence determine the dislocation density. An attempt was made to calculate the dislocation densities in bainitic ferrite from the $\{110\}_\alpha$, $\{200\}_\alpha$, $\{211\}_\alpha$, $\{220\}_\alpha$ and $\{310\}_\alpha$ peak line broadening (after subtraction of the contribution from instrumental broadening) using the Williamson–Hall (W–H) method [27,28]. Unfortunately, the W–H values for dislocation densities were unphysically high for both alloys, almost certainly because tetragonality-induced peak splitting was present. However, the W–H plots did show a strong linear dependence on sin θ, indicating that strain broadening and not crystallite size was the major constituent. The degree of influence of vanadium on the dislocation density in bainitic ferrite was estimated by comparing the angular width at half maximum of the $\{011\}_\alpha$ peaks from both alloys at the end of the isothermal hold (Figure 5a,b). The $\{222\}_\alpha$ reflection, whose angular width is less sensitive to tetragonality than $\{011\}_\alpha$, would be a better choice for this. However, as there is almost perfect superposition between $\{222\}_\alpha$ and $\{311\}_\gamma$ reflections, these peaks cannot be separated. Based on the trends seen in the W–H plots, it is reasonable to interpret Figure 5b as a qualitative plot of the variation of dislocation density with transformation temperature. For both steels, the dislocation density was highest during transformation at 375 °C and linearly decreased as the temperature increases. From Figure 5b the absolute difference in the $\{011\}_\alpha$ FWHM of Ref and Ref+V was never more than 1.7% at any transformation temperature (measured after 600 s). As this is comparable with the uncertainty in the FWHM measurement, it is not possible to demonstrate that there is any significant difference in the dislocation density of bainitic ferrite due to vanadium additions.

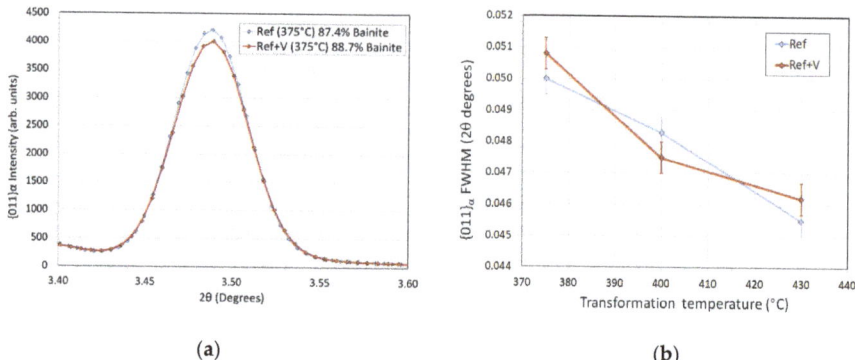

Figure 5. (a) HEXRD $\{011\}_\alpha$ reflection for Ref and Ref+V transformed for 600 s at 375 °C. (b) Comparison of $\{011\}_\alpha$ peak broadening for Ref and Ref+V transformed for 600 s at 375 °C, 400 °C and 430 °C.

3.4. TEM—Microstructure

Thin foil samples from all of the specimens listed in Table 3 were examined in the TEM. Consistent with Figure 1, the matrix phase was lath bainite (Figure 6a) with thin (<100 nm) films of retained austenite situated between the bainite laths (Figure 6b). Some martensite was detected at higher transformation temperatures, usually associated with the RA films (Figure 6c). The linear defects visible in the austenite lath of Figure 6c are most likely epsilon martensite. Other areas showed plate or granular bainite (Figure 7). Overall, very few cementite precipitates were observed, in agreement with the HEXRD findings.

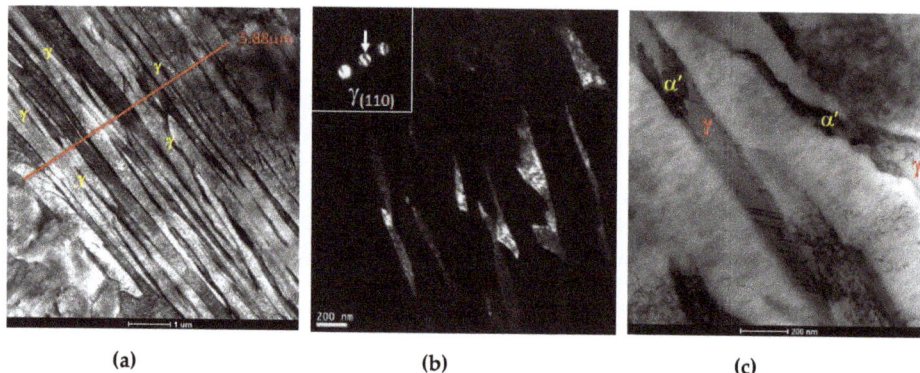

Figure 6. (a) Bright field STEM image of the Ref alloy transformed at 430 °C; (b) dark field STEM image showing austenite films ($\gamma_{(110)}$ reflection) from Ref+V transformed at 430 °C and (c) bright field STEM image of Ref transformed at 430 °C showing RA partially transformed to martensite.

The bainitic lath size was estimated by selecting blocks with lath boundaries that were normal to the foil surface and measuring the average distance between the lath boundaries (Figure 6a). The results can be seen in Table 4 and in Figure 7. The heterogeneous nature of these microstructures makes it very difficult to determine a reliable value for the lath width and it is impossible to say from the data that there is any significant difference between the Ref and Ref+V specimens. The bainite lath width actually appears to increase slightly with decreasing transformation temperature for the Ref alloy, which is contrary to current theory [25]. However, the scatter in the measurement is large and the trend could be misleading.

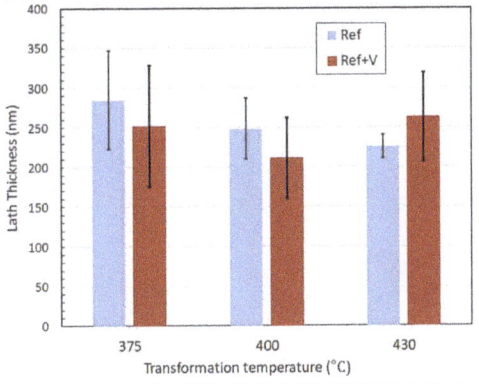

Figure 7. STEM measurement of bainite lath widths.

Figure 8a shows a bright field image of plate bainite containing RA films in Ref+V transformed at 400 °C. Figure 8b is the same area superimposed with a vanadium EDX chemical map and Figure 8c shows a carbon EDX map of the same region. Even although EDX carbon maps are not quantitative, carbon enrichment in retained austenite films is qualitatively obvious. Some coarse and fine precipitates are visible in Figure 8a. The chemical map in Figure 8b confirms that the majority of these particles contain vanadium. Interestingly, there is little evidence from carbon or nitrogen (not shown) maps that these are classical carbide or nitride precipitates. The largest of these particles are aligned, probably along austenite grain boundaries formed during hot rolling as their position is clearly uncorrelated with either the bainitic structure or the retained austenite films. In fact, the V precipitate density in retained austenite appears to be quite low, compared to the surrounding ferrite.

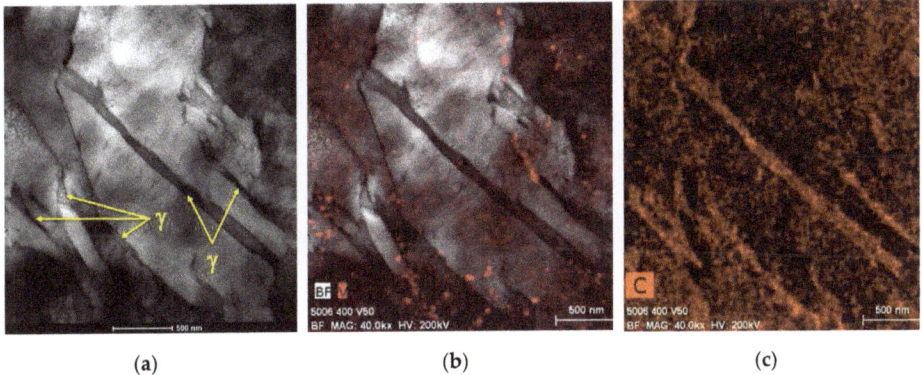

Figure 8. (a) Bright field STEM image of Ref+V transformed at 400 °C; (b) same as Figure 8a with a superimposed energy dispersive X-ray analysis (EDX) vanadium map (red) and (c) EDX carbon map (qualitative) from the same area.

Most of the particles observed in Ref+V specimens transformed at 400 °C and 430 °C contained vanadium with small amounts of carbon and nitrogen being occasionally detected, especially for the larger precipitates (Figure 9e). However, at 375 °C some of the particles contained V+Mo (Figure 9a–d). An equilibrium calculation (not shown) predicted that Mo enrichment of V(C,N) is only significant at temperatures below 800 °C, suggesting that these particles formed in ferrite.

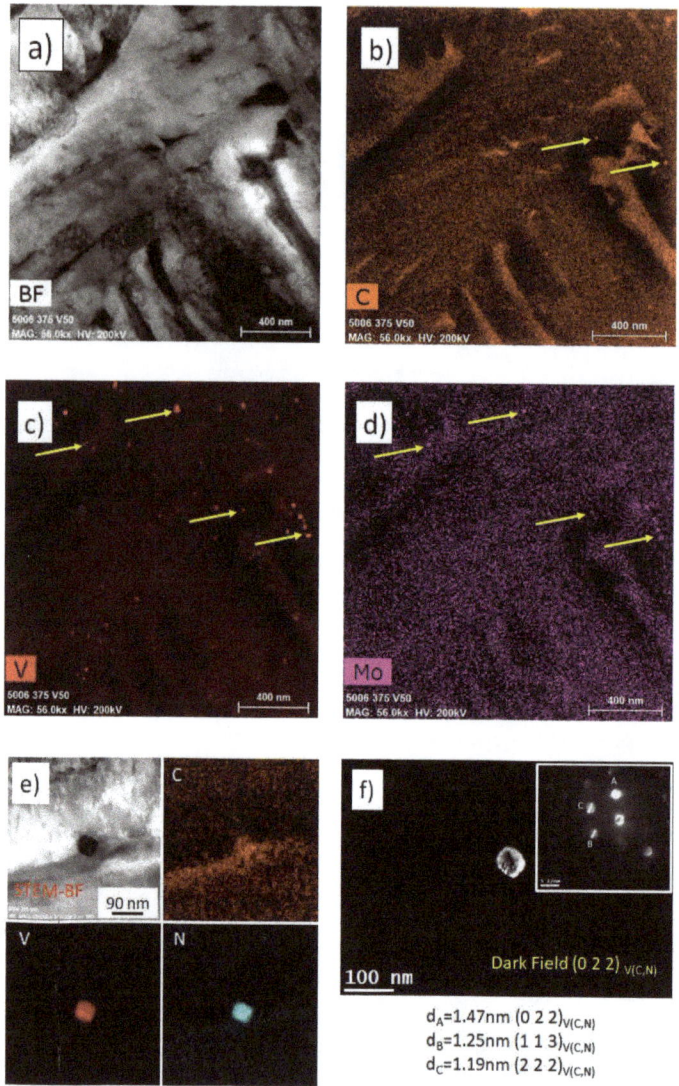

Figure 9. Ref+V alloy transformed at 375 °C, (**a**) STEM bright field image, (**b**) EDX carbon map, (**c**) EDX vanadium map, (**d**) EDX molybdenum map from the same region, (**e**) STEM bright field image and EDX maps of a V(C,N) particle and (**f**) STEM dark field imaging and convergent beam electron diffraction (CBED) pattern confirming the crystal structure of the particle in e).

Measuring the Local Carbon Contents

The local carbon content, C_γ, in a number of retained austenite and martensite islands was measured using a standards-based EELS technique [18]. At a carbon concentration level of 0.8 wt. % the EELS measurement precision was ±4% and the detection limit for solute carbon was 0.03 wt. %. The ex-situ EELS results for C_γ were averaged and compared in Figure 10a with the in-situ HEXRD synchrotron data from Figure 4b and Table 5. Note that the error in the HEXRD carbon measurement (±0.011 wt. %) is too small to show in Figure 10a. The error bars in the EELS data are not measurement errors but rather represent the standard deviation in the carbon contents obtained from different austenite islands. Clearly, there is a wide island-to-island scatter in C_γ, as previously noted for TRIP steels [18].

From Figure 10a, the global HEXRD measurement and the local EELS values both predict the same trend of decreasing C_γ with increasing transformation temperature. For Ref+V (yellow and brown points), the mean EELS data show very close agreement with HEXRD, especially bearing in mind the scatter in RA island carbon contents. The agreement for Ref (light blue and dark blue points) was slightly less good, but the overall trend was the same. In Figure 10b the mean carbon content in martensite $C_{\alpha'}$ is compared with C_γ from the EELS data. As expected, martensite islands contained less carbon than stable retained austenite. However, it is interesting to note that on average Ref+V martensite contained less carbon than Ref, i.e., Ref+V austenite appeared to be more stable, especially at lower transformation temperatures, and both alloys show $C_{\alpha'}$ (375 °C) < $C_{\alpha'}$ (400 °C). These observations clearly show that RA stability is not solely determined by carbon content [29].

If all the available carbon diffused into retained austenite, then the lever rule would predict significantly higher concentrations, C_γ than those found in Table 5. In the absence of any significant precipitation this implies that bainitic ferrite is supersaturated in carbon. EELS analyses were therefore carried out to try to measure C_b, the bainitic ferrite carbon content. These measurements were made in regions of bainitic ferrite where there were no obvious signs of carbide precipitation. It was observed that C_b was heterogeneous—in areas close to austenite laths usually no carbon was detected (i.e., C_b < 0.03 wt. %). The experimental values for C_b given in Table 5 and plotted in Figure 10b were determined by averaging EELS measurements from bainitic ferrite far from austenite laths. The experimental C_b results can be compared with values calculated from a simple carbon mass balance using the phase fraction data in Table 3 and the experimental values for C_γ and $C_{\alpha'}$ in Table 5. The two show very good agreement (Table 5). Evidently, bainitic ferrite remained highly supersaturated in carbon for both alloys, hence the tetragonality problem encountered with the W–H analysis. C_b increased as the transformation temperature decreased, in accordance with theory [30]. There appeared to be a slight influence of vanadium on C_b at 375 °C and 400 °C but the difference was within the experimental error. The carbon mass balance calculations in Table 5 suggest that vanadium additions increased C_b. More experimental work is required to confirm this.

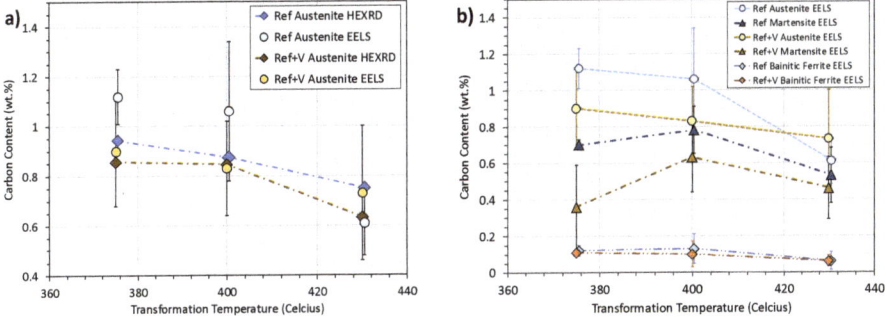

Figure 10. (a) Comparison of HEXRD and electron energy loss spectroscopy (EELS) carbon contents in austenite and (b) EELS carbon contents of austenite, martensite and bainitic ferrite. The error in the HEXRD carbon measurements is ±0.011 wt. % The error bars in the EELS data show the standard deviation in the range of the measured values.

Table 5. Final carbon contents in austenite, martensite and bainitic ferrite.

Alloy	Transformation Temperature	C_g HEXRD (wt. %)	C_g EELS (wt. %)	$C_{a'}$ EELS (wt. %)	C_b Calculated from Mass Balance (wt. %)	C_b Measured from EELS (wt. %)
Ref	375 °C	0.94	1.12	0.7	0.12	0.12
Ref	400 °C	0.87	1.06	0.78	0.11	0.13
Ref	430 °C	0.75	0.61	0.53	0.08	0.06
Ref+V	375 °C	0.86	0.9	0.36	0.14	0.11
Ref+V	400 °C	0.85	0.83	0.63	0.11	0.1
Ref+V	430 °C	0.63	0.73	0.46	0.09	0.06

4. Discussion

The yield stress σ_e of a single phase fully bainitic structure can be expressed as the sum of contributions:

$$\sigma_e = \sigma_0 + \sigma_{ss} + \sigma_d + \sigma_p + \sigma_L \quad (2)$$

where σ_0 is the Peierls stress for pure iron, σ_{ss} is the contribution from solid solution hardening, σ_d is the strengthening due to forest dislocations, σ_p is due to precipitation strengthening and σ_L is the hardening due to grain or lath size dependence. As the interest here lies in explaining the difference in strength between Ref and Ref+V shown in Figure 3, the first two terms on the right side of Equation (1) can be neglected by assuming there is no difference in C_b and that small amounts of vanadium in solid solution have a negligible influence so that:

$$\Delta\sigma_e = \Delta\sigma_d + \Delta\sigma_p + \Delta\sigma_L \quad (3)$$

Figure 5a,b show that the dislocation term $\Delta\sigma_d$ is very small. This leaves two terms to evaluate; the contribution from precipitation strengthening, $\Delta\sigma_p$ and a size dependent term, $\Delta\sigma_L$. The precipitation term consists of two distinct components; strengthening from (a) precipitates inherited from prior processing steps and (b) precipitates formed at the isothermal transformation temperature. The two cases are considered separately below.

4.1. Strengthening from V Precipitation in Austenite and Ferrite

The presence of 1.8 wt. % silicon strongly inhibits cementite formation in these steels [12]. This was confirmed by the lack of evidence for Fe_3C in the TEM images and the absence of any cementite reflections in the HEXRD data. In view of the scarcity of these precipitates, there is no evidence to suggest that vanadium additions have a significant influence on their size, volume fraction or distribution or that they make a measurable contribution to the mechanical properties of these alloys. Therefore, cementite precipitation can be omitted from further discussion.

In order to restrict austenite grain growth the choice was made to austenitize at 900 °C, which is below the calculated solution temperature of 1067 °C for V(C,N) in Ref+V. Consequently, some undissolved vanadium-containing particles are present before the bainitic transformation occurs (Figures 8 and 9). In those images a majority of rather coarse (20–40 nm) vanadium particles can be seen, along with some finer (5–10 nm) precipitates. It must be emphasized that many areas showed little or no evidence of V precipitation, so that the average particle density observed in the TEM was actually quite low. Further, a close examination of the carbon and nitrogen maps suggests that many of these vanadium-rich regions are not carbides or nitrides and in fact could be due to local segregation.

An anisothermal kinetic precipitation calculation was carried out with the MatCalc 6 code in order to determine the amount of V(C,N) remaining undissolved before the bainitic transformation. Figure 11a shows the thermal path of the simulation. Starting at 1220 °C, hot rolling is represented by a uniform cooling rate of 10 °C/s in the austenite domain. This is decreased to 2 °C/s to represent air cooling in ferrite from 700 °C to room temperature. The material is then reheated to 900 °C at a constant rate of 3 °C/s and held for 120 s before quenching. Two types of vanadium precipitates were simulated; the first class nucleates in austenite on grain boundaries and triple points (assuming a 100 µm austenite grain size), the second nucleates in ferrite on dislocations and defects. The dislocation density in ferrite is assumed to be low, 1×10^{12} m^{-2} due to the slow air cooling after finish rolling. No attempt was made to include strain induced precipitation in austenite as previous experimental studies indicate that very little vanadium is actually precipitated before this point [3,4]. Figure 11b shows the evolution of the phase fractions, Figure 11c the number density and Figure 11d the mean radii of the two types of precipitate.

At the end of re-austenitization, the mean diameter of the GB V(C,N) precipitates was predicted to be 120 nm with a very low number density of 1×10^{15}/m^3. The particles nucleated in ferrite were much smaller with a mean diameter of 9 nm and higher number density of 1.6×10^{19}/m^3. The total amount of V and C precipitated was actually very small: 0.0006 wt. % V and 0.0002 wt. % C. Compared to the TEM images, the calculated GB particle size appeared to be overestimated. However, the size of the dislocation nucleated precipitates correlated well with the thin foil images. Given their large size and scarcity, the strengthening effect of the GB particles can be neglected.

The contribution to strengthening made by the smaller precipitates can be estimated using the modified Ashby–Orowan equation proposed by Gladman [31]:

$$\sigma_p = 0.538 Gb \left(\frac{f^{0.5}}{X} \right) \ln\left(\frac{X}{2b} \right) \quad (4)$$

where G is the shear modulus (80,000 MPa), b is the Burgers vector (2.5×10^{-4} µm), f is the volume fraction of precipitates and X is the mean particle diameter in µm. Substituting the appropriate values from Figure 11 gives $\sigma_p = 8.5$ MPa. Thus, the strength contribution from undissolved vanadium precipitates formed before the end of the re-austenitization step is expected to be very small.

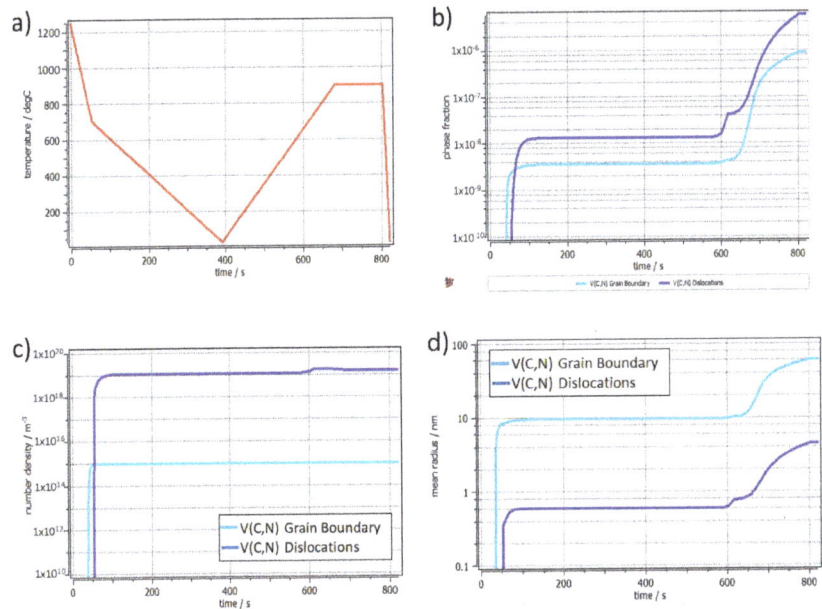

Figure 11. (a) Simulation of the thermal path for V(C,N) precipitation in austenite and ferrite. (b) Calculated phase fractions of V(C,N) nucleated on GB and dislocation sites. (c) Precipitate number densities. (d) Mean radii.

4.2. Strengthening from V Precipitation in Bainite

Here the question of whether or not new vanadium precipitates can nucleate and grow during the fast cooling and isothermal bainitic transformation stage is addressed. It can be assumed that any further growth of the pre-existing vanadium particles is limited, as this would require low temperature long-range diffusion of vanadium across the (presumably) depleted zones surrounding those precipitates. Figure 12a is a calculated precipitation-time-temperature (PTT) diagram for V(C,N) formation in Ref+V calculated using the MatCalc 6 code [32] assuming a constant dislocation density of $1 \times 10^{14}/m^2$ in ferrite. The blue dots show the cooling path of the Ref+V alloy. The high defect density decreased the nucleation time for V(C,N) by several orders of magnitude, but it is still not possible to nucleate vanadium particles in any ferrite that might form during cooling.

A PTT section from a lower temperature region corresponding to the isothermal bainitic transformations is shown in Figure 12b. Here the dislocation density was increased to $5 \times 10^{14}/m^2$, which is a reasonable value for lower bainite [1]. The simulation predicts that no V(C,N) could nucleate during the transformation times tested here. In fact, the temperature needs to be raised to almost 500 °C before any detectable precipitation can occur. The evidence points to the conclusion that no new vanadium precipitation occurred during the bainitic transformations studied here and that the precipitates observed in Figures 8 and 9 were all formed prior to bainite. The $\Delta\sigma_p$ strengthening term is thus a constant for all three transformation temperatures.

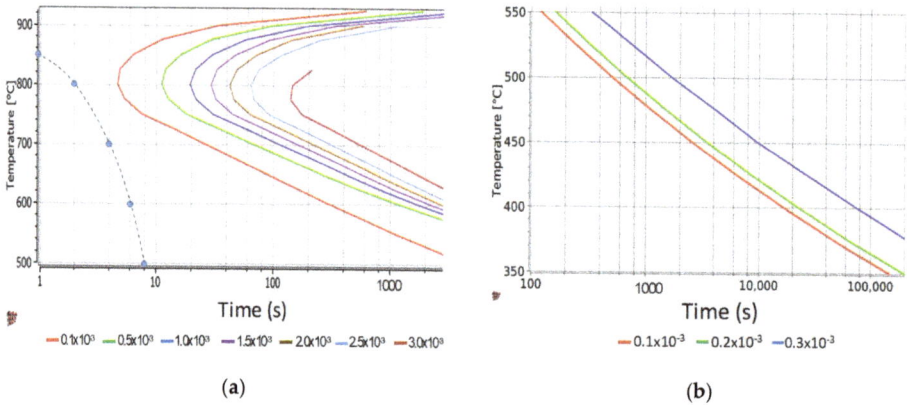

Figure 12. (a) V(C,N) PTT in Ref+V ferrite containing $1 \times 10^{14}/m^2$ dislocations and (b) low temperature PTT for V(C,N) in Ref+V bainitic ferrite containing $5 \times 10^{14}/m^2$ dislocations.

Even if classical precipitates do not form, the possibility that vanadium could influence the mechanical properties by segregating or clustering at a finer scale during the bainitic transformation is not precluded. Sourmail et al. [3] could not find any vanadium segregation at all in a medium carbon forging steel transformed at 450 °C using 3D tomographic atom probe reconstructions. However, many vanadium-rich areas that did not appear to contain either C or N were observed and these could well be an indication of clustering or segregation to boundaries and defects. Further investigations, beyond the scope of this work, are required.

4.3. Size-Dependence Effects

There is no doubt that the bainite matrix of both alloys becomes significantly stronger as the transformation temperature decreases from 430 to 375 °C (Figure 3). Following the current literature, a combination of packet and/or block size and/or lath refinement should be observed as the transformation temperature decreases. There are many arguments as to what is the best effective length L to use and whether yielding has a Hall–Petch or simple reciprocal type dependence on L [23]. The EBSD analysis of low angle ($\theta < 5°$) misorientations representing lath boundaries and the TEM analysis of lath sizes (Table 4) did not show the expected monotonic temperature-dependence (although it should be stated that the size variations were subtle and the trends might be being distorted by experimental error). It can be concluded that, based on the experimental data, the behavior shown in Figure 3 could not be explained by variations in the $\Delta\sigma_L$ term.

4.4. Strengthening from Minority Phases

The hardness measurements in Figure 3 involved plastic deformation and hence depended to some extent on the work hardening behavior of the alloys. The latter is greatly influenced by the minority retained austenite and martensite phases. In particular, the stability of retained austenite islands will play an important role. Tables 3–5 show that vanadium additions introduced four microstructural changes in Ref+V that may have a bearing on this:

1. Retained austenite in Ref+V contained less carbon, although the volume fraction of stable austenite was the same in both alloys.
2. Both OM and EBSD data show that MA islands in Ref+V were refined at the two lowest transformation temperatures.
3. Thermal martensite in Ref+V contained less carbon, especially at lower transformation temperatures.
4. At higher transformation temperatures (430 °C) more thermal martensite formed in Ref+V.

Points 1–3 indicate that retained austenite in Ref+V transformed at low temperatures must be being stabilized by a strong size effect, i.e., the RA stability is not governed uniquely by the local chemistry [29]. This is very striking at 375 °C where the mean Ref+V thermal martensite carbon content was much lower than the Ref alloy (Figure 10b). From this together with point 4 two different transformation temperature dependent strengthening mechanisms to qualitatively explain the hardness response in Figure 3 can be proposed:

- At 375 °C the retained austenite in Ref+V was significantly stabilized by size refinement. This should result in a higher work hardening coefficient and hence higher hardness, mitigated by the fact that any strain induced martensite will contain less carbon.
- At 400 °C minority phase strengthening was at a minimum and only weak precipitation strengthening was apparent.
- At 430 °C there was no size effect operating so the austenite stability should actually reverse. However, this was compensated by more thermal martensite formation in Ref+V at this transformation temperature.

5. Conclusions

The influence of vanadium additions on isothermally formed bainite in medium carbon steels containing retained austenite was investigated using in-situ HEXRD and ex-situ electron microscopy techniques. In the range of transformation temperatures investigated (375–430°C) the main conclusions are:

- Vanadium did not change the final fractions of bainite or of retained austenite.
- There was no significant effect on the bainite reaction rate.
- The presence of bainitic ferrite tetragonality hindered the quantitative dislocation density analysis. Qualitatively, there was no detectable effect of vanadium on the dislocation density in bainitic ferrite.
- No clear influence of vanadium on bainite lath widths could be observed.
- No significant vanadium precipitation occurred in bainite.
- No direct evidence of bainitic ferrite carbon supersaturation levels showing a vanadium dependence was found. Indirect calculations (from carbon mass balance) suggest that C_b was slightly raised.

- At low transformation temperatures, vanadium additions provided blocky MA and RA refinement.
- Retained austenite carbon content was decreased. This is not believed to be due to competition with vanadium precipitation.
- At low transformation temperatures (375 °C), retained austenite stability was enhanced due to a strong size effect.
- More thermal martensite forms during cooling, especially after transformation at 430 °C.
- Vanadium additions resulted in a modest increase in strength levels.

It is the author's opinion that the main benefit of adding vanadium to these steels is to be found in the refinement of martensite and austenite islands at low transformation temperatures. This provides a strong size effect stabilization of RA that could result in a significant increase in the ductility. Further, refining of blocky martensite should improve the fracture toughness behavior. If more strengthening is required then a combination of vanadium with higher nitrogen levels should be studied.

Author Contributions: Conceptualization, C.S.; methodology, G.G., S.Y.P.A., C.S., F.F; formal analysis, I.P.; investigation, I.P., B.S.-A., G.G., S.Y.P.A., M.S.; writing—original draft preparation, I.P.; writing—review and editing, C.S., S.Y.P.A.; project administration, F.F. All authors have read and agreed to the published version of the manuscript.

Funding: This research was funded by the Canadian Office of Energy Research and Development (OERD) and the LABEX DAMAS (ANR-11-LABX-008-01) from Lorraine, France.

Acknowledgments: The authors would like to express their gratitude to the Canadian Office of Energy Research and Development (OERD) and the LABEX DAMAS (ANR-11-LABX-008-01) from Lorraine, France for their support. Finally, we wish to acknowledge the help and professionalism of all the staff at CanmetMATERIALS who participated in the study and in particular Jian Li for aid with the EBSD analysis.

Conflicts of Interest: The authors declare no conflict of interest.

References

1. Fazeli, F.; Amirkhiz, B.S.; Scott, C.P.; Arafin, M.; Collins, L. Kinetics and microstructural change of low-carbon bainite due to vanadium microalloying. *Mater. Sci. Eng. A* **2018**, *720*, 248–256. [CrossRef]
2. Garcia-Mateo, C.; Morales-Rivas, L.; Caballero, F.G.; Milbourn, D.; Sourmail, T. Vanadium Effect on a Medium Carbon Forging Steel. *Metals* **2016**, *6*, 130. [CrossRef]
3. Sourmail, T.; Garcia-Mateo, C.; Caballero, F.G.; Cazottes, S.; Epicier, T.; Danoix, F.; Milbourn, D. The Influence of Vanadium on Ferrite and Bainite Formation in a Medium Carbon Steel. *Metall. Mater. Trans. A* **2017**, *48*, 3985–3996. [CrossRef]
4. Wang, Z.; Hui, W.; Chen, Z.; Zhang, Y.; Zhao, X. Effect of vanadium on microstructure and mechanical properties of bainitic forging steel. *Mater. Sci. Eng. A* **2020**, *771*, 138653. [CrossRef]
5. Scott, C.P.; Fazeli, F.; Amirkhiz, B.S.; Pushkareva, I.; Allain, S.Y.P. Structure-properties relationship of ultra-fine grained V-microalloyed dual phase steels. *Mater. Sci. Eng. A* **2017**, *703*, 293–303. [CrossRef]
6. Rijkenberg, R.A.; Blowey, A.; Bellina, P.; Wooffindin, C. Advanced High Stretch-Flange Formability Steels for Chassis & Suspension Applications. In Proceedings of the 4th International Conference on Steels in Cars and Trucks (SCT2014), Braunschweig, Germany, 15–19 June 2014; pp. 426–433.
7. Ioannidou, C.; Arechabaleta, Z.; Navarro-Lopez, A.; Rijkenberg, R.A.; Dalgliesh, R.M.; Kolling, S.; Bliznuk, V.; Pappas, C.; Sietsma, J.; Van Well, A.A.; et al. Interaction of precipitation with austenite-to-ferrite phase transformation in vanadium micro-alloyed steels. *Acta Mater.* **2019**, *181*, 10–24. [CrossRef]
8. Scott, C.P.; Remy, B.; Collet, J.-L.; Cael, A.; Bao, C.; Danoix, F.; Malard, B.; Curfs, C. Precipitation strengthening in high manganese austenitic TWIP steels. *Int. J. Mater. Res.* **2011**, *102*, 538–549. [CrossRef]
9. Gwon, H.; Kima, J.-K.; Shina, S.; Choa, L.; De Cooman, B.C. The effect of vanadium micro-alloying on the microstructure and the tensile behavior of TWIP steel. *Mater. Sci. Eng. A* **2017**, *696*, 416–428. [CrossRef]
10. Lagneborg, R.; Siwecki, T.; Zajac, S.; Hutchinson, B. The Role of Vanadium in Microalloyed steels. *Scand. J. Metall.* **1999**, *28*, 186–241.
11. Siwecki, T.; Eliasson, J.; Lagneborg, R.; Hutchinson, B. Vanadium microalloyed bainitic hot strip steels. *ISIJ Int.* **2010**, *50*, 760–767. [CrossRef]

12. Matsumura, O.; Sakuma, Y.; Takechi, H. Enhancement of Elongation by Retained Austenite in Intercritical Annealed 0.4C-1.5Si-0.8Mn Steel. *Trans. ISIJ* **1987**, *27*, 570–579. [CrossRef]
13. Perrard, F.; Scott, C. Vanadium Precipitation during Intercritical Annealing in Cold Rolled TRIP Steels. *ISIJ Int.* **2007**, *47*, 1168–1177. [CrossRef]
14. Allain, S.Y.P.; Gaudez, S.; Geandier, G.; Hell, J.-C.; Gouné, M.; Danoix, F.; Soler, M.; Aoued, S.; Poulon-Quintin, A. Internal stresses and carbon enrichment in austenite of Quenching and Partitioning steels from high energy X-ray diffraction experiments. *Mater. Sci. Eng. A* **2018**, *710*, 245–250. [CrossRef]
15. Available online: http://www.esrf.eu/computing/scientific/FIT2D/ (accessed on 18 March 2020).
16. Rodriguez-Carvajal, J. Recent advances in magnetic structure determination by neutron powder diffraction. *Phys. B* **1993**, *192*, 55–69. [CrossRef]
17. Cheng, L.; Bottger, A.; de Keijser, T.H.; Mittemeijer, E.J. Lattice parameters of iron-carbon and iron-nitrogen martensites and austenites. *Scr. Metall. Mater.* **1990**, *24*, 509–514. [CrossRef]
18. Scott, C.P.; Drillet, J. A study of the carbon distribution in retained austenite. *Scr. Mater.* **2007**, *56*, 489–492. [CrossRef]
19. Available online: https://www.matcalc-engineering.com/index.php/matcalc-software/matcalc-6 (accessed on 18 March 2020).
20. Fazeli, F.; Scott, C.P.; Amirkhiz, B.S.; Pushkareva, I. Challenges with Design and Processing of Carbide-free Bainitic AHSS Sheets. In Proceedings of the International Symposium on New Developments in Advanced High-Strength Sheet Steels, Keystone, CO, USA, 30 May 2017; pp. 135–145.
21. Venugopalan, D.; Kirkaldy, J. New Relations for Predicting the Mechanical Properties of Quenched and Tempered Low Alloy Steels. In Proceedings of the Hardenability Concepts with Application to Steel, Chicago, IL, USA, 26–28 October 1977; pp. 249–268.
22. Andrews, K.W. Empirical Formulae for the Calculation of Some Transformation temperatures. *JISI* **1965**, *203*, 721–727.
23. Zhu, K.; Bouaziz, O.; Oberbillig, C.; Huang, M. An approach to define the effective lath size controlling yield strength of bainite. *Mater. Sci. Eng. A* **2010**, *527*, 6614–6619. [CrossRef]
24. Pavlina, E.J.; Van Tyne, C.J. Correlation of Yield Strength and Tensile Strength with Hardness for Steels. *J. Mater. Eng. Perform.* **2008**, *17*, 888–893. [CrossRef]
25. Bhadeshia, H.K.D.H. *Bainite in Steels*; Maney Publishing: Leeds, UK, 2015.
26. Takebayashi, S.; Kunieda, T.; Yoshinaga, N.; Ushioda, K.; Ogata, S. Comparison of the Dislocation Density in Martensitic Steels Evaluated by Some X-ray Diffraction Methods. *ISIJ Int.* **2010**, *50*, 875–882. [CrossRef]
27. Williamson, G.K.; Hall, W.H. X-ray line broadening from filed aluminum and wolfram. *Acta Metall.* **1953**, *1*, 22–31. [CrossRef]
28. Zak, A.K.; Majid, W.A.; Abrishami, M.E.; Yousefi, R. X-ray analysis of ZnO nanoparticles by Williamson–Hall and size–strain plot methods. *Solid State Sci.* **2011**, *13*, 251–256.
29. He, S.H.; He, B.B.; Zhu, K.Y.; Ding, R.; Chen, H.; Huang, M.X. Revealing the role of dislocations on the stability of retained austenite in a tempered bainite. *Scr. Mater.* **2019**, *168*, 23–27. [CrossRef]
30. Hulme-Smith, C.N.; Peet, M.J.; Lonardelli, I.; Dippel, A.C.; Bhadeshia, H.K.D.H. Further evidence of tetragonality in bainitic ferrite. *Mater. Sci. Technol.* **2015**, *31*, 254–256. [CrossRef]
31. Gladman, T. *The Physical Metallurgy of Microalloyed Steels*; The Institute of Materials: London, UK, 1997.
32. Available online: https://www.matcalc.at/wiki/doku.php?id=tutorials:t18 (accessed on 18 March 2020).

© 2020 by the authors. Licensee MDPI, Basel, Switzerland. This article is an open access article distributed under the terms and conditions of the Creative Commons Attribution (CC BY) license (http://creativecommons.org/licenses/by/4.0/).

MDPI
St. Alban-Anlage 66
4052 Basel
Switzerland
Tel. +41 61 683 77 34
Fax +41 61 302 89 18
www.mdpi.com

Metals Editorial Office
E-mail: metals@mdpi.com
www.mdpi.com/journal/metals

www.ingramcontent.com/pod-product-compliance
Lightning Source LLC
LaVergne TN
LVHW070739100526
838202LV00013B/1268